METHODS IN MOLECULAR BIOLOGY

For further volumes:
http://www.springer.com/series/7651

For over 35 years, biological scientists have come to rely on the research protocols and methodologies in the critically acclaimed Methods in Molecular Biology series. The series was the first to introduce the step-by-step protocols approach that has become the standard in all biomedical protocol publishing. Each protocol is provided in readily-reproducible step-by step fashion, opening with an introductory overview, a list of the materials and reagents needed to complete the experiment, and followed by a detailed procedure that is supported with a helpful notes section offering tips and tricks of the trade as well as troubleshooting advice. These hallmark features were introduced by series editor Dr. John Walker and constitute the key ingredient in each and every volume of the Methods in Molecular Biology series. Tested and trusted, comprehensive and reliable, all protocols from the series are indexed in PubMed.

Manuscript Instructions https://drive.google.com/file/d/1K4Az5vpAFVzYadJ3jKx2-QSo7Dqa_SNA/view?usp=sharing

Quantum Mechanics Applications in Drug Design

Edited by

Alexander Heifetz

Sygnature Discovery, Nottingham, UK

Humana Press

Editor
Alexander Heifetz
Sygnature Discovery
Nottingham, UK

ISSN 1064-3745 ISSN 1940-6029 (electronic)
Methods in Molecular Biology
ISBN 978-1-0716-5419-4 ISBN 978-1-0716-5420-0 (eBook)
https://doi.org/10.1007/978-1-0716-5420-0

This Humana imprint is published by the registered company Springer Science+Business Media, LLC, part of Springer Nature.
The registered company address is: 1 New York Plaza, New York, NY 10004, U.S.A.

This book is dedicated to the memory of my mother, Asia Heifetz, whose love, strength, and generosity shaped my life and taught me that love has no limits. Her warmth, courage, and quiet wisdom continue to guide me, and her presence remains a lasting source of inspiration in all that I do.

To my father, Gennady Heifetz, for his boundless love, strength, and wisdom; for his guidance, patience, and steadfast support through every stage of my life; and for his unwavering belief in me, which has given me confidence to pursue my path, to question, to learn, and to keep moving forward.

And to my children, Hasia, Naomi, Michael, and Jonathan, whose curiosity reminds me that discovery is not only science but also wonder, love, and imagination.

Preface

This book builds on the foundation laid by the 2020 Springer volume *Quantum Mechanics in Drug Discovery*. Whereas that earlier work provided a broad overview of the role of quantum mechanics (QM) in pharmaceutical research, the present volume focuses specifically on practical applications of QM methods for designing new drug candidates and for supporting decision-making in modern drug design.

Over the past decade, computational chemistry has become an integral component of contemporary drug design pipelines. Analyses of published projects suggest that programs supported by computational methods are approximately 16–20% more likely to succeed and may require 30–40% fewer compounds during hit-to-lead and lead-optimization stages compared with projects that do not systematically employ such approaches. These trends highlight the growing importance of physics-based modeling as a key enabler of efficiency, mechanistic understanding, and innovation in medicinal chemistry.

The adoption of QM in drug design has expanded significantly in recent years, although it is not yet fully mainstream across the industry. This expansion has been driven by rapid advances in high-performance computing (HPC), cloud infrastructure, and workflow automation, together with increasing integration of QM with artificial intelligence (AI) and machine learning (ML). These developments are reshaping computational drug design, enabling researchers to address increasingly complex biological systems and chemical modalities with improved accuracy and predictive power.

QM methods contribute to drug design in multiple complementary ways. They enable detailed characterization of protein–ligand, protein–water, and protein–protein interactions; support structure-based drug design (SBDD); enhance molecular dynamics simulations; provide mechanistic rationalization of structure–activity relationships; and serve as high-accuracy scoring approaches in docking and virtual screening. Among the available methods, the Fragment Molecular Orbital (FMO) approach has emerged as a leading QM framework for structure-based design. Over the past decade, major technical challenges—including fragmentation strategies, treatment of solvation and polarization, geometry optimization, and computational speed—have been addressed. As a result, FMO has evolved into increasingly automated and user-friendly platforms capable of supporting high-throughput workflows and industrial drug design applications.

FMO and related QM methodologies are now routinely applied to protein–small molecule systems and are rapidly extending to emerging therapeutic modalities, including proteolysis-targeting chimeras (PROTACs), protein–protein interaction (PPI) modulators, molecular glue degraders, antibody–drug conjugates (ADCs), peptides and peptidomimetics, and nucleic-acid-targeting therapeutics. Beyond these applications, QM approaches support the design of enzyme inhibitors, the interpretation of mutagenesis data, and the refinement of ligand binding hypotheses. In predictive toxicology and pharmacokinetics, QM methods contribute to ADMET modeling, including evaluation of metabolic pathways, reactivity, and potential liabilities such as hERG-related cardiotoxicity. QM also plays an important role in understanding the contribution of water molecules to ligand binding and residence time, clarifying GPCR activation mechanisms, and analyzing conformational preferences of drug candidates.

This volume brings together contributions from leading researchers across academia and industry to provide a comprehensive and practical overview of QM methods in drug design. The chapters cover methodological developments, integration with molecular dynamics and AI, applications to diverse target classes and modalities, and perspectives on automation and high-throughput deployment. The book is intended for a broad readership, including computational and medicinal chemists, structural biologists, pharmacologists, and researchers working at the interface of physics, chemistry, and drug design.

Our goal is to provide both an authoritative reference and a practical guide for scientists seeking to harness quantum mechanical methods to accelerate and improve the design of new therapeutics.

Oxfordshire, UK *Alexander Heifetz*

Contents

Preface .... *vii*
Contributors .... *xi*

1 Visualization of Properties in the Fragment Molecular Orbital Method Using Facio 1
Masahiko Suenaga and Dmitri G. Fedorov

2 Perspectives on Hybridizing Quantum Mechanics and Artificial Intelligence for Drug Design 25
Alexander Heifetz, Girinath G. Pillai, Mussa Quareshy, Tarun Jain, Louise Birch and Colin Sambrook Smith

3 Applications of Integrated Molecular Dynamics and Fragment Molecular Orbital (MD+FMO) Calculations in Drug Discovery and Structural Biology ... 39
Kaori Fukuzawa

4 Advancing Medicinal Chemistry Through FMO: Sygnature's Platform 49
Alexander Heifetz, Girinath G. Pillai, Anna Cavalleri, Maryamdokht Taimoory, Louise Birch and Colin Sambrook Smith

5 Protein–Protein Interactions with the Fragment Molecular Orbital Method ... 81
Dustin C. Woods, Stefania Monteleone and Stéphanie Labouille

6 Beyond Small Molecules: Applying Fragment Molecular Orbital Sygnature Platform (FMO-SP) to Emerging Modalities 91
Alexander Heifetz, Girinath G. Pillai, Maryamdokht Taimoory, Louise Birch and Colin Sambrook Smith

7 Exploring Nucleic Acid Structure and Function with Molecular Dynamics and Fragment Molecular Orbital Methods 115
Tatsuya Ohyama, Hisae Tateishi-Karimata, Naoki Sugimoto and Teruki Honma

8 Protein–Protein Interactions Analyzed by the Fragment Molecular Orbital Method 133
Shigenori Tanaka

9 Untangling the Complexity of Binding Across Biomolecular Interfaces with Interaction Network Models 153
Vladimir Sladek and Dmitri G. Fedorov

10 Hidden Interactions: Quantum Perspectives in Drug Design 187
Taisiia Feoktistova and Tahsin Kellici

11 Quantum Chemistry in Drug Design: Applications, Challenges, and the Need for Speed 223
Christofer S. Tautermann, Matthias Degroote and Benjamin Ries

12 Drug Censorship at Scale: The Practicalities of Automation and Parallelism for Fragment Molecular Orbital (FMO) in *GAMESS* 233
Jintao Long

13 Quantum Methods in Drug Discovery: An Introduction to the Theoceptor (Theoretical Receptor) Approach 241
Nikeel Cull, Yixin He, Sylvia Kaempf, Andrew G. Leach, Mauno Lius and James Osbiston

14 Quantum-Driven High-Throughput Docking: Advances in Molecular Scoring and Drug Discovery 259
Nicolás Arrupe and Claudio N. Cavasotto

15 Integration of GPCR Modeling with QM for Drug Design: Academic and Industry Collaboration 271
Alexander Heifetz and Andrea Townsend-Nicholson

16 HessFit: Deriving Molecule-Specific Force Fields from Quantum Mechanical Hessians 283
Emanuele Falbo and Antonio Lavecchia

17 Fragment Molecular Orbital-Based Molecular Dynamics (FMO-MD): Current Status and Challenges 301
Yuto Komeiji, Shu Ohno, Shuhei Miyakawa and Kaori Fukuzawa

18 How Can QM Methods Be Used to Guide Synthesis Planning in Drug Discovery? 311
Jemima Haque

19 Fragment Molecular Orbital Drug Design Consortium (FMODD): Bridging Academia and Industry Through Prospective FMO-Driven Drug Design 323
Alexander Heifetz and Kaori Fukuzawa

20 Integrating AI-Based Protein Modeling and Fragment Molecular Orbital Analysis for Structure-Based Drug Design 333
Alessio Atzori, Louise Birch, Colin Sambrook Smith and Alexander Heifetz

Index 347

Contributors

NICOLÁS ARRUPE • *Computational Drug Design and Biomedical Informatics Laboratory, Instituto de Investigaciones en Medicina Traslacional (IIMT), Universidad Austral-CONICET, Pilar, Buenos Aires, Argentina; Austral Institute for Applied Artificial Intelligence, Universidad Austral, Pilar, Buenos Aires, Argentina*

ALESSIO ATZORI • *Sygnature Discovery, BioCity Nottingham, Nottingham, UK*

LOUISE BIRCH • *Sygnature Discovery, BioCity Nottingham, Nottingham, UK*

ANNA CAVALLERI • *Sygnature Discovery, BioCity Nottingham, Nottingham, UK*

CLAUDIO N. CAVASOTTO • *Computational Drug Design and Biomedical Informatics Laboratory, Instituto de Investigaciones en Medicina Traslacional (IIMT), Universidad Austral-CONICET, Pilar, Buenos Aires, Argentina; Austral Institute for Applied Artificial Intelligence, Universidad Austral, Pilar, Buenos Aires, Argentina; Facultad de Ingeniería, Universidad Austral, Pilar, Buenos Aires, Argentina; Facultad de Ciencias Biomédicas, Universidad Austral, Pilar, Buenos Aires, Argentina*

NIKEEL CULL • *Faculty of Biology Medicine and Health, University of Manchester, Manchester, UK*

MATTHIAS DEGROOTE • *Quantum Lab, Boehringer Ingelheim, Ingelheim am Rhein, Germany*

EMANUELE FALBO • *Dipartimento di Farmacia, "Drug Discovery Laboratory", Università degli Studi di Napoli "Federico II", Napoli, Italy*

DMITRI G. FEDOROV • *Materials DX Research Center (MDX), National Institute of Advanced Industrial Science and Technology (AIST), Tsukuba, Japan*

TAISIIA FEOKTISTOVA • *Modeling and Informatics, Merck & Co., Inc., West Point, PA, USA*

KAORI FUKUZAWA • *Graduate School of Pharmaceutical Sciences, The University of Osaka, Suita, Osaka, Japan*

JEMIMA HAQUE • *Evotec (UK) Ltd., Abingdon, Oxfordshire, UK*

YIXIN HE • *Faculty of Biology Medicine and Health, University of Manchester, Manchester, UK*

ALEXANDER HEIFETZ • *Sygnature Discovery, BioCity Nottingham, Nottingham, UK*

TERUKI HONMA • *RIKEN Center for Integrative Medical Sciences, Yokohama, Kanagawa, Japan*

TARUN JAIN • *Sygnature Discovery, BioCity Nottingham, Nottingham, UK*

SYLVIA KAEMPF • *Faculty of Biology Medicine and Health, University of Manchester, Manchester, UK*

TAHSIN KELLICI • *Modeling and Informatics, Merck & Co., Inc., West Point, PA, USA*

YUTO KOMEIJI • *Graduate School of Pharmaceutical Sciences, The University of Osaka, Suita, Osaka, Japan*

STÉPHANIE LABOUILLE • *Evotec (UK) Ltd., Abingdon, Oxfordshire, UK*

ANTONIO LAVECCHIA • *Dipartimento di Farmacia, "Drug Discovery Laboratory", Università degli Studi di Napoli "Federico II", Napoli, Italy*

ANDREW G. LEACH • *Faculty of Biology Medicine and Health, University of Manchester, Manchester, UK*

MAUNO LIUS • *Faculty of Biology Medicine and Health, University of Manchester, Manchester, UK*

JINTAO LONG • *Evotec (UK) Ltd., Abingdon, Oxfordshire, UK*
SHUHEI MIYAKAWA • *Graduate School of Pharmaceutical Sciences, The University of Osaka, Suita, Osaka, Japan*
STEFANIA MONTELEONE • *Evotec (UK) Ltd., Abingdon, Oxfordshire, UK*
SHU OHNO • *Graduate School of Pharmaceutical Sciences, The University of Osaka, Suita, Osaka, Japan*
TATSUYA OHYAMA • *RIKEN Center for Integrative Medical Sciences, Yokohama, Kanagawa, Japan*
JAMES OSBISTON • *Faculty of Science and Engineering, University of Manchester, Manchester, UK*
GIRINATH G. PILLAI • *Sygnature Discovery, BioCity Nottingham, Nottingham, UK*
MUSSA QUARESHY • *Sygnature Discovery, BioCity Nottingham, Nottingham, UK*
BENJAMIN RIES • *Department of Medicinal Chemistry, Boehringer Ingelheim Pharma GmbH & Co KG, Biberach, Germany*
VLADIMIR SLADEK • *Institute of Chemistry, Slovak Academy of Sciences, Bratislava, Slovakia*
COLIN SAMBROOK SMITH • *Sygnature Discovery, BioCity Nottingham, Nottingham, UK*
MASAHIKO SUENAGA • *Department of Advanced Materials Science and Engineering, Faculty of Engineering Sciences, Kyushu University, Kasuga City, Fukuoka, Japan*
NAOKI SUGIMOTO • *Frontier Institute for Biomolecular Engineering Research (FIBER), Konan University, Kobe, Hyogo, Japan*
MARYAMDOKHT TAIMOORY • *Sygnature Discovery, Montréal, QC, Canada*
SHIGENORI TANAKA • *Molecular Photoscience Research Center, Kobe University, Kobe, Japan*
HISAE TATEISHI-KARIMATA • *Frontier Institute for Biomolecular Engineering Research (FIBER), Konan University, Kobe, Hyogo, Japan*
CHRISTOFER S. TAUTERMANN • *Department of Medicinal Chemistry, Boehringer Ingelheim Pharma GmbH & Co KG, Biberach, Germany*
ANDREA TOWNSEND-NICHOLSON • *Research Department of Structural & Molecular Biology, Division of Biosciences, University College London, London, UK*
DUSTIN C. WOODS • *Evotec (US) Inc, Princeton, NJ, USA*

Check for updates

Chapter 1

Visualization of Properties in the Fragment Molecular Orbital Method Using Facio

Masahiko Suenaga and Dmitri G. Fedorov

Abstract

For evaluating binding in biochemical systems, the fragment molecular orbital method is used to calculate interaction and binding energies at various levels of theory. The graphical user interface program Facio is used to prepare input files and visualize fragment properties, supporting rational design of new molecular agents. The protein-ligand binding in the complex of HIV protease 6X mutant with an AB-2 ligand (PDB: 2HC0) is studied with various analyses.

Keywords Fragment molecular orbital method, FMO, Facio, Visualization, Protein–ligand binding, Pair interaction energy, PIEDA, Partition analysis, Subsystem analysis, Fluctuation analysis, HIV protease, GAMESS, DFTB3, MP2, Hotspot analysis, Rational drug design

1 Introduction

Molecular simulations provide valuable insight into the binding of biochemical molecules, delivering a detailed picture of binding with contributions of individual residues that can only indirectly be measured in experiment (e.g., in mutation studies). By combining simulations and experiments, very powerful methods are born, such as structure determination in X-Ray and NMR spectroscopies, or rational drug discovery.

Traditionally, parametrized force fields based on molecular mechanics have been used in biomolecular simulations. Recently, the use of potentially exact quantum mechanical (QM) simulations is becoming commonplace in biochemical studies [1], assisted by artificial intelligence and machine learning [2].

To facilitate large scale QM simulations, various low-scaling methods have been developed [3–6], including the fragment molecular orbital (FMO) method [7–9]. The use of fragmentation helps to identify hotspots, facilitating fragment-based drug design (FBDD) [10].

Alexander Heifetz (ed.), *Quantum Mechanics Applications in Drug Design*, Methods in Molecular Biology, vol. 3061,
https://doi.org/10.1007/978-1-0716-5420-0_1,

FMO is available in various programs, ABINIT-MP [11], PAICS [12], OpenFMO [13], and GAMESS [14]. FMO has been used for many biochemical simulations [15–17]. For visualizing FMO calculations, several GUI can be used, Biostation [18], PAICSView [19], FU [20], and Facio [21, 22]. As a text-based tool for making input files, FragIt [23] has attractive features for an automatic fragmentation of various materials, whereas the PyProGA plugin [24] for PyMol can be used for hotspot identification and residue interaction network models. Noncovalent interactions (NCI) can be visualized via density-based models [25].

Several books and dedicated chapters [26–29] provide guidelines for performing FMO calculations. However, the description of the Facio functionality for FMO is outdated [22] and does not cover much of recent development. In this chapter, it is shown how Facio can be used to prepare input files for various kinds of FMO calculations and visualize the results, demonstrated on a protein-ligand complex.

2 Summary of FMO Properties

In FMOn, the energy and other properties for a system divided into N fragments are obtained from a many-body expansion for $n=2$ or 3 [30],

$$E^{\mathrm{FMO2}}=\sum_{I}^{N} E'_{I}+\sum_{I>J}^{N} \Delta E_{IJ} \tag{1}$$

$$E^{\mathrm{FMO3}}=\sum_{I}^{N} E'_{I}+\sum_{I>J}^{N} \Delta E_{IJ}+\sum_{I>J>K}^{N} \Delta E_{IJK} \tag{2}$$

Most commonly FMO2 is used, and FMO3 is employed for systems with large charge transfer (such as coordinated metal cations). E'_I, ΔE_{IJ}, and ΔE_{IJK} are the energies of fragments (monomers), pair interaction energies (PIE), and PIE couplings, respectively.

The auxiliary polarization (AP) method [31] may be used with caps (C), denoted by APC [32, 33], for systems with covalent bonds at fragment boundaries. For molecular clusters AP and APC are the same, but for covalently connected fragments the treatment of boundaries differs (no caps in AP vs hydrogen caps in APC). The main objective of AP(C) is to use large diffuse basis sets in FMO, that are otherwise very difficult to employ. The energy in APC is given by (a similar form is used for AP),

$$E^{\mathrm{APC}}=E(\mathrm{BS1},\mathrm{V})-E(\mathrm{BS1},\mathrm{C})+E(\mathrm{BS2},\mathrm{C}) \tag{3}$$

The decomposition in Eq. 1 is used for each of the three energies on the right-hand side of Eq. 3. BS1 and BS2 denote two basis sets (BS), the smaller and the larger ones, respectively. V means the regular embedding, and C denotes that the embedding is replaced by hydrogen caps.

Pair interaction energies in FMOn can be split in energy decomposition analyses (EDAn, n=2 or 3) [34–39]. EDA2 is also known as the PIE decomposition analysis (PIEDA). The decomposition terms vary depending on the QM method. In this chapter, single reference closed shell methods are employed, although EDAn may be used for open-shell methods as well [40].

For second order Møller-Plesset perturbation theory (MP2) or coupled cluster (CC) in FMO2, PIEDA is formulated as

$$\Delta E_{IJ} = \Delta E_{IJ}^{\mathrm{ES}} + \Delta E_{IJ}^{\mathrm{EX}} + \Delta E_{IJ}^{\mathrm{CT+MIX}} + \Delta E_{IJ}^{\mathrm{RC+DI}} + \Delta E_{IJ}^{\mathrm{BS}} + \Delta E_{IJ}^{\mathrm{solv}} \tag{4}$$

EDAn components are electrostatic (ES), exchange-repulsion (EX), charge transfer, MIX (coupling), remainder correlation (RC), dispersion (DI), basis set correction (BS), and solvent screening (SOLV). A list of decompositions is provided in Table 1. For Hartree-Fock (HF), there is no correlation (RC) term. For density functional theory (DFT), RC and DI may be split. For density-functional tight-binding (DFTB), EX+RC is represented by the 0-order Hamiltonian contribution. DFTB is not used with AP(C) because currently there is only one minimum basis set available, so there is no BS term. The CT and MIX terms may be split [41].

There are three variations of EDAn/AP(C), with (1) a separation of polarization (PL), (2) a separation of a BS correction, and (3) an incorporation of PL and BS into other terms. In this chapter, the second approach is used, whereby the BS term is obtained as $E(\mathrm{BS2}, \mathrm{C}) - E(\mathrm{BS1}, \mathrm{C})$ and all other terms come from a decomposition of $E(\mathrm{BS1}, V)$ in Eq. 4.

In FMO3, ΔE_{IJK} is decomposed. For a reduction in complexity, Eq. 2 can be rewritten in terms of effective pair interactions, i.e., two-body values corrected with three-body terms

$$E^{\mathrm{FMO3}} = \sum_{I}^{N} E'_I + \sum_{I>J}^{N} \Delta \widetilde{E}_{IJ} \tag{5}$$

where for MP2

$$\Delta \widetilde{E}_{IJ} = \Delta E_{IJ}^{\mathrm{ES}} + \Delta \widetilde{E}_{IJ}^{\mathrm{EX}} + \Delta \widetilde{E}_{IJ}^{\mathrm{CT+MIX}} + \Delta \widetilde{E}_{IJ}^{\mathrm{RC+DI}} + \Delta \widetilde{E}_{IJ}^{\mathrm{BS}} + \Delta \widetilde{E}_{IJ}^{\mathrm{solv}} \tag{6}$$

Table 1
Interaction energy decomposition terms[a]

Method	Electrostatic	Exchange-repulsion	Remainder correlation	Dispersion	Charge transfer+mix	Basis set	Solvation
HF	$\Delta E_{IJ}^{\mathrm{ES}}$	$\Delta E_{IJ}^{\mathrm{EX}}$		$\Delta E_{IJ}^{\mathrm{DI}}$	$\Delta E_{IJ}^{\mathrm{CT+MIX}}$	$\Delta E_{IJ}^{\mathrm{BS}}$	$\Delta E_{IJ}^{\mathrm{solv}}$
MP2,CC	$\Delta E_{IJ}^{\mathrm{ES}}$	$\Delta E_{IJ}^{\mathrm{EX}}$	$\Delta E_{IJ}^{\mathrm{RC+DI}}$		$\Delta E_{IJ}^{\mathrm{CT+MIX}}$	$\Delta E_{IJ}^{\mathrm{BS}}$	$\Delta E_{IJ}^{\mathrm{solv}}$
DFT	$\Delta E_{IJ}^{\mathrm{ES}}$	$\Delta E_{IJ}^{\mathrm{EX}}$	$\Delta E_{IJ}^{\mathrm{RC}}$	$\Delta E_{IJ}^{\mathrm{DI}}$	$\Delta E_{IJ}^{\mathrm{CT+MIX}}$	$\Delta E_{IJ}^{\mathrm{BS}}$	$\Delta E_{IJ}^{\mathrm{solv}}$
DFTB	$\Delta E_{IJ}^{\mathrm{ES}}$	$\Delta E_{IJ}^{\mathrm{H0}}$		$\Delta E_{IJ}^{\mathrm{DI}}$	$\Delta E_{IJ}^{\mathrm{CT+ES}}$		$\Delta E_{IJ}^{\mathrm{solv}}$

[a]The BS term appears only in Hartree-Fock with three corrections (HF-3c) [42] or AP(C). H0 stands for the 0-order Hamiltonian contribution in DFTB, describing EX, RC, and REP (repulsion)

Some terms are identical in FMO2 and FMO3 (indicated by no tilde in Eq. 6): BS and DI in empirical models, and ES in all cases.

In the partition analysis (PA) [43, 44] applied to M segments, the energy is

$$E^{\mathrm{PA}} = \sum_{i}^{M} E_i + \sum_{i>j}^{M} \Delta E_{ij} \tag{7}$$

where E_i and ΔE_{ij} are the energies of segments and pair interactions, respectively. Atomic charges and electron density from FMOn are used as input data for PA. The same form of Eq. 7 is used in PA for FMOn (n=2 and 3), but the values of effective E_i and ΔE_{ij} depend on n (in some sense, these energies are analogous to the effective values $\Delta \widetilde{E}_{IJ}$ of PIEs in FMO, where 3-body interactions are integrated). PA was applied to biochemical systems [45, 46].

The decomposition in PA is simpler than in EDAn (compare to Eq. 4)

$$\Delta E_{ij} = \Delta E_{ij}^{\mathrm{ES}} + \Delta E_{ij}^{\mathrm{DI}} + \Delta E_{ij}^{\mathrm{solv}} \tag{8}$$

because the equivalent of CT+MIX+EX+RC is integrated into monomer energies E_i due to PA using relaxed electron densities after interfragment CT takes place.

The solvent screening is decomposed into electrostatic (es) and nonelectrostatic (nones) components,

$$\Delta E_{ij}^{\mathrm{solv}} = \Delta E_{ij}^{\mathrm{es}} + \Delta E_{ij}^{\mathrm{nones}} \tag{9}$$

The subsystem analysis (SA) [47] is applied to a binding process or a chemical reaction. For a common addition,

$$A + B \rightarrow AB \tag{10}$$

$$\Delta E^{\mathrm{bind}} = E_{AB} - E_A - E_B \tag{11}$$

The binding energy ΔE^{bind} in SA is decomposed by applying Eq. 1 (FMO) or Eq. 7 (PA) for each term in Eq. 11. In SA, it is possible either to (a) separate or (b) neglect the deformation energy, or (c) merge it into other terms. In this chapter, it is neglected. The binding energy is decomposed into polarization (PL), desolvation (desolv), and interation (INT). Here, desolv should not be confused with solv. The former (desolv) shows the amount of energy required to remove solvent from the active site of A and B when the complex is formed. In contrast, the latter (solv) is the solvent screening of interactions in the complex.

$$\Delta E^{\text{bind}} = \Delta E_A^{\text{PL}} + \Delta E_B^{\text{PL}} + \Delta E_A^{\text{desolv}} + \Delta E_B^{\text{desolv}} + \Delta E_{AB}^{\text{INT}} \tag{12}$$

where PL and desolv terms can be split into fragment contributions for $X = A$ or B,

$$\Delta E_X^{\text{PL}} = \sum_{I \in X} \Delta E_I^{\text{PL}} \tag{13}$$

$$\Delta E_X^{\text{desolv}} = \sum_{I \in X} \Delta E_I^{\text{desolv}} \tag{14}$$

The INT term is a sum of pair contributions. In FMO2, the interaction energy is

$$\Delta E_{AB}^{\text{INT}} = \sum_{I \in A} \sum_{J \in B} \Delta E_{IJ} \tag{15}$$

and in FMO3, the effective interaction energies $\Delta \widetilde{E}_{IJ}$ are used instead. The binding energy is

$$\Delta E = \sum_{I \in AB} (\Delta E_I^{\text{PL}} + \Delta E_I^{\text{desolv}}) + \sum_{I \in A} \sum_{J \in B} \Delta E_{IJ} \tag{16}$$

which can be modified as

$$\Delta E = \sum_{I \in AB} \Delta E_I^{\text{BIND}} \tag{17}$$

where

$$\Delta E_I^{\text{BIND}} = \Delta E_I^{\text{PL}} + \Delta E_I^{\text{desolv}} + \Delta E_I^{\text{INT}} \tag{18}$$

$$\Delta E_I^{\text{INT}} = \sum_{J \in B} w_{IJ}^A \Delta E_{IJ}, \quad I \in A \tag{19}$$

$$\Delta E_I^{\text{INT}} = \sum_{J \in A} w_{IJ}^B \Delta E_{IJ}, \quad I \in B \tag{20}$$

where w is a weighing factor to avoid double counting.

Two common choices can be made for $I \in A$, $J \in B$: (1) $w_{IJ}^A = 1$ and $w_{IJ}^B = 0$ (denoted as 1:0) and (2) $w_{IJ}^A = w_{IJ}^B = 1/2$ uniformly (denoted as 0.5:0.5). The former may be used for protein (A) and ligand (B) binding, when residue-ligand interactions are counted in the binding energy of residues. The latter is suitable when A and B are both macromolecules, for example, in protein-protein or protein-DNA complexes. In either case, $\Delta E_I^{\mathrm{BIND}}$ is a convenient measure of the importance (score) of fragment I to the total binding. The weights are chosen to add up to one ($w_{IJ}^A + w_{IJ}^B = 1$), leaving the sum of fragment binding energies in Eq. 17 invariant of the definition of weights.

The fluctuation analysis (FA) [36] is performed by averaging the energy in Eq. 1 along an MD trajectory (denoted by <>). It is commonly done by adding and subtracting a reference energy for the global minimum in the trajectory (denoted by 0), defining thermal fluctuations (deviations of the average from the reference).

$$\langle E \rangle = E^0 + \langle E - E^0 \rangle \tag{21}$$

Both the energy of each conformation (E) and of the global minimum (E^0) are decomposed according to Eq. 1.

$$\begin{aligned} \langle E \rangle = & \sum_{I}^{N} E_I^0 + \sum_{I>J}^{N} \Delta E_{IJ}^0 + \sum_{I>J>K}^{N} \Delta E_{IJK}^0 \\ & + \sum_{I}^{N} \langle \Delta E_I \rangle + \sum_{I>J}^{N} \langle \Delta\Delta E_{IJ} \rangle + \sum_{I>J>K}^{N} \langle \Delta\Delta E_{IJK} \rangle \end{aligned} \tag{22}$$

where the polarization effect of thermal fluctuations is

$$\Delta E_I = E_I - E_I^0 \tag{23}$$

and fluctuations of pair and triple interaction terms are

$$\Delta\Delta E_{IJ} = \Delta E_{IJ} - \Delta E_{IJ}^0 \tag{24}$$

$$\Delta\Delta E_{IJK} = \Delta E_{IJK} - \Delta E_{IJK}^0 \tag{25}$$

The averaged pair interaction energy in MD is (and likewise the effective value in FMO3 can be defined)

$$\langle \Delta E_{IJ} \rangle = \Delta E_{IJ}^0 + \langle \Delta\Delta E_{IJ} \rangle \tag{26}$$

In addition to the potential energies above, kinetic energies are also averaged in FA. Although the total kinetic energy in NVT or NPT MD is normalized to maintain a constant global temperature T, the kinetic energy per fragment varies, defining a local temperature T_I of fragment I. These values may be useful in understanding the flexibility of residues in the complex.

3 Computational Details

All FMO computations were performed with FMO in GAMESS [14, 48] parallelized with generalized distributed data interface (GDDI) [49, 50]. MP2 was done with the 6–31 G* basis set. All DFTB calculations were performed at the third order level (DFTB3) with 3ob parameters [51]. In all QM calculations, polarizable continuum model (PCM) was used at the one-body level PCM<1> [52]. In NVT MD at 298 K, the time step of 1 fs was used with RATTLE constraints. Of the 40 ps trajectory, the initial step of 20 ps was done for pre-equilibration, and in the final 20 ps a production averaging in FA was conducted. In the energy plots identifying hotspots, the threshold of −3.0 kcal/mol was used to filter out unimportant contributions.

HIV protease 6X mutant in complex with an AB-2 ligand ([1-((1S,2R)-1-benzyl-2-hydroxy-3-{isobutyl[(4-methoxyphenyl)sulfonyl]amino}propyl)-1H-1,2,3-triazol-4-yl]methyl (1R,2R)-2-hydroxy-2,3-dihydro-1h-inden-1-yl carbamate) was used for all calculations, PDB: 2hc0. There are 99 residues per chain, two chains, and a ligand, 3214 atoms in total. All atoms are treated quantum-mechanically on equal footing.

In PA, small letters (a, b etc) in residue names denote chains. For Cα-C fragment boundaries, a dash is inserted in residue names (as in Asp-25), whereas for C-N boundaries, no dash is used. For example, bAsp25 means conventional residue Asp25 of chain **b** for C-N boundaries. In segment names, the residue numbers stored in the PDB are preserved so that, for example, the first residue in chain **a** is aPro1001 and in chain **b** it is bPro2101. In Facio, fragments are numbered from one consequently in both chains together.

The general rule is that when total properties in FMO are of primary interest, for example, the total energy and its analytic gradient in a geometry optimization or MD, Cα-C boundaries (and no S-S boundaries) are used as the more accurate setup. On the other hand, for discussing properties of individual fragments in FMO, C-N, and S-S boundaries are used to define conventional residues as fragments.

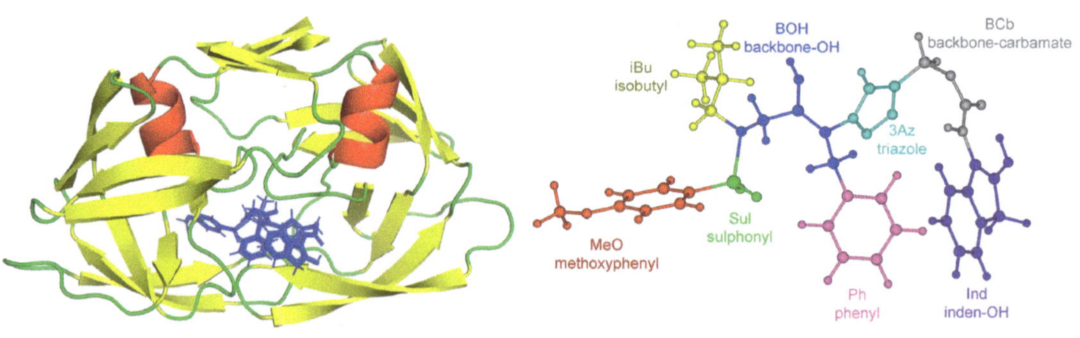

Fig. 1 Protein-ligand complex: structure and ligand segmentation

PA is typically done by taking FMO properties obtained with Cα-C (no S-S) boundaries and partitioning the energy for segments corresponding to C-N and S-S boundaries. PA can also be used for finer segments, such as side chains and backbone, or functional groups in a ligand. It may be noted that some ligands may also be fragmented in FMO, but in this chapter, it is not done.

In PA, the ligand is treated in two ways, as one big segment (84 atoms) and as eight segments (by defining functional groups as

Table 2

A section of the protein-ligand complex PDB, showing a selection of ligand atoms and their assignment to segments[a]

HETATM	1	C	MeO	3501	53.188	23.704	6.188	1	0	C
HETATM	2	H	MeO	3501	52.604	24.613	6.353	1	0	H
HETATM	3	H	MeO	3501	53.528	23.66	5.144	1	0	H
HETATM	4	O	MeO	3501	54.304	23.759	7.075	1	0	O
HETATM	5	C	MeO	3501	55.066	22.648	7.26	1	0	C
HETATM	6	C	MeO	3501	56.091	22.75	8.222	1	0	C
HETATM	7	H	MeO	3501	56.241	23.673	8.777	1	0	H
HETATM	8	C	MeO	3501	56.91	21.654	8.461	1	0	C
HETATM	9	H	MeO	3501	57.694	21.714	9.205	1	0	H
HETATM	10	C	MeO	3501	56.718	20.452	7.76	1	0	C
HETATM	11	C	MeO	3501	55.684	20.349	6.815	1	0	C
HETATM	12	H	MeO	3501	55.52	19.415	6.284	1	0	H
HETATM	13	C	MeO	3501	54.851	21.44	6.568	1	0	C
HETATM	14	H	MeO	3501	54.062	21.346	5.83	1	0	H
HETATM	15	O	Sul	3502	58.559	19.256	9.3	1	0	O
HETATM	16	O	Sul	3502	57.099	17.828	7.815	1	0	O
HETATM	17	N	BOH	3503	59.249	19.008	6.621	1	0	N
HETATM	18	C	iBu	3504	58.821	19.505	5.311	1	0	C
HETATM	19	H	iBu	3504	59.107	20.559	5.157	1	0	H
HETATM	20	C	iBu	3504	59.402	18.61	4.165	1	0	C
…										
HETATM	74	H	Ind	3508	64.44	15.868	12.089	1	0	H
HETATM	75	S	Sul	3502	57.835	19.078	8.054	1	0	S
…										

[a]3501, 3502, 3503, and 3504 are unique segment IDs. All atoms with the same segment ID are put into the same segment. Atoms in the same segment do not have to be listed consequently, an intermittent list is also acceptable (see Sul here, which is made of atoms 15, 16, and 75)

shown in Fig. 1). The assignment of segments was accomplished by a manual editing of the PDB and modifying the residue ID column for the atoms in the ligand, assigning the same ID to residues in a given functional group (Table 2). The following segments were defined for the ligand: methoxyphenyl (MeO), sulphonyl (Sul), backbone-OH (BOH), isobutyl (iBu), phenyl (Ph), backbone-carbamate (BCb), and inden-OH (Ind). The ligand is made from formally neutral functional groups, but because PA allows for charge transfer, the actual charge on segments is not zero.

4 Overview of Facio

Development of Facio was originally started in 1995 for a computational chemistry class, because there was no freely available GUI with all required functionality. Facio was named after a Latin verb *facio* ("I make"), meaning "I make an original molecular modeling software for my class." The first stage of development was implementation of (a) a pre-processor (an input file maker) for commonly used computational chemistry software, such as Gaussian, GAMESS US, Firefly (called PC GAMESS at that time), MOPAC, TINKER, and MSMS, and (b) a post-processor for visualization of outputs, such as molecular orbitals, normal modes of vibration, and energy changes during structure optimization. In addition to the pre- and post-processors, utilities and functions for molecular modeling were added and Facio became a full-fledged computational chemistry environment by ~2005.

2007 was a turning point for Facio, when Prof. Kitaura, the pioneer of the FMO method, requested development of GUI for FMO/GAMESS. In response to his requests, basic functions for FMO were implemented into Facio, such as an automatic fragmentation of polypeptides, preparation of input for FMO calculations, and local structure viewers used to check the structure of each fragment in a large system. In 2008, visualization of PIEDA was implemented. Since then, Facio was extended to add various functionality in response to requests from all over the world. These years are the second stage of development.

2024 was the beginning of the third stage of development to correspond to the many novel features of FMO, such as new fragmentation scheme (conventional residues) and AP. Further development is currently underway, expected to be completed in 2025. In this chapter, version 27.1.1 is described.

Although there are many (over than 150) functions in Facio, a small subset of this functionality related to FMO is briefly explained in this chapter. In Facio, almost all file types can be read by a drag-and-drop to the Facio's main window. Thus, there is no need to worry about selecting a file format (such as structure or a calculation output). For example, one can drag-and-drop an output of FMO

```
GAMESS FMO Subsystem Analysis by fmosa.py (ver.16July2024)

initial input for optimization of (pro + lig) : KDELR4_CKHDEL_R169_opt1.inp
base name for SA : KDELR4_CKHDEL_R169_opt1
 total fragments : 213
ligand fragments : 6
 $OPTFMO OPTTOL=1E-4

-----------------------------------------------------------------------------------
 Direct Binding Energy (deltaE = sum(Ebind)) :  -352.425 kcal/mol
 Deformation Energy of Protein (deltaE_defP) :    12.560 kcal/mol
     EP : -2645805.2234433456 a.u.        (in the complex geometry)
    EP~ : -2645817.783600707 a.u.        (in the isolated energy-minimum state)
  Deformation Energy of Ligand (deltaE_defL) :     8.715 kcal/mol
     EL : -81707.38082147202 a.u.        (in the complex geometry)
    EL~ : -81716.09625510934 a.u.        (in the isolated energy-minimum state)
-----------------------------------------------------------------------------------
 Binding Energy = deltaE + deltaE_defP + deltaE_defL :   -331.149 kcal/mol
###################################################################################
Fragment-wise subsystem partition summary
   I       Name    Sys   Epart-complex    PIE      Epart-isolated  delta-Epart     Ebind
-----------------------------------------------------------------------------------
   1      MET-1    1   -15.15925099   -83.759   -15.22990857   44.33830332    -39.42
   2      ASN-2    1   -20.55776671   -38.619   -20.58676987   18.19975780    -20.42
   3      ILE-3    1   -20.21809404    20.055   -20.21205942   -3.78678326     16.27
   4      PHE-4    1   -24.54499246    -0.513   -24.54616706    0.73707391      0.22
```

Fig. 2 Example of an output from fmosa.py

calculation to Facio main window. If there is PIEDA data in the output file, a PIEDA panel will automatically pop up, in addition to the molecular structure shown in the main window.

Recently, a Python script fmosa.py was developed, which performs all steps in FMO/SA for a binary complex *AB*, namely, it optimizes the complex *AB* and isolated *A* and *B*, then it performs single point FMO calculations of *AB*, *A*, and *B*, while partitioning properties (needed to define each component in Eq. 12), and finally, the binding energy is decomposed in SA.

The script is executed as "python3 fmosa.py job 1 $N_{AB}N_B$," where N_B and N_{AB} is the number of ligands in *B* (typically, ligand) and *AB*, respectively. Before running this script, an input file job1.inp should be prepared, setting up a geometry optimization of complex *AB*. The number "1" may be replaced by 2 etc (i.e., job2.inp is executed as "python3 fmosa.py job 2 $N_{AB}N_B$"), if a geometry restart is required (although it may be easier to set the maximum number of steps to avoid having to restart). An example of the script output is shown in Fig. 2.

5 Making Input Files for FMO

Initial preparations are described in Table 3. When a structure (PDB) is read into Facio, it is quite common to have a problem of missing or extraneous chemical bonds (e.g., a H atom may be bound to two oxygens). It may be unfortunately quite tedious to resolve this problem in Facio for a system containing thousands of atoms.

It helps if the structure is already optimized (e.g., with a force field). Getting rid of the bond problem is an issue outside the scope of this chapter. If this problem is not resolved, Facio will not be able to assign fragment charges, and it may confuse fragmentation too.

As a guide for biochemical applications, a set of typical FMO calculations was performed for input files prepared in Facio. A summary of all calculations is given in Table 4.

First, let us look at making an FMO2-DFTB3/D3(BJ)PCM<1> input file for geometry optimizations (here and below, the protein-ligand complex is used). The sequence of operations is shown in Fig. 3. There are eight steps. **1**,**2**, and **3** are clicked consequently. Note that here Cα-C fragmentation is chosen because *ConventionalPeptide* is not selected. S-S bonds are also left intact (*Break S-S* is not selected). After step **3** is pressed, Facio will print a summary in the small Worksheet screen. If there is a problem (e.g., a H atom is bound to two O atoms), then there may be a warning that fragment charges are wrong. If that happened, most likely one will have to manually remove or add missing bonds, and repeat everything starting from step **1**.

After doing step **3**, the second half of the panel will be revealed. There, RUNTYP=OPTIMIZE is chosen (**4**), solvent model is selected (**5**), DFTB3/3ob-3-1 is set (**6**), D3(BJ) dispersion is turned on as *Grimme, v4* (**7**). Finally, by pressing **8**, an input file is made.

As a second example, an input for SA/PIEDA with HF-3c/PCM<1> is made. Note that if two input files are made consequently, Facio will remember all selected options and they may have to be reset. Here, the sequence of actions is shown in Fig. 4 for a fresh start of Facio when all options are at the default. Steps **1**–**5** set up C-N and S-S fragmentation, as suitable for an analysis.

Table 3
Initial preparation steps of the protein-ligand complex

Step	Description
PDB download and pruning	Using RCSB database, http://www.rcsb.org (Br ions and water were removed by hand).
Protein protonation	Using pdb2pqr tool, [a]http://server.possonboltzmann.org
Ligand protonation	Using FU for an initial protonation, followed by a manual addition of H in Facio using Molview, which can reconstruct structure from a full chemical name, http://molview.org
Optimization of added hydrogen atoms in the ligand	At the DFTB level using GAMESS, freezing heavy atoms to their position stored in the PDB
Concatenation of the protonated protein and ligand	Using text editor

[a]The protonation state of bAsp25 assigned to be neutral by pdb2pqr was manually changed to anionic, making the two chains identical. Although pdb2pqr should in principle protonate ligands, it did not work due to an unknown problem

Table 4
Summary of FMO calculations

Type	Level[a]	Boundary[b]	Structures	Timing[c]
Geometry optimization	FMO2-DFTB3/3ob	Cα-C	1358	9.6 h
PA/FMO	FMO2-DFTB3/3ob	Cα-C C-N (S-S)	1	7.3 min
SA/PIEDA	FMO2-HF-3c/MINIX	C-N (S-S)	2[d]	128.8 min
PIEDA	FMO2-MP2/6-31G*	C-N (S-S)	1	15.9 h
FA	FMO2-DFTB3/3ob20+ 20 ps NVT MD	Cα-C	40,000	10.7 days

[a]MINIX is the basis set used in HF-3c. 3ob are the DFTB parameters used for biochemical applications. For all methods, solvent is treated with PCM<1>.
[b]PA uses FMO with Cα-C fragments to determine electron densities, followed by a decomposition using segments with C-N boundaries. In this system, there are no S-S boundaries
[c]On 1 dual 3.1 GHz Xeon node (40 cores total)
[d]Technically, three structures (AB, A, and B), but the count is set to two corresponding to the cost of two (AB and A+B) calculations

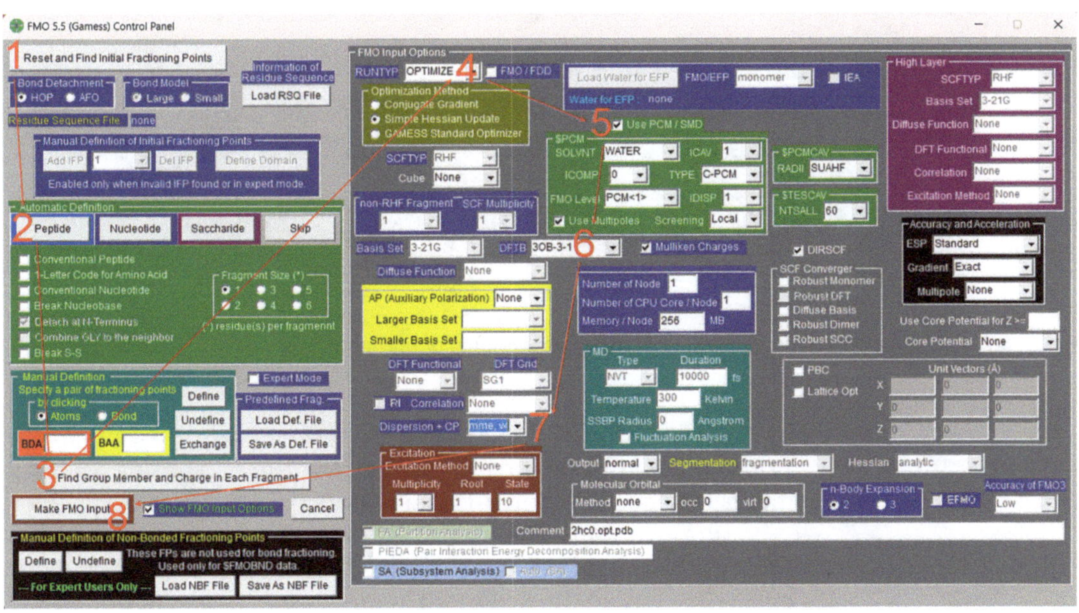

Fig. 3 Making input file for a geometry optimization with FMO2-DFTB3/D3(BJ)/PCM<1>

If SA is to be set up, however, one further step is necessary, after step **5**. SA requires that every fragment is assigned to a subsystem. By default, all fragments are in subsystem 1. In this complex, only ligand (the last fragment $I = 199$) should be assigned to subsystem 2. For that, a Local Structure Viewer #1 in the *FMO* menu is opened (Fig. 5a), *Define Local Structure* is set to *by FMO Fragment* (**1**), fragment 199 is chosen in the pull-down menu of *From Frag #* (**2**),

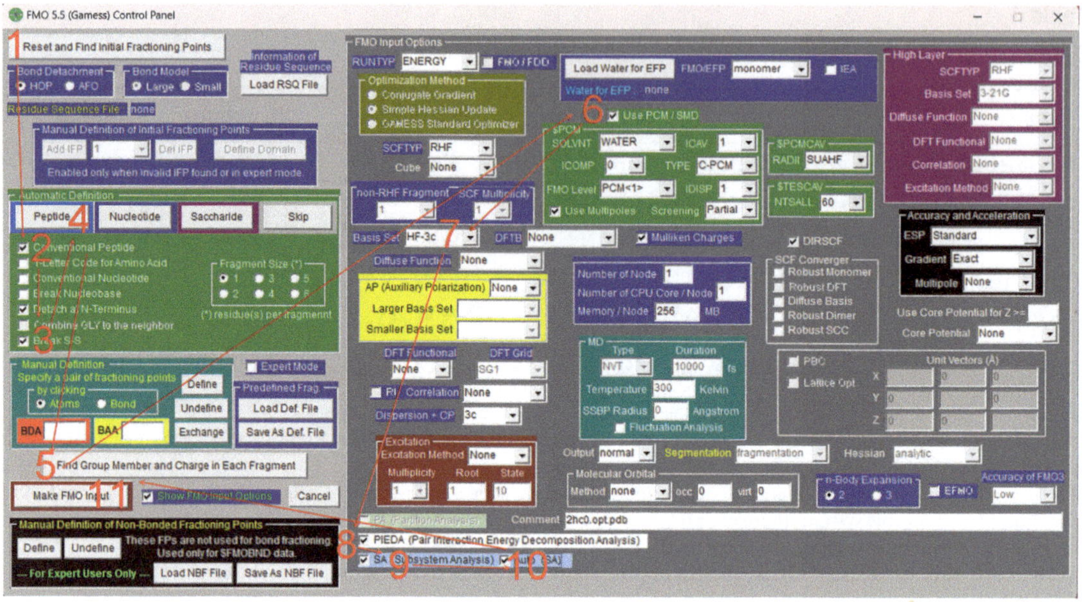

Fig. 4 Making input file for SA/PIEDA with FMO2-HF-3c/PCM<1>

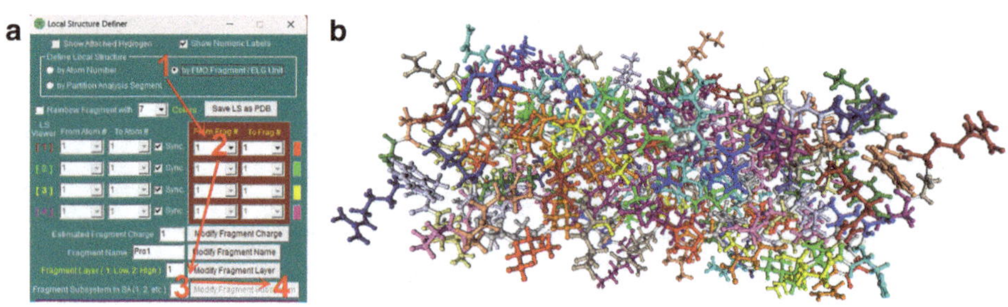

Fig. 5 (**a**) Local structure viewer, and (**b**) rainbow representation of fragments.

its subsystem is set to 2 as *Fragment Subsystem in SA* (**3**), and *Modify Fragment Subsystem* is pressed (**4**). This finishes up setting SA. Fragments may be plotted in different colors (Fig. 5b).

Auto SA means that *AB*, *A*, and *B* (in Eq. 10) are calculated in one run by providing geometry of the complex *AB* (to study the effect of deformation, neglected in this chapter, separately optimized structures of *A* and *B* should be manually inserted in the input file as $ISAXYZ and $ISBXYZ, respectively). If the *auto SA* option is not selected, then only *AB* is calculated with a subsystem partitioning of energies. If the chemical process differs from Eq. 10, for example, it is $A+B+C \rightarrow ABC$, then 4 calculations (of *ABC*, *A*, *B*, and *C*) have to be done separately one by one without selecting *Auto SA*.

Then, steps **6**–**10** are performed. HF-3c should not have any dispersion model selected (3c already includes an internal dispersion model). If one only wants PIEDA and no SA, then this step of setting subsystems is skipped. Finally, an input file is made by pressing **11**.

6 Visualization of Interactions

6.1 Structure Optimization

An overlay of the experimental (solid state crystal) and FMO-DFTB optimized (aqueous solution) structures is shown in Fig. 6. Some part of the difference may be due to the packing effect in the experimental structure, affecting mainly the protein surface

6.2 Energy Decomposition Analysis

The results of PIEDA calculations can be visualized by placing a log and its input file into one folder and dropping the log file onto the main Facio window (Fig. 7a). As a demonstration, the following sequence of actions was performed: step **1** to choose fragment 199 (the ligand), steps **2–5** to set upper and lower bounds for adding fragment labels on the plot for pairs with interactions stronger than 1 kcal/mol, and step **6** to change the scale (to 2).

To check the structure of a pair of interest, local structure viewers from the *FMO* menu are opened, first #1, with the setting *Define Local Structure* set to *by FMO fragment* (Fig. 8a). Then a similar viewer #2 is opened (two viewers are needed to show a pair). Then on the PIEDA panel, 7 is checked, and by pressing the longest bar in step **8**, the main hotspot (the residue with the strongest interaction) is selected. The structure of the dimer is shown in the main window (viewer #1 shows a fragment (199) in red, whereas viewer #2 shows a fragment (25) in green.

The whole set of interactions can be plotted as a heat map (Fig. 8), either the total values or individual components (accomplished by

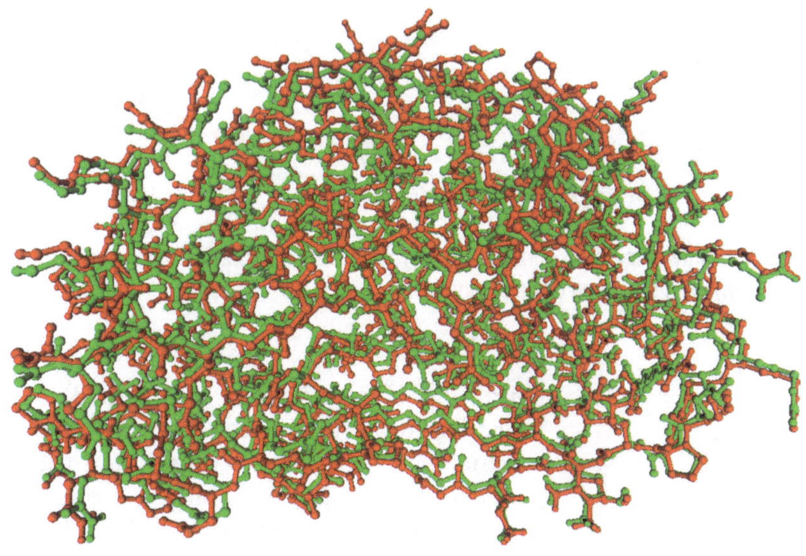

Fig. 6 Experimental (solid state, X-ray with the resolution of 1.30Å, shown in red) and optimized (solution, FMO-DFTB, green) structures (only heavy atoms). The RMSD is 0.934Å

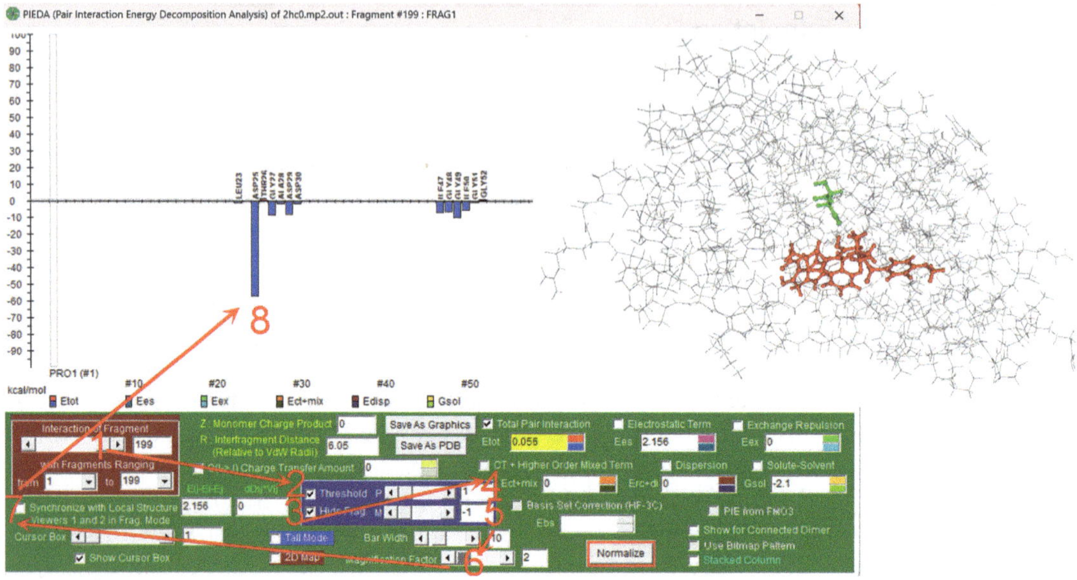

Fig. 7 MP2 interaction energies in 2hc0: PIEDA panel showing ΔE_{IJ} for fixed $I = 199$ (ligand) with varied J (residue). On the insert on the right, the pair of the ligand (red) and the main hotspot aAsp1025 (green) is shown.

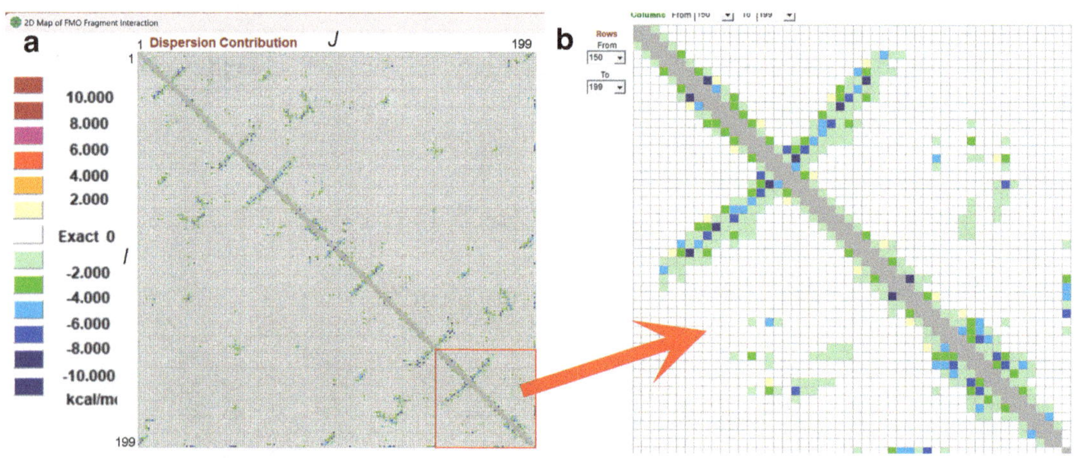

Fig. 8 Mainly dispersion contribution $\Delta E_{IJ}^{\mathrm{RC+DI}}$(MP2) in 2hc0 shown as a heat map for (**a**) all fragment pairs *IJ* and (**b**) for a selection of fragments (from 150 to 199).

pressing *2D Map* on the PIEDA panel). Here, dispersive interactions are displayed (the component to plot can be chosen from the PIEDA panel). The rightmost column and the lowest row represent the ligand-residue dispersion interactions. Other data points describe residue-residue interactions. Although such a heat map may not be very relevant for a single protein-ligand complex, it may be useful for understanding conformation stability when comparing different complex structures.

PIEDA/MP2 results are shown in Fig. 9. For many residues the main interaction type is electrostatic (ES), reduced by the solvent screening (es). In some pairs, dispersion (RC+DI) makes a large contribution (aIle1050, etc).

6.3 Partition Analysis

PA results for the ligand treated as one large segment are shown in Fig. 10. The most important residue is aAsp1025, followed by bAsp2125. Both of them are tightly bound to the ligand, so that the interaction is barely screened by the solvent. For other residues the solvent screening (es) cancels off a large part of the solute electrostatic interaction ES. Dispersion plays a big role, in some cases, it is the dominant interaction (aAla1028, bIle2184, etc). The non-electrostatic solvent screening (its nature is exchange-repulsion and dispersion interactions between solute and solvent) is small, as usual (the largest is for bArg2108, −1.3 kcal/mol). Both chains **a** and **b** interact with the ligand with a similar total strength although for chain **a** the interactions are more localized.

PA results for the ligand divided into functional groups are shown in Fig. 11. It is clear that the main hotspot segment is BOH,

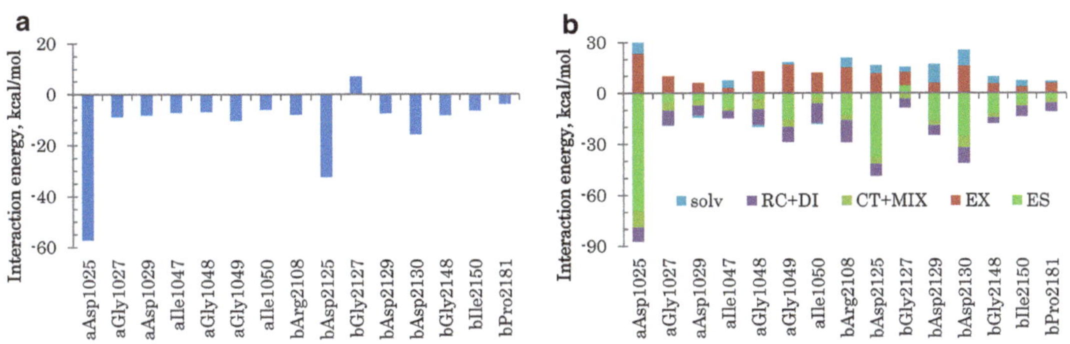

Fig. 9 PIEDA/MP2 interaction energies for main ligand-residue pairs in 2hc0: (**a**) total energies, (**b**) components

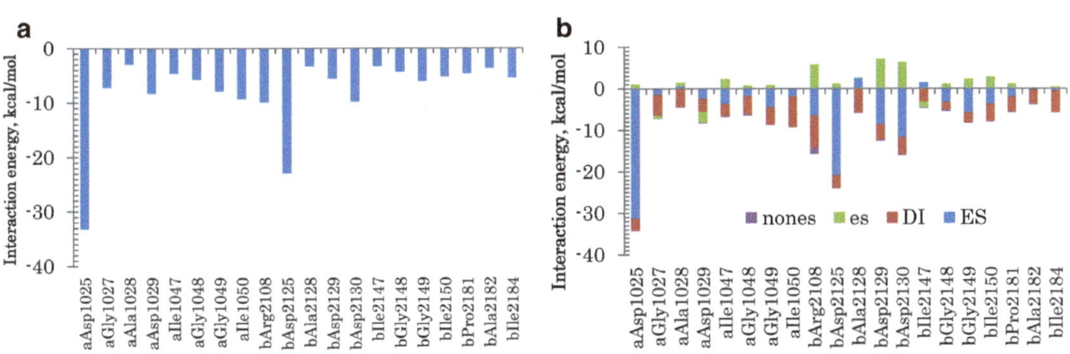

Fig. 10 PA/DFTB3 interaction energies for main ligand-residue pairs in 2hc0: (**a**) total energies, (**b**) components (whole ligand)

presumably because of the OH group in BOH that forms two very strong interactions with aAsp1025 and bAsp2125. The third in importance is the inden-OH segment (Ind), strongly bound to bArg2108. Another important functional group is methoxyphenyl (MeO), bound to bAsp2130. However, every functional group in the ligand is busy being bound to some residue with the energy of at least −3.8 kcal/mol. Because the structure was optimized, there are no repulsions.

6.4 Subsystem Analysis

The results are shown in Table 5. The interaction energies are strongly attractive (exothermic), and the polarization is overall conducive to binding but the desolvation penalty cancels most of the attraction. The total interaction energy is −152.9 kcal/mol, but the binding energy is −28.7 kcal/mol. The neglected deformation energy may further decrease the binding energy.

Detailed contributions of individual residues to binding are shown in Fig. 12. Similarly to PA/DFTB3, aAsp1025, and bAsp2125 have large interaction energies, but they also have large desolvation penalties. The ligand has a remarkably large desolvation penalty, indicating that it is happily solvated in water when isolated, whereas forming a complex results in a large loss of this ligand-water attraction.

One explanation why some polarization energies are negative is that binding the ligand removes some water from the protein surface decreasing the repulsive solvent screening of residue-residue interactions in the protein, thus lowering the energy of the protein in the complex compared to its isolated state in solvation. In other residue pairs, removing the solvent can decrease attractive screening. Also, there is a direct mutual polarization effect of the electronic states of the ligand and protein, which is usually repulsive for complexation in

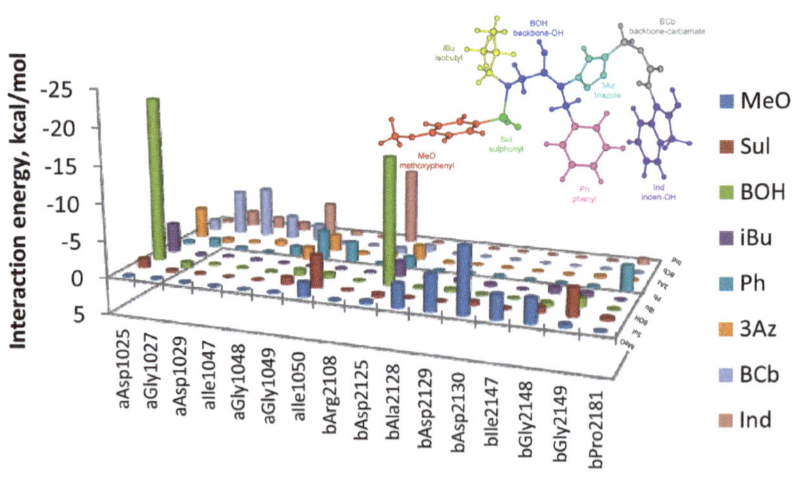

Fig. 11 PA/DFTB3 interaction energies for functional groups in the ligand and important residues (2hc0)

Table 5
Subsystem analysis applied to the binding in 2hc0 at the level of HF-3c (kcal/mol)

Contribution	value
$\Delta E^{PL}_{protein}$	−29.0
ΔE^{PL}_{ligand}	8.2
$\Delta E^{desolv}_{protein}$	84.7
$\Delta E^{desolv}_{ligand}$	60.3
$\Delta E^{INT}_{protein-ligand}$	−152.9
ΔE^{bind}	−28.7

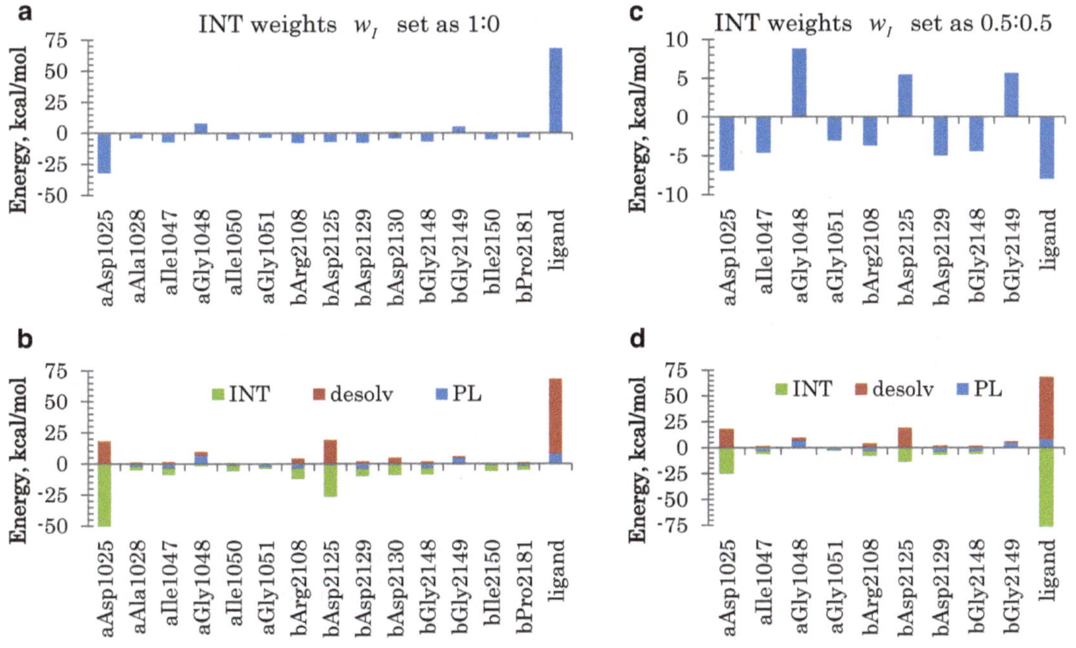

Fig. 12 Components in the subsystem analysis at the level of HF-3c (2hc0): (**a**) total for 1:0, (**b**) components for 1:0, (**c**) total for 0.5:0.5, and (**d**) components for 0.5:0.5

vacuum, but in solution it may have both signs. The PL components describe all these effects, which can be decomposed if desired [48].

AB interaction energies can be split between *A* and *B* in two ways, 1:0 and 0.5:0.5 (see Eqs. 19 and 20) in the definition of ΔE_I^{BIND}. For the former setup, the ligand features a large binding energy arising from its desolvation penalty. For the latter, one half of the residue-ligand interaction energies is added to ΔE_I^{BIND} as the INT term, largely canceling out the desolvation. It is natural that

INT and desolv components cancel out—their sum describes the replacement of solute-solvent interactions in the isolated state by solute-solute (e.g., protein-ligand) interactions in the complex, so that the binding is a differential process rather than an addition.

The 0.5:0.5 division results in much more tractable contributions of residues on the order of 10 kcal/mol. Some residues such as aGly1048a, bAsp2125, and bGly2149 act as *anti-hotspots*, meaning that they prefer solvent to the company of the ligand. Two ways to improve the drug efficacy emerge: **increasing the affection of hotspots and reducing the aversion of anti-hotspots**.

6.5 Fluctuation Analysis

Fragment temperatures are shown in Fig. 13. They represent the kinetic energy possessed by atoms in a given residue. There is a substantial variation: some residues are hot ($T_I > 350$ K) and some are cold ($T_I < 260$ K). The temperature of residues in the active site may be related to the retention time of the ligand.

Interaction energies at the global minimum in MD and averaged fluctuations are shown in Fig. 14. Compared to the optimized structure (starting from the X-ray structure obtained for a crystal), the

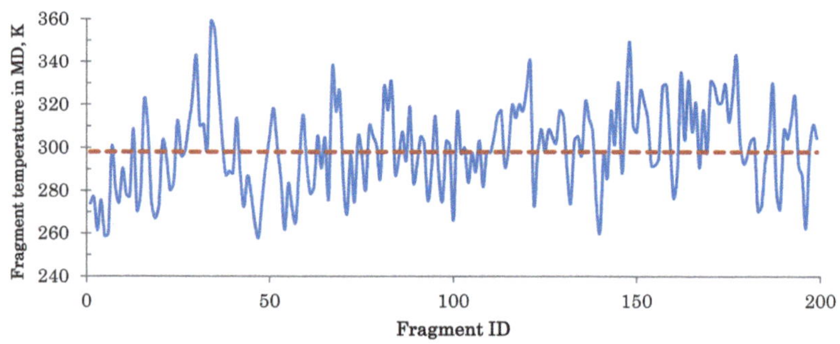

Fig. 13 Fragment temperatures T_I in 2hc0 from FMO2-DFTB3 NVT MD. The global temperature of 298 K is shown as the red dashed line

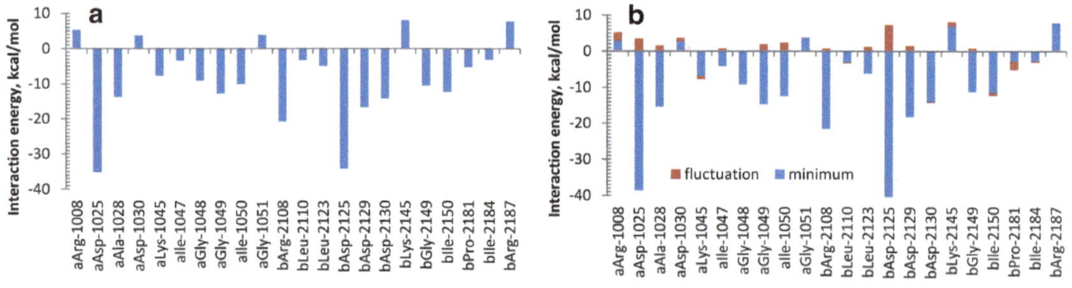

Fig. 14 Ligand-residue interaction energies in 2hc0 from FMO2-DFTB3 NVT MD at 298 K: **(a)** $\langle \Delta E_{IJ} \rangle$ and **(b)** their decomposition into ΔE_{IJ}^0 (minimum) and $\Delta\Delta E_{IJ}$ (fluctuation)

interaction energy to bAsp2125 is higher, making the two Asp residues of chains **a** (aAs1025) and **b** (bAsp2125) almost equally important. Thermal fluctuations, as usual, typically weaken binding (the contributions $\langle \Delta\Delta E_{IJ} \rangle$ are positive), because the ligand and residues adopt less favorable orientations due to thermal fluctuations.

7 Comparative Remarks

In this chapter, various QM methods are used, each with its own strength and weakness. Let us compare PIEDA/MP2 (Fig. 9a) with PA/DFTB (Fig. 10a) and SA/HF-3c (Fig. 12c). Conceptually, PIEDA and PA are similar. SA, however, stands quite apart from PIEDA and PA, because SA deals with binding energies compared to interactions in the other two.

In the MP2 plot, there is a small repulsion (bGly2127) missing in DFTB. This may be because the geometry is optimized with DFTB, removing repulsions at the DFTB level. Overall, there is a very good agreement between MP2 and DFTB, especially among all principal hotspots, with a few minor spots different. Overall, MP2 has larger PIEs, in agreement with the general trend [53].

SA, however, has a very different picture, probably more realistic in terms of relating to experiment because desolvation (missing in PIEDA and PA) is an important factor that removes the overestimation of energies. Among the residues, the main hotspot (aAsp1025) is consistent with PA and PIEDA. One striking difference of SA (0.5:0.5) and the other methods is that bAsp2125 is repulsive (an anti-hotspot), because its desolvation penalty is larger than its attraction to the ligand (counted as 1/2, with the remaining half count as the ligand binding energy).

8 Conclusions and Outlook

The recent development of the FMO functionality in Facio has been summarized, describing many new features that have recently become available, such as the support of segments, and various formulations of energy decompositions using the fragment molecular orbital method.

A summary of the FMO method is provided for completeness, followed by a description of how to use Facio to prepare input files and visualize the results. All calculations are done on the protein-ligand complex 2hc0. It is shown that, although pair decomposition energies can pinpoint the hotspots, the absolute values may be overestimated, and a way to correct them is paved by the subsystem analysis. A concept of anti-hotspots was introduced denoting residues that like the solvent more than the ligand. Temperature may also be a factor not to be neglected, in particular, as a way to locate a low-energy structure, denoted here as the global minimum in the MD trajectory.

Because full QM-based MD simulations of biochemical size (several thousands of atoms) are still somewhat expensive and a meager 40 ps MD simulation is reported here (3214 QM atoms), barely enough to probe the vicinity of one conformation, although 1–2 ns MD could be done for smaller systems (~300 atoms) [54]. On the other hand, geometry optimizations of protein-ligand complexes can be done instead of MD, with a reasonable cost (~10 h on 1 node here).

It is hoped that in the future, FMO/MD can be accelerated into a production level of at least 1 ns for complete protein-ligand complexes. For more reliable energy evaluations, coupled cluster [55] and quantum computing [56, 57] may be used. Another desirable feature is the inclusion of entropy, which can in principle be done with the partition analysis of vibrational energies (PAVE) [58] or free energy decomposition analysis (FEDA) [59], not covered here.

References

1. Manathunga M, Götz AW, Merz KM Jr (2022) Computer-aided drug design, quantum-mechanical methods for biological problems. Curr Opin Str Biol 5:102417
2. Gupta R, Srivastava D, Sahu M, Tiwari S, Ambasta RK, Kumar P (2021) Artificial intelligence to deep learning: machine intelligence approach for drug discovery. Mol Divers 25:1315–1360
3. Gordon MS, Pruitt SR, Fedorov DG, Slipchenko LV (2012) Fragmentation methods: a route to accurate calculations on large systems. Chem Rev 112:632–672
4. Kussmann J, Beer M, Ochsenfeld C (2013) Linear-scaling self-consistent field methods for large molecules. WIREs: Comput Mol Sci 3:614–636
5. Akimov AV, Prezhdo OV (2015) Large-scale computations in chemistry: a bird's eye view of a vibrant field. Chem Rev 115:5797–5890
6. Liu J, He X (2023) Recent advances in quantum fragmentation approaches to complex molecular and condensed-phase systems. WIREs Comput Mol Sci 13:e1650
7. Kitaura K, Ikeo E, Asada T, Nakano T, Uebayasi M (1999) Fragment molecular orbital method: an approximate computational method for large molecules. Chem Phys Lett 313:701–706
8. Fedorov DG (2017) The fragment molecular orbital method: theoretical development, implementation in GAMESS, and applications. WIREs. Comput Mol Sci 7:e1322
9. Fukuzawa K, Tanaka S (2022) Fragment molecular orbital calculations for biomolecules. Curr Opin Struct Biol 72:127–134
10. Xu W, Kang C (2025) Fragment-based drug design: from then until now, and toward the future. J Med Chem 68:5000–5004
11. Komeiji Y, Inadomi Y, Nakano T (2004) PEACH 4 with ABINIT-MP: a general platform for classical and quantum simulations of biological molecules. Comput Biol Chem 28:155–161
12. Ishikawa T, Ishikura T, Kuwata K (2009) Theoretical study of the prion protein based on the fragment molecular orbital method. J Comput Chem 30:2594–2601
13. Takami T, Maki J, Ooba J, Inadomi Y, Honda H, Susukita R, Inoue K, Kobayashi T, Nogita R, Aoyagi M (2007) Multi-physics extension of OpenFMO framework. AIP Conf Proc 963:122–125
14. Barca GMJ, Bertoni C, Carrington L, Datta D, Silva ND, Deustua JE, Fedorov DG, Gour JR, Gunina AO, Guidez E et al (2020) Recent developments in the general atomic and molecular electronic structure system. J Chem Phys 152:154102
15. Ibrahim GFPE, Zuccotto F, Zachariae U, Gilbert I, Bodkin M (2024) Accurate prediction of dynamic protein-ligand binding using P-score ranking. J Comput Chem 45:1762–1778
16. Handa Y, Okuwaki K, Kawashima Y, Hatada R, Mochizuki Y, Komeiji Y, Tanaka S, Furuishi T, Yonemochi E, Honma T, Fukuzawa K (2024) Prediction of binding pose and affinity of nelfinavir, a SARS-CoV-2 main protease repositioned drug, by combining docking, molecular dynamics, and fragment molecular orbital calculations. J Phys Chem B 128:2249–2265

17. Chuntakaruk H, Hengphasatporn K, Shigeta Y, Aonbangkhen C, Lee VS, Khotavivattana T, Rungrotmongkol T, Hannongbua S (2024) FMO-guided design of darunavir analogs as HIV-1 protease inhibitors. SC Rep 14:3639
18. Kato A BioStation viewer. https://fmodd.jp/biostationviewer-dl/. Accessed 25 Mar 2025
19. PAICS View. http://www.paics.net/paics_view_e.html. Accessed 25 Mar 2025
20. Fedorov DG, Kitaura K (2017) Modeling and visualization for the fragment molecular orbital method with the graphical user interface FU, and analysis of protein-ligand binding. In: Gordon MS (ed) Fragmentation: toward accurate calculations on complex molecular systems. Wiley, Hoboken, pp 119–139
21. Suenaga M (2005) Facio: new computational chemistry environment for PC GAMESS. J Comput Chem Jpn 4:25–32
22. Suenaga M (2008) Development of GUI for GAMESS/FMO calculation. J Comput Chem Jpn 7:33–54
23. Steinmann C, Ibsen MW, Hansen AS, Jensen JH (2012) FragIt: a tool to prepare input files for fragment based quantum chemical calculations. PLoS One 7:e44480
24. Sladek V, Yamamoto Y, Harada R, Shoji M, Shigeta Y, Sladek V (2021) pyProGA – a PyMOL plugin for protein residue network analysis. PLoS One 16:1–17
25. Fedorov DG, Inostroza D, Courbiere B, Guegan F, Contreras-García J, Mori S (2025) Decomposition analysis for visualization of non-covalent interactions based on the fragment molecular orbital method. J Chem Theory Comput 21:4435–4446
26. Mochizuki, Y, Tanaka, S, Fukuzawa, K, 2021. Recent advances of the fragment molecular orbital method. Springer, Singapore
27. Fedorov DG (2024) Quantum-chemical analyses of interactions for biochemical applications. In: Poongavanam V, Ramaswamy V (eds.). Computational drug discovery: methods and applications. Wiley, pp 183–210
28. Heifetz A (2020) Quantum mechanics in drug discovery, Methods in molecular biology, vol 2114. Springer, New York
29. Fedorov DG (2023) Complete guide to the fragment molecular orbital method in GAMESS. World Scientific, Singapore
30. Fedorov DG, Asada N, Nakanishi I, Kitaura K (2014) The use of many-body expansions and geometry optimizations in fragment-based methods. Acc Chem Res 47:2846–2856
31. Fedorov DG, Kitaura K (2014) Use of an auxiliary basis set to describe the polarization in the fragment molecular orbital method. Chem Phys Lett 597:99–105
32. Fedorov DG (2024) Use of caps in the auxiliary basis set formulation of the fragment molecular orbital method. J Comput Chem 45:1540–1551
33. Fedorov DG (2025) Importance of charge balance for the embedding of zwitterionic solutes in the fragment molecular orbital method. J Phys Chem A 129:2325–2336
34. Fedorov DG, Kitaura K (2007) Pair interaction energy decomposition analysis. J Comput Chem 28:222–237
35. Fedorov DG (2020) Three-body energy decomposition analysis based on the fragment molecular orbital method. J Phys Chem A 124:4956–4971
36. Fedorov DG, Kitaura K (2018) Pair interaction energy decomposition analysis for density functional theory and density-functional tight-binding with an evaluation of energy fluctuations in molecular dynamics. J Phys Chem A 122:1781–1795
37. Fedorov DG, Kitaura K (2012) Energy decomposition analysis in solution based on the fragment molecular orbital method. J Phys Chem A 116:704–719
38. Fedorov DG (2019) Solvent screening in zwitterions analyzed with the fragment molecular orbital method. J Chem Theory Comput 15:5404–5416
39. Fedorov DG (2025) Polarization and basis set superposition error in interaction energies in the fragment molecular orbital method. J Chem Phys 163:144108
40. Green MC, Fedorov DG, Kitaura K, Francisco JS, Slipchenko LV (2013) Open-shell pair interaction energy decomposition analysis (PIEDA): Formulation and application to the hydrogen abstraction in tripeptides. J Chem Phys 138:074111
41. Fedorov DG (2025) Analyzing many-body charge transfer effects with the fragment molecular orbital method. J Comput Chem 46:e70128
42. Fedorov DG, Kromann JC, Jensen JH (2018) Empirical corrections and pair interaction energies in the fragment molecular orbital method. Chem Phys Lett 706:328–333
43. Fedorov DG (2020) Partition analysis for density-functional tight-binding. J Phys Chem A 124:10346–10358
44. Fedorov DG (2024) Partition analysis of dipole moments in solution applied to functional groups in polypeptide motifs. Phys Chem Chem Phys 26:18614–18628
45. Sladek V, Fedorov DG (2022) The importance of charge transfer and solvent screening in the interactions of backbones and functional groups in amino acid residues and nucleotides. Int J Mol Sci 23:13514

46. Sladek V, Artiushenko PV, Fedorov DG (2024) Effect of direct and water-mediated interactions on the identiffcation of hotspots in biomolecular complexes with multiple subsystems. J Chem Inf Model 64:7602–7615
47. Fedorov DG, Kitaura K (2016) Subsystem analysis for the fragment molecular orbital method and its application to protein-ligand binding in solution. J Phys Chem A 120:2218–2231
48. Fedorov DG, Kitaura K (2004) The importance of three-body terms in the fragment molecular orbital method. J Chem Phys 120:6832–6840
49. Fedorov DG, Olson RM, Kitaura K, Gordon MS, Koseki S (2004) A new hierarchical parallelization scheme: generalized distributed data interface (GDDI), and an application to the fragment molecular orbital method (FMO). J Comput Chem 25:872–880
50. Fedorov DG, Pham BQ (2023) Multi-level parallelization of quantum-chemical calculations. J Chem Phys 158:164106
51. Gaus M, Goez A, Elstner M (2013) Parametrization and benchmark of DFTB3 for organic molecules. J Chem Theory Comput 9:338–354
52. Nagata T, Fedorov DG, Li H, Kitaura K (2012) Analytic gradient for second order Møller-Plesset perturbation theory with the polarizable continuum model based on the fragment molecular orbital method. J Chem Phys 136:204112
53. Morao I, Fedorov DG, Robinson R, Southey M, Townsend-Nicholson A, Bodkin MJ, Heifetz A (2017) Rapid and accurate assessment of GPCR-ligand interactions using the fragment molecular orbital-based density-functional tight-binding method. J Comput Chem 38:1987–1990
54. Vuong VQ, Nishimoto Y, Fedorov DG, Sumpter BG, Niehaus TA, Irle S (2019) The fragment molecular orbital method based on long-range corrected density-functional tight-binding. J Chem Theory Comput 15:3008–3020
55. Fedorov DG, Kitaura K (2005) Coupled-cluster theory based upon the fragment molecular-orbital method. J Chem Phys 123:134103
56. Lim H, Kang DH, Kim J, Pellow-Jarman A, Mcfarthing S, Pellow-Jarman R, Jeon HN, Oh B, Rhee JKK, No KT (2024) Fragment molecular orbital-based variational quantum eigensolver for quantum chemistry in the age of quantum computing. SC Rep 14:2422
57. Sugisaki K, Nakano T, Mochizuki Y (2024) Size-consistency and orbital-invariance issues revealed by VQE-UCCSD calculations with the FMO scheme. J Comput Chem 45:2204–2213
58. Fedorov DG (2021) Partitioning of the vibrational free energy. J Phys Chem Lett 12:6628–6633
59. Fedorov DG, Nakamura T (2022) Free energy decomposition analysis based on the fragment molecular orbital method. J Phys Chem Lett 13:1596–1601

Check for updates

Chapter 2

Perspectives on Hybridizing Quantum Mechanics and Artificial Intelligence for Drug Design

Alexander Heifetz, Girinath G. Pillai, Mussa Quareshy, Tarun Jain, Louise Birch and Colin Sambrook Smith

Abstract

Drug design is one of the most resource-intensive and failure-prone stages of pharmaceutical R&D. This chapter presents the perspective of computational chemists at Sygnature Discovery on the growing need to integrate quantum mechanics (QM) with artificial intelligence (AI) to improve the efficiency and effectiveness of modern drug design. Published retrospective analyses and case studies indicate that programs supported by computational chemistry can progress more efficiently through the design phase while reducing experimental burden, including synthetic effort. Structure-based drug design (SBDD) remains the dominant computational approach; however, its accuracy is constrained by simplified scoring functions. QM, particularly the Fragment Molecular Orbital (FMO) method, addresses some of these limitations by providing residue-level, experimentally aligned energetic insight. AI methods have expanded rapidly but remain constrained by data quality, interpretability, and the practical challenge of experimentally validating large numbers of generated designs. In this chapter, we discuss the perspectives and opportunities of combining AI and QM within SBDD to create more predictive, efficient, and mechanistically informed drug design workflows.

Keywords Computational chemistry, Drug design, Quantum mechanics, Fragment molecular orbital, Structure-based drug design, Artificial intelligence, Machine learning, Hit-to-lead, Lead optimization, Design–Make–Test–Analyze cycle, High-performance computing, Hybrid AI–QM workflows

1 Introduction

This chapter presents the authors' perspectives on the growing need to integrate artificial intelligence (AI) with quantum mechanics (QM) within computational chemistry (CC) to advance modern drug design [1]. Drug design is one of the most critical and resource-intensive stages of the pharmaceutical pipeline [2–5], focused on optimizing small molecules or biologics for potency, selectivity, pharmacokinetics, and safety. Unlike early discovery, which aims to identify and validate new therapeutic targets, drug design refines the structures

Alexander Heifetz (ed.), *Quantum Mechanics Applications in Drug Design*, Methods in Molecular Biology, vol. 3061, https://doi.org/10.1007/978-1-0716-5420-0_2,

of candidate molecules to maximize therapeutic benefit while minimizing risk [4].

Computational chemistry methods are now widely recognized as an integral component of modern drug design [1, 6–10]. Numerous retrospective and prospective studies have demonstrated that the use of CC can improve decision-making, reduce attrition, and increase efficiency across hit-to-lead (H2L) and lead optimization (LO) stages [1, 6–13]. Although much industrial work remains confidential, trends reported in the public literature consistently support the conclusion that computational support enhances progression through the design phase and reduces experimental burden, contributing to lower costs and shorter development timelines [1, 6–10, 12, 13].

Structure-based drug design (SBDD) remains the dominant computational approach and is frequently associated with the most consistent success in progressing compounds beyond the design stage [14, 15]. However, classical SBDD relies on simplified scoring functions that often fail to capture important enthalpic and entropic contributions to binding [14]. Its performance can be substantially enhanced through the use of QM methods, such as the Fragment Molecular Orbital (FMO) approach, which provides near experiment-level accuracy and enables residue-level pair-interaction energy decomposition [16–21]. By overcoming many of the intrinsic limitations of conventional scoring functions, FMO offers deeper mechanistic insight into molecular recognition.

To support client projects, Sygnature Discovery has developed its own FMO-based platform, FMO-SP, which delivers residue-resolved quantitative information on interaction energies (in kcal/mol), the chemical nature of interactions (e.g., electrostatic, dispersion, or hydrophobic contributions), and key energetic terms such as desolvation and depolarization. As described in Chaps. 4 and 6 of this volume, the integration of SBDD with FMO represents a practical and highly efficient strategy for guiding medicinal chemistry decisions.

Ligand-based drug design (LBDD) [22–27] represents the second most widely used computational paradigm. LBDD methods can be highly effective when rich and reliable structure–activity relationship (SAR) data are available, but their applicability is limited in situations where chemical space is sparsely explored or where extrapolation beyond known chemistry is required.

AI and data-driven methods [28–32] have expanded rapidly over the past decade, evolving from a relatively minor role to a substantial component of contemporary drug design workflows [33, 34]. These methods are now applied across many stages of the pipeline, including target identification, virtual screening, lead optimization, ADMET prediction, repurposing strategies, and selected aspects of clinical trial design [35]. Despite this rapid growth, AI has not yet fundamentally transformed drug design practice [30, 32, 36]. Progress remains constrained by data quality and availability, the scarcity of well-curated negative data, bias and noise in training sets, limited interpretability of black-box models, and ongoing regulatory uncertainties.

In practice, the impact of AI-driven approaches is further limited by synthetic feasibility and the finite number of compounds that can be synthesized and experimentally evaluated. These constraints help explain why reported success rates remain below those achieved by more established physics-based methods [28, 30, 32]. Much of the recent increase in AI-focused publications, therefore, reflects methodological innovation and proof-of-concept studies rather than widespread industrial deployment [28–33, 36–38]. While a small number of AI-designed or AI-repurposed molecules have entered clinical development [34], broader impact will require improved data standards, mechanistically informed and multimodal models, and deeper integration of AI with experimental and physics-based methodologies [33].

Hybrid workflows that combine SBDD, LBDD, QM, and AI are now beginning to emerge in the literature [39–44]. Although these approaches currently represent a smaller fraction of reported studies, they consistently demonstrate improvements in efficiency, robustness, and predictive power and are increasingly recognized as a promising direction for the field. This chapter focuses on one particularly important aspect of this hybridization: the integration of AI and QM within the Design–Make–Test–Analyze (DMTA) cycle and, potentially, in virtual screening workflows.

2 The Role of Quantum Mechanics in Drug Design

QM represents the accuracy benchmark of computational chemistry because, unlike molecular mechanics (MM), it models electrons explicitly and therefore captures the fundamental drivers of molecular recognition that force-field (FF) methods often miss [16, 17, 20, 21, 45]. Even when high-resolution crystal structures are available, classical MM methods often fail to capture the full complexity of protein–ligand interactions. This makes QM a powerful and reliable basis for understanding structure–activity relationships (SAR), although its higher computational cost traditionally limited its routine application. The FMO method overcomes this barrier by dividing large biomolecular complexes into fragments and performing QM calculations on each fragment and their interactions. This strategy enables near-linear scaling, provides residue-level pair-interaction energies (PIEs) and pair interaction energy decomposition analysis (PIEDA) and delivers QM precision within a SBDD framework [46–50]. FMO is implemented in several widely used QM software platforms, including GAMESS and ABINIT-MP, which have helped establish QM as a practical tool in both academia and industry [46–50].

QM now plays a targeted but essential role in drug design, particularly where classical force fields struggle, such as charged or highly polar binding sites, metal coordination, covalent mechanisms, atypical protonation states, and subtle SAR driven by polarization or charge transfer. Production-level QM calculations for binding-site analysis are typically performed with tools such as Gaussian [51], Q-Chem [52], ORCA [53], Jaguar [54], and Spartan [55], which provide robust single-point energies, geometry optimizations, and conformational analysis, pK_a/tautomer evaluation, and electrostatic potential maps. These tools help correct docking poses and refine scoring in systems that are difficult for molecular mechanics. QM/MM platforms [54], including QSite and Jaguar QM/MM, are routinely used to model covalent inhibitors and enzymatic reaction mechanisms, capturing bond-making and bond-breaking events at the active site while keeping the rest of the system at MM resolution to maintain computational tractability. For conformational analysis, dihedral scans and quantification of bioactive strain energy, workflows based on MacroModel combined with Jaguar QM refinements, MOE torsional exploration with local DFT minimization [56] and Spartan's conformational and energy-profile tools are widely used to map torsional landscapes and estimate the energetic penalty of adopting the bioactive conformation. Together, these QM-based methods address specific weaknesses of classical approaches and are increasingly integrated with SBDD and AI to deliver physically grounded, mechanistically interpretable guidance to medicinal chemistry.

Compared with free-energy perturbation approaches (FEP/FEP+), which offer predictive estimates of relative binding free energies but remain constrained by FF accuracy and limited interpretability, FMO provides mechanistic, residue-level insight into the electrostatics, polarization and charge-transfer components of binding. FEP predicts, while FMO explains, and the two methods together form a powerful dual strategy for H2L and LO stages of drug design. In contrast, MM methods are fast and well suited for exploring conformational dynamics and entropic effects, but their accuracy is limited by the underlying force fields. QM/MM hybrids offer a practical compromise, enabling reactive or electronically complex regions to be treated quantum mechanically, while keeping the remainder of the system computationally efficient.

The overarching message is that no single computational method is sufficient for modern drug design. QM and FMO deliver accuracy and mechanistic detail; MM and FEP provide speed and optimization. When combined thoughtfully, these approaches form a complementary and highly effective toolkit for guiding SBDD.

3 The Role of AI in Drug Design

In the last decade, AI has moved rapidly from hype to practical application in drug design [28–32, 35–38, 57]. Although its standalone impact remains modest, AI consistently adds value across several areas, particularly in property prediction, generative molecular design and accelerating the DMTA cycle. AI is not a replacement for physics-based methods; rather, it complements them by providing scale and speed, while methods such as QM deliver accuracy and mechanistic insight.

In current practice, AI is most widely used for predicting molecular properties through QSAR models, machine-learning (ML) regressors, and deep learning (DL) architectures; for generative design through techniques such as variational autoencoders, generative adversarial networks, and reinforcement learning and graph-based models; and for profiling ADMET and drug-likeness. AI also enables more efficient DMTA optimization by using uncertainty estimates, active learning, and Bayesian optimization to prioritize experiments and reduce iteration time. A growing ecosystem of platforms, including Makya (Iktos), REINVENT (AstraZeneca) [58], Chemprop, DeepChem [59], AutoQSAR [60], Chemistry42 [61], and Schrödinger's generative modeling tools, is becoming increasingly embedded in pharmaceutical pipelines. However, their effectiveness in real-world projects is still constrained by synthetic feasibility and the limited number of compounds that can be experimentally tested [33].

AlphaFold [62–64] has transformed SBDD by providing high-quality protein structure prediction for targets that lack experimental data. The AlphaFold2 architecture uses DL on multiple sequence alignments and evolutionary information to predict 3D protein structures with near experimental accuracy for many domains, and large public resources, such as the AlphaFold Protein Structure Database [62], now cover most of the human proteome and many key drug targets.

Despite its promise, AI faces several well-recognized challenges. Industrial datasets are often noisy, sparse, or biased, which limits the robustness and transferability of models. Many modern architectures remain difficult to interpret, making it unclear why a model recommends a particular molecule.

AI can generate far more designs than any organization can realistically synthesize and test, making it essential to prioritize candidates using physics-based methods. Current AI workflows often incorporate synthetic-feasibility filters and physicochemical property predictors, which help identify molecules that can be made in the laboratory [33]. However, these filters provide little insight into how well such molecules are likely to bind or interact with the target protein. To address this gap, docking combined with quantum-mechanical methods, such as FMO, can re-score and refine AI-generated designs, enabling the selection of candidates with both favorable interaction energetics and realistic synthetic accessibility. This reinforces the importance of integrating AI with physics-based approaches, such as QM and SBDD, to create predictive, experimentally grounded workflows.

4 AI for Drug Design at Sygnature Discovery

Sygnature Discovery has integrated generative AI (GenAI) and machine learning (ML) into its discovery workflows to accelerate the design and optimization of high-quality lead compounds (for more details, see Sygnature's website: https://www.sygnaturediscovery.com). AI now supports multiple stages of drug discovery, from target assessment and hit generation to lead optimization and translational design. Modern platforms combine generative models for molecule creation with predictive ML for potency and ADMET, bioinformatics and LLMs for target evaluation, and structure prediction tools to guide mechanism-based design. The greatest impact, however, arises when AI is embedded within a human-in-the-loop (HITL) framework, where domain expertise, medicinal chemistry judgment and contextual feedback steer algorithms toward solutions that perform robustly across the DMTA cycle. This section outlines Sygnature's AI-enabled workflows, supported by practical case studies and aligned with current best practice.

4.1 Generative AI for De Novo Design and Lead Optimization

Diffusion-, transformer-, VAE-, and GAN-based generative models are used to propose novel chemical structures and guide optimization through reinforcement learning or multi-objective strategies. At Sygnature, platforms such as SygDesign, Makya (Iktos), and REINVENT are routinely applied for core hopping, linker design, macrocyclization and library enumeration. These models interface with potency and ADMET predictors to address multi-parameter optimization and align designs with Target Product Profiles.

4.2 Predictive ML Models (QSAR/QSPR) in Closed-Loop DMTA

Predictive models built on Sygnature's proprietary project data, augmented with selected public-domain datasets, provide potency and ADMET estimates for properties such as solubility, clearance, and CNS penetration. Integration via KNIME hub and uncertainty-aware active learning enables rapid refinement of chemical ideas, faster DMTA cycles, and reduced experimental burden.

4.3 Case Study: AI-Enabled CNS Drug Design

4.3.1 Phase 1: Generative Ideation at Scale

In an AI-driven CNS discovery program (Fig. 1), Makya and REINVENT were used to explore chemical space and propose diverse scaffolds through controlled substitution and core diversification. This step produced more than 27,000 chemically realistic designs within a few hours, establishing a broad foundation for data-driven prioritization.

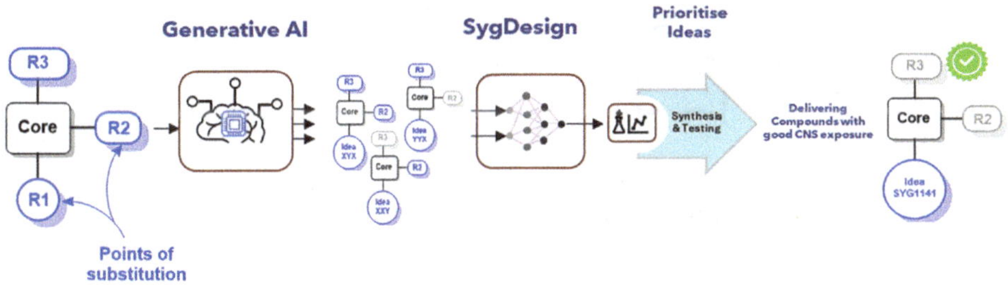

Fig. 1 An illustration of Sygnature's GenAI workflow, where Makya and REINVENT explore chemical space to generate diverse scaffolds and substitution patterns (R1–R3). More than 27,000 novel ideas can be generated within hours

4.3.2 Phase 2: Data-Guided Design and Prioritization

Sygnature's SygDesign platform evaluated the virtual library using predictive models trained on proprietary and public data to assess potency, ADME, and CNS exposure (Fig. 2). Multi-parameter optimization condensed the 27,000 designs into a few hundred high-value candidates. Explainable ML ensured that selected compounds aligned with project objectives and balanced potency, metabolic stability, permeability, and CNS penetration.

4.3.3 Experimental Validation

Shortlisted candidates progressed to synthesis and experimental evaluation. Unlike traditional high-throughput strategies, only a small number of compounds were synthesized, each demonstrating high CNS exposure and favorable ADME characteristics. The final optimized molecule achieved the desired affinity and pharmacokinetic profile, confirmed through in vivo studies (Fig. 3).

4.4 Deliverables

This integrated workflow delivered several concrete benefits. First, the generative models produced thousands of design ideas within hours, providing a broad and chemically diverse starting point. Second, ML-driven prioritization and careful triage led to a substantial reduction in synthesis workload while simultaneously improving CNS exposure outcomes. Third, the seamless linkage of generative AI, ML-based ranking, and experimental validation ensured that only the most promising candidates advanced to the laboratory. As a result, the program delivered a validated CNS-active compound more quickly and with fewer resources than would have been possible using conventional approaches.

Sygnature Discovery's integration of generative AI platforms (Makya, REINVENT), bespoke machine-learning models deployed through KNIME hub, LLM-enabled bioinformatics workflows, and AlphaFold/OmegaFold-based structural prediction provides a practical example of this end-to-end, human-in-the-loop design philosophy. This approach is already influencing real projects, enabling faster, more informed, and mechanistically grounded drug design.

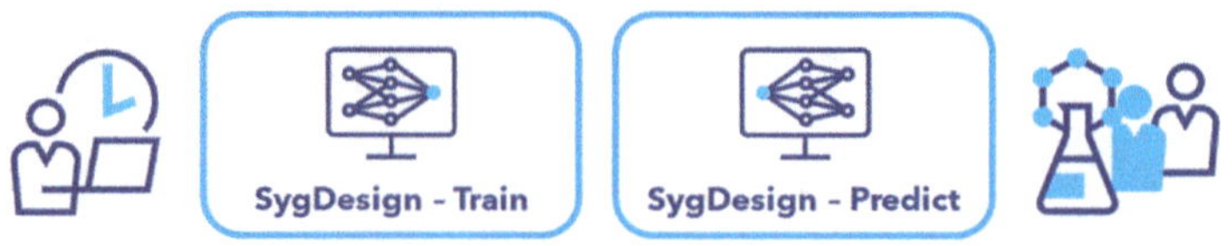

Fig. 2 This image shows the data-guided design process implemented through the SygDesign platform

Fig. 3 A presentation of the full workflow from AI-based design to in vivo confirmation of CNS penetration

5 Integrating AI and QM (AI-QM)

The most powerful advances in computational drug design are emerging from the integration of AI and QM [34, 65]. AI provides breadth by rapidly exploring vast regions of chemical space, while QM contributes depth through experiment-like accuracy and mechanistic insight. Together, these complementary capabilities have begun to reshape the DMTA cycle, allowing ideation at scale and validation at a level of physical realism previously unattainable.

The integration of AI within SBDD has already shown promising results. One notable example is Pocket2Mol [66], a geometry-driven generative framework that constructs ligand structures directly from the three-dimensional shape and chemical features of a protein binding pocket. Pocket2Mol uses graph neural networks and diffusion-based sampling to generate molecules that match the pocket topology, recover key pharmacophoric elements, and provide scaffold diversity beyond traditional library-based approaches. Despite these advantages, pocket-based generative models still depend on simplified scoring functions or surrogate machine-learning predictors, which limit their ability to capture complex interaction energetics and to discriminate between physically plausible and unstable designs.

QM methods, particularly FMO, can overcome these limitations by re-scoring and ranking Pocket2Mol or docking-generated protein–ligand complexes using residue-level QM decomposition. This enables early identification of the most energetically favorable and synthetically tractable candidates for progression. Such AI-QM hybrid workflows retain the creative breadth of generative AI while grounding decision-making in physics-based accuracy, making them one of the most promising directions for future SBDD.

Integration between AlphaFold and FMO can be a highly powerful addition to modern drug-design workflows (see Chap. 20 for more details). AlphaFold has transformed structural biology by enabling accurate prediction of protein 3D structures directly from sequence, including previously intractable targets with no experimental data. These AI-generated models provide reliable starting points for downstream tasks, such as binding-site assessment, docking, and molecular-dynamics refinement. The combination with FMO is particularly impactful: AlphaFold supplies high-quality structural frameworks and putative binding pockets, while FMO delivers residue-level pair-interaction energies (PIEs) that highlight key stabilizing contacts, energetic hot spots and opportunities for ligand optimization. In practice, AlphaFold structures can be docked with candidate ligands and then evaluated with FMO to quantify PIEs patterns that classical scoring functions often fail to capture. This integration enables SBDD on targets that previously lacked structural information and represents a natural interface between AI-driven structure prediction and QM-driven mechanistic insight in next-generation drug-design pipelines.

An increasing number of studies now provide concrete examples of successful integration of AI with QM for drug design [67–71]. One example is the use of symmetry-adapted atomic-orbital features derived from low-cost semiempirical QM calculations, which are then supplied to machine-learning models trained to reproduce density functional theory (DFT) energies for drug-like molecules. These hybrid models reach near DFT-level accuracy at a fraction of the computational cost and can, therefore, support large-scale virtual screening and scoring. Similar approaches have been applied in property prediction, where QM-derived descriptors [72, 73] such as partial charges [69], orbital energies and DFTB components improve accuracy and interpretability compared with models built only on classical descriptors. Together, these examples illustrate how QM information can be used to inform, strengthen and regularize AI models within prospective drug design workflows.

An additional complementary direction is the development of QM-based ML force fields [74], where AI models are trained directly on large quantum chemical datasets and then used as fast surrogates for QM in design workflows. Neural-network potentials such as ANI-1 [75], OrbNet [76], and related machine-learned force fields can deliver DFT-like accuracy at close to MM computational cost, enabling routine use of QM-level energetics for conformational sampling, molecular dynamics, and scoring within AI-driven design pipelines. Commercial platforms are also beginning to industrialize these methods. Rowan's quantum-accurate neural network potentials focus on bio-relevant chemistry, and Pharmacelera's QM-based 3D descriptors [65] provide enhanced ligand profiling and virtual screening. These QM-informed FF offer a practical route to embedding physics-based accuracy inside scalable AI engines for structure-based design, property prediction, and virtual screening.

Although QM excels in accuracy and interpretability, it remains relatively slow and unsuitable for large-scale library screening. AI addresses this limitation by enabling fast, large-scale exploration and property prediction. Conversely, AI alone often produces far more designs than can be synthesized or experimentally evaluated. QM can bridge this gap by filtering AI-generated structures, enriching training sets with physics-based descriptors and reducing initial design libraries to a tractable number of high-quality candidates. Without QM, AI risks overwhelming experimental capacity; without AI, QM lacks scalability. The synergy of the two, therefore, transforms AI from an idea generator into a practical, experiment-aligned design engine.

Real-world use highlights several lessons. AI models must be integrated directly into DMTA workflows, with QM validating their outputs and improving interpretability [28–32, 36–38]. Data quality remains critical, and QM-derived descriptors can strengthen biased or sparse datasets. Iterative learning loops, where QM refines AI predictions and AI guides subsequent QM evaluations, improve SAR understanding and accelerate optimization. The current bottleneck is computational speed: AI produces predictions in seconds, whereas QM calculations require minutes to hours. For now, QM is most effective as a post-design filter or biasing tool, but advances in HPC and GPU-accelerated QM, linear-scaling FMO, and quantum computing may close this timescale gap. Emerging AI-SBDD approaches, such as geometry-based ligand generation, still rely on simplified scoring and stand to benefit from QM-based residue-level accuracy. SBDD remains the anchor method in computational chemistry. AI extends its reach into generative design, and QM methods such as FMO correct its scoring limitations. Together they form a next-generation paradigm: AI-driven, QM-validated drug design (Fig. 4).

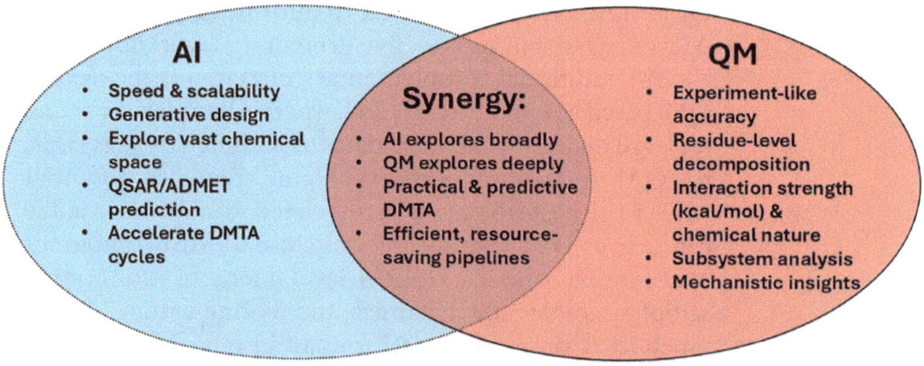

Fig. 4 Complementary strengths of artificial intelligence (AI) and quantum mechanics (QM) in drug design. AI provides speed, scalability, and broad chemical exploration, while QM delivers experiment-like accuracy, residue-level decomposition, and mechanistic insight. Their synergy enables predictive, resource-efficient DMTA cycles by combining AI's breadth with QM's depth

6 Conclusion

The integration of AI, QM, and SBDD is emerging as one of the most promising directions for future drug design. Each method brings a complementary strength: SBDD provides structural context, AI delivers rapid exploration of chemical space and QM contributes the experiment-like accuracy needed to evaluate interactions that classical models miss. Used together, they offer a pathway to more predictive, efficient, and mechanistically informed design workflows.

AI can generate and prioritize thousands of potential molecules, but only a small fraction can be synthesized and tested. QM, particularly through fragment-based approaches such as FMO, provides a critical filtering step by identifying realistic, stabilizing interactions at the residue level and highlighting energetically meaningful hot-spots for optimization. This synergy allows AI-driven ideation to be grounded in physics-based validation, reducing attrition and focusing medicinal chemistry effort on compounds with the highest likelihood of success. Chaps. 4 and 6 illustrate how this hybrid approach is already being applied at Sygnature Discovery.

The same integration offers powerful opportunities for virtual screening. AI models can propose diverse, synthetically feasible molecules or perform rapid pocket-based ligand generation, while SBDD provides structural placement through docking or geometry-driven methods. QM evaluation then refines these poses, resolves ambiguous scoring results and identifies the most credible candidates. Such AI–SBDD–QM hybrid screening could substantially improve hit quality, reduce false positives from classical scoring functions and unlock targets that were previously inaccessible due to a lack of high-quality structural data.

Looking ahead, advances in HPC and GPU-accelerated quantum chemistry and linear-scaling QM methods will continue to narrow the timescale gap between AI and QM, enabling more seamless integration within design loops. As these technologies mature, AI–QM–SBDD workflows have the potential not only to support drug design but to transform it, making the entire DMTA cycle faster, leaner, and more predictive.

References

1. Sadybekov AV, Katritch V (2023) Computational approaches streamlining drug discovery. Nature 616:673–685
2. Waring MJ, Arrowsmith J, Leach AR et al (2015) An analysis of the attrition of drug candidates from four major pharmaceutical companies. Nat Rev Drug Discov 14:475–486
3. Kola I, Landis J (2004) Can the pharmaceutical industry reduce attrition rates?. Nat Rev Drug Discov 3:711–716
4. Hughes J, Rees S, Kalindjian S et al (2011) Principles of early drug discovery. Br J Pharmacol 162:1239–1249
5. Pammolli F, Magazzini L, Riccaboni M (2011) The productivity crisis in pharmaceutical R&D. Nat Rev Drug Discov 10:428–438
6. Ouma RBO, Ngari SM, Kibet JK (2024) A review of the current trends in computational approaches in drug design and metabolism. Discov Public Health 21:108

7. Lin X, Li X, Lin X (2020) A review on applications of computational methods in drug screening and design. Molecules 25:1375
8. Neelam AKN (2025) Advancing drug discovery: the role of computer-aided design and development in modern pharmaceuticals. Discov Pharm Sci 1:8
9. Mihai DP, Nitulescu GM (2025) Computer-aided drug design and drug discovery. Pharmaceuticals 18:436
10. Sabe VT, Ntombela T, Jhamba LA et al (2021) Current trends in computer aided drug design and a highlight of drugs discovered via computational techniques: a review. Eur J Med Chem 224:113705
11. Gurung AB, Ali MA, Lee J et al (2021) An updated review of computer-aided drug design and its application to COVID-19. BioMed Res Int 2021:8853056
12. Wu Z, Chen S, Wang Y et al (2024) Current perspectives and trend of computer-aided drug design: a review and bibliometric analysis. Int J Surg 110:3848–3878
13. Medina-Franco JL (2021) Grand Challenges of computer-aided drug design: the road ahead. Front Drug Discov 1:728551
14. Schneuing A, Harris C, Du Y et al (2024) Structure-based drug design with equivariant diffusion models. Nat Comput Sci 4:899–909
15. Amzel LM (1998) Structure-based drug design. Curr Opin Biotechnol 9:366–369
16. Takaya D (2024) Computer-aided drug design using the fragment molecular orbital method: current status and future applications for SBDD. Chem Pharm Bull (Tokyo) 72:781–786
17. Fukuzawa K, Tanaka S (2022) Fragment molecular orbital calculations for biomolecules. Curr Opin Struct Biol 72:127–134
18. Kawashita N (2024) Interaction analysis by fragment molecular orbital method for drug discovery research. Chem Pharm Bull (Tokyo) 72:787–793
19. Ginex T, Vázquez J, Estarellas C et al (2024) Quantum mechanical-based strategies in drug discovery: finding the pace to new challenges in drug design. Curr Opin Struct Biol 87:102870
20. Niazi SK (2025) Quantum mechanics in drug discovery: a comprehensive review of methods, applications, and future directions. Int J Mol Sci 26:6325
21. Arodola OA, Soliman ME (2017) Quantum mechanics implementation in drug-design workflows: does it really help? Drug Des Devel Ther 11:2551–2564
22. Shim J, MacKerell AD Jr (2011) Computational ligand-based rational design: role of conformational sampling and force fields in model development. MedChemComm 2:356
23. Vázquez J, López M, Gibert E et al (2020) Merging ligand-based and structure-based methods in drug discovery: an overview of combined virtual screening approaches. Molecules 25:4723
24. Rajagopalan M, Balasubramanian S, Ramaswamy A et al (2013) Pharmacophore based 3D-QSAR modeling and free energy analysis of VEGFR-2 inhibitors. J Enzyme Inhib Med Chem 28:1236–1246
25. Yang S-Y (2010) Pharmacophore modeling and applications in drug discovery: challenges and recent advances. Drug Discov Today 15:444–450
26. Acharya C, Coop AE, Polli J et al (2011) Recent advances in ligand-based drug design: relevance and utility of the conformationally sampled pharmacophore approach. Curr Comput Aided-Drug Des 7:10–22
27. Lee C-H, Huang H-C, Juan H-F (2011) Reviewing ligand-based rational drug design: the search for an ATP synthase inhibitor. Int J Mol Sci 12:5304–5318
28. (2023) AI's potential to accelerate drug discovery needs a reality check. Nature 622:217–217
29. Bender A, Cortés-Ciriano I (2021) Artificial intelligence in drug discovery: what is realistic, what are illusions? Part 1: ways to make an impact, and why we are not there yet. Drug Discov Today 26:511–524
30. Walters WP, Barzilay R (2021) Critical assessment of AI in drug discovery. Expert Opin Drug Discov 16:937–947
31. Ocana A, Pandiella A, Privat C et al (2025) Integrating artificial intelligence in drug discovery and early drug development: a transformative approach. Biomark Res 13:45
32. Duran T, Chaudhuri B (2024) Where might artificial intelligence be going in pharmaceutical development?. Mol Pharm 21:993–995
33. Pitt WR, Bentley J, Boldron C et al (2025) Real-world applications and experiences of AI/ML deployment for drug discovery. J Med Chem 68:851–859
34. Ali S, Tian X, Chen H et al (2025) A new era of artificial intelligence (AI): transforming drug discovery and development. J Med Chem 68:23643–23652
35. Heifetz A, (2022) Artificial intelligence in drug design. Springer US, New York, NY
36. Segall M (2024) Beyond the hype: the reality of implementing AI to drive drug discovery success. GEN. Biotechnol 3:338–340
37. Bender A, Cortes-Ciriano I (2021) Artificial intelligence in drug discovery: what is realistic, what are illusions? Part 2: a discussion of chemical and biological data. Drug Discov Today 26:1040–1052

38. Goldman B, Kearnes S, Kramer T et al (2022) Defining levels of automated chemical design. J Med Chem 65:7073–7087
39. Ghiandoni GM, Evertsson E, Riley DJ et al (2024) Augmenting DMTA using predictive AI modelling at AstraZeneca. Drug Discov Today 29:103945
40. Ocana A, Pandiella A, Privat C et al (2025) Integrating artificial intelligence in drug discovery and early drug development: a transformative approach. Biomark Res 13:45
41. Vázquez J, López M, Gibert E et al (2020) Merging ligand-based and structure-based methods in drug discovery: an overview of combined virtual screening approaches. Molecules 25:4723
42. Gill J, Sharma A (2022) Prospects of halofuginone as an antiprotozoal drug scaffold. Drug Discov Today 27:2586–2592
43. Lillich FF, Willems S, Ni X et al (2021) Structure-based design of dual partial peroxisome proliferator-activated receptor γ Agonists/soluble epoxide hydrolase inhibitors. J Med Chem 64:17259–17276
44. Schneuing A, Harris C, Du Y et al (2024) Structure-based drug design with equivariant diffusion models. Nat Comput Sci 4:899–909
45. Fedorov DG (2017) The fragment molecular orbital method: theoretical development, implementation in GAMESS, and applications. WIREs Comput Mol Sci 7:e1322
46. Fukuzawa K, Tanaka S (2022) Fragment molecular orbital calculations for biomolecules. Curr Opin Struct Biol 72:127–134
47. Tanaka S, Tokutomi S, Hatada R et al (2021) Dynamic cooperativity of ligand-residue interactions evaluated with the fragment molecular orbital method. J Phys Chem B 125:6501–6512
48. Heifetz A, Trani G, Aldeghi M et al (2016) Fragment molecular orbital method applied to lead optimization of novel interleukin-2 inducible T-cell kinase (ITK) inhibitors. J Med Chem 59:4352–4363
49. Fedorov DG (2021) Electron density from the fragment molecular orbital method combined with density-functional tight-binding. Chem Phys Lett 780:138900
50. Fedorov DG, Nagata T, Kitaura K (2012) Exploring chemistry with the fragment molecular orbital method. Phys Chem Chem Phys 14:7562
51. McGann MR, Almond HR, Nicholls A et al (2003) Gaussian docking functions. Biopolymers 68:76–90
52. Epifanovsky E, Gilbert ATB, Feng X et al (2021) Software for the frontiers of quantum chemistry: An overview of developments in the Q-Chem 5 package. J Chem Phys 155:084801
53. Neese F, Wennmohs F, Becker U et al (2020) The ORCA quantum chemistry program package. J Chem Phys 152:224108
54. Molani F, Cho AE (2024) Accurate protein-ligand binding free energy estimation using QM/MM on multi-conformers predicted from classical mining minima. Commun Chem 7:247
55. Hehre W, Klunzinger P, Deppmeier B et al (2019) Efficient protocol for accurately calculating[13] C chemical shifts of conformationally flexible natural products: scope, assessment, and limitations. J Nat Prod 82:2299–2306
56. Habgood M, James T, Heifetz A (2020) Conformational searching with quantum mechanics. In: Heifetz A (ed) Quantum mechanics in drug discovery. Springer US, New York, NY, pp 207–229
57. Hysmith H, Foadian E, Padhy SP et al (2024) The future of self-driving laboratories: from human in the loop interactive AI to gamification. Digit Discov 3:621–636
58. Blaschke T, Arús-Pous J, Chen H et al (2020) REINVENT 2.0: an AI tool for de novo drug design. J Chem Inf Model 60:5918–5922
59. Wu Z, Ramsundar B, Feinberg EN et al (2018) MoleculeNet: a benchmark for molecular machine learning. Chem Sci 9:513–530
60. Dixon SL, Duan J, Smith E et al (2016) AutoQSAR an automated machine learning tool for best-practice quantitative structure–activity relationship modeling. Future Med Chem 8:1825–1839
61. Ivanenkov YA, Polykovskiy D, Bezrukov D et al (2023) Chemistry42: an AI-driven platform for molecular design and optimization. J Chem Inf Model 63:695–701
62. Varadi M, Anyango S, Deshpande M et al (2022) AlphaFold protein structure database: massively expanding the structural coverage of protein-sequence space with high-accuracy models. Nucleic Acids Res 50:D439–D444
63. Tunyasuvunakool K, Adler J, Wu Z et al (2021) Highly accurate protein structure prediction for the human proteome. Nature 596:590–596
64. Jumper J, Evans R, Pritzel A et al (2021) Highly accurate protein structure prediction with AlphaFold. Nature 596:583–589
65. Ginex T, Vázquez J, Estarellas C et al (2024) Quantum mechanical-based strategies in drug discovery: finding the pace to new challenges in drug design. Curr Opin Struct Biol 87:102870
66. Peng X, Luo S, Guan J et al (2022) Pocket2Mol: efficient molecular sampling based on 3D protein pockets. https://arxiv.org/abs/2205.07249
67. Smith JS, Isayev O, Roitberg AE (2017) ANI-1: an extensible neural network potential with DFT accuracy at force field computational cost. Chem Sci 8:3192–3203
68. Unke OT, Chmiela S, Sauceda HE et al (2021) Machine learning force fields. Chem Rev 121:10142–10186

69. Bleiziffer P, Schaller K, Riniker S (2018) Machine learning of partial charges derived from high-quality quantum-mechanical calculations. J Chem Inf Model 58:579–590
70. Smith DGA, Lolinco AT, Glick ZL et al (2021) Quantum chemistry common driver and databases (QCDB) and quantum chemistry engine (QCE NGINE): automation and interoperability among computational chemistry programs. J Chem Phys 155:204801
71. Baiardi A, Reiher M (2020) The density matrix renormalization group in chemistry and molecular physics: Recent developments and new challenges. J Chem Phys 152:040903
72. Ramakrishnan R, Dral PO, Rupp M et al (2014) Quantum chemistry structures and properties of 134 kilo molecules. Sci Data 1:140022
73. Smith DGA, Lolinco AT, Glick ZL et al (2021) Quantum chemistry common driver and databases (QCDB) and quantum chemistry engine (QCE NGINE): automation and interoperability among computational chemistry programs. J Chem Phys 155:204801
74. Unke OT, Chmiela S, Sauceda HE et al (2021) Machine learning force fields. Chem Rev 121:10142–10186
75. Smith JS, Isayev O, Roitberg AE (2017) ANI-1: an extensible neural network potential with DFT accuracy at force field computational cost. Chem Sci 8:3192–3203
76. Qiao Z, Welborn M, Anandkumar A et al (2020) OrbNet: deep learning for quantum chemistry using symmetry-adapted atomic-orbital features. J Chem Phys 153:124111

Chapter 3

Applications of Integrated Molecular Dynamics and Fragment Molecular Orbital (MD+FMO) Calculations in Drug Discovery and Structural Biology

Kaori Fukuzawa

Abstract

By integrating the fragment molecular orbital (FMO) method, a quantum chemical calculation approach, with classical molecular dynamics (MD) simulations, the MD+FMO framework extends interaction analysis beyond a single static structure to enable quantitative structure–activity relationship evaluation across diverse scaffolds, pose estimation for complexes with unknown structures, and investigation of molecular recognition mechanisms in mutant proteins. This chapter summarizes current evaluation strategies using the MD+FMO approach and integration with diverse experimental techniques, highlighting its value as an analytical method that complements experimental studies in molecular biology.

Keywords Fragment molecular orbital (FMO) method, Classical molecular dynamics (MD) simulation, Structure-based drug design (SBDD), Quantum mechanics (QM), Molecular mechanics (MM)

1 Introduction

Applications of the fragment molecular orbital (FMO) method [1] have historically been limited to static structure analysis because of the high computational cost of quantum chemical calculations. Single-point calculations using molecular structures determined by X-ray crystallography and cryo-electron microscopy (cryo-EM) have proven useful for quantitative assessment of intermolecular interactions in protein–ligand and protein–protein interactions (PPIs), evaluating inter-fragment interaction energies (IFIEs) and their energy components. The scope of the FMO method extends beyond post-structure interpretation [2]. It has been used to validate interactions during X-ray crystallographic structure refinement [3, 4], and it serves as foundational data for comprehensive FMO databases [5, 6]. Molecular recognition in living systems is inherently dynamic, whereas single-point calculations capture only a snapshot.

Alexander Heifetz (ed.), *Quantum Mechanics Applications in Drug Design*, Methods in Molecular Biology, vol. 3061,
https://doi.org/10.1007/978-1-0716-5420-0_3,

A practical approach to incorporate dynamics is to combine classical molecular dynamics (MD) simulations with FMO calculations. In this workflow, MD simulations characterize molecular motions under conditions approximating the biological environment (e.g., 310 K in aqueous solution). Representative structures are then sampled from the resulting trajectories for FMO calculations. Statistical averaging of interaction energies across these samples yields a dynamic description of intermolecular interactions. Dynamic structure analysis using FMO method typically requires more than 100-fold greater computational resources than static analysis, which has historically constrained its use. For example, even on the K computer in Japan, which operated until 2019, such sampling calculations required approximately 3 months [7]. With the advent of the supercomputer Fugaku, comparable calculations became feasible within a day, accelerating the adoption of this approach as a practical method. Fugaku began operation in 2020, coinciding closely with the onset of the COVID-19 pandemic. Early in the pandemic, the crystal structure of the SARS-CoV-2 main protease (Mpro) in complex with the N3 inhibitor was reported (PDB ID: 6LU7). Dynamical interaction analysis combining MD and FMO on this complex was reported shortly thereafter using Fugaku [8]. This example illustrates how advances in computational infrastructure can rapidly expand the practical scope of high-cost quantum chemical analyses and enable methodological shifts as technology evolves.

2 Combination of Classical MD and FMO Methods

The approach that combines classical MD simulations with FMO calculations is termed the MD+FMO method. In this framework, multiple molecular structures are sampled by MD simulations, FMO calculations are performed on the sampled structures, and interaction energies are averaged to enable quantitative time-resolved analysis. To date, several applications of MD+FMO have been reported, including studies focused on computational protocols.

In one study predicting binding affinities between cyclin-dependent kinase 2 (CDK2) and seven ligands, FMO calculations were performed for 100 structures of each CDK2–ligand complex sampled from MD trajectories initiated from X-ray crystal structures to estimate dynamic ligand binding energies [7]. Compared with static binding energies derived directly from the crystal structures, the dynamic analysis showed substantially improved correlation with experimental ΔG values ($R^2 = 0.75$ to 0.99). One interpretation is that calculations initiated from distinct co-crystal structures for each ligand can incorporate ligand-specific structural distortions, introducing noise because of differences in experimental conditions and resolution. In contrast, dynamic averaging enables binding affinity evaluation with reduced sensitivity to these distortions. This strategy is well-suited to ligand structure changes in structure–activity relationship studies and broadens the applicability of the FMO method to problems such as predicting binding properties of protein variants and compounds with distinct scaffolds.

When applied to protein–ligand docking models in the absence of co-crystal structures, the method can migrate residue-specific interaction biases characteristic of individual docking poses. MD sampling redistributes these interactions and, after averaging, yields interaction patterns more stably accommodated within the binding pocket, thereby supporting more reliable structure-based design. For example, an MD+FMO analysis of the anti-HIV-1 inhibitor nelfinavir illustrates its use in drug repositioning for SARS-CoV-2 Mpro (Fig. 1) [9]. In the absence of X-ray co-crystal data, docking calculations using a template PDB structure generated multiple candidate poses. For each pose, MD simulations were performed, followed by dynamical FMO analysis of 100 sampled structures. In the docking models, nelfinavir formed a strong hydrogen bond with a single nearby residue, whereas in the MD-sampled ensembles, it formed stable attractive interactions with multiple residues within the pocket, yielding a configuration that better reflects the biological environment. In addition, pair interaction energy decomposition analysis (PIEDA) [10] distinguished functional groups that contribute to binding from those that do not, providing actionable guidance for subsequent molecular design.

Similarly, MD+FMO calculations have been applied to elucidate the mechanism of action of remdesivir, an approved inhibitor identified through drug repositioning that targets the SARS-CoV-2

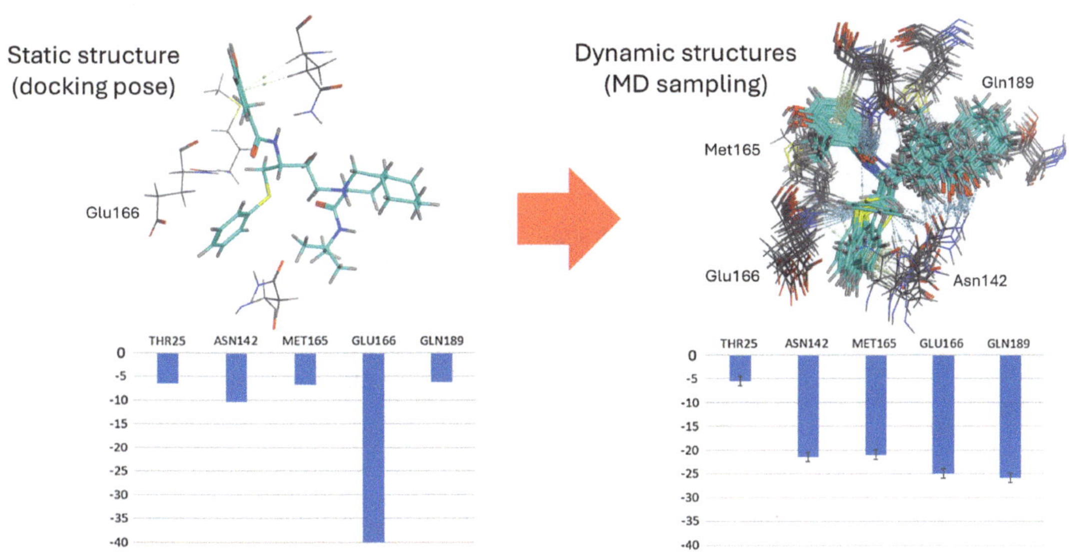

Fig. 1 Interactions between Mpro and nelfinavir in docking poses and MD sampling structure

RNA-dependent RNA polymerase (RdRp). Following initial static FMO analysis based on the available cryo-EM structure [11], subsequent work used dynamical FMO analysis to investigate the inhibition mechanism while accounting for structural fluctuations [12]. Remdesivir binds at the +1 position at the end of the primer RNA strand and inhibits RNA elongation at the −3 position after incorporation of three additional nucleotides. Static analysis evaluated interactions among remdesivir, RNA bases, and RdRp residues in a complex in which remdesivir was covalently incorporated at the +1 position. As a nucleotide analog, remdesivir was suggested to perturb canonical double-stranded RNA interactions by forming additional hydrogen bonds with intrastrand bases and oblique π–π stacking interactions with interstrand bases, driven by its chemically modified functional group, in addition to base-pair hydrogen bonding and intrastrand stacking. In the dynamical analysis, remdesivir was modeled at four distinct RNA positions, followed by MD simulations and FMO analysis across sampled ensembles (Fig. 2). This approach indicated that binding at positions such as −1 does not markedly perturb RNA structure, whereas remdesivir at the −3 position forms strong interactions with RdRp Lys593, thereby inhibiting RNA elongation. Together, these examples illustrate how the combined MD and FMO framework can clarify molecular recognition mechanisms under

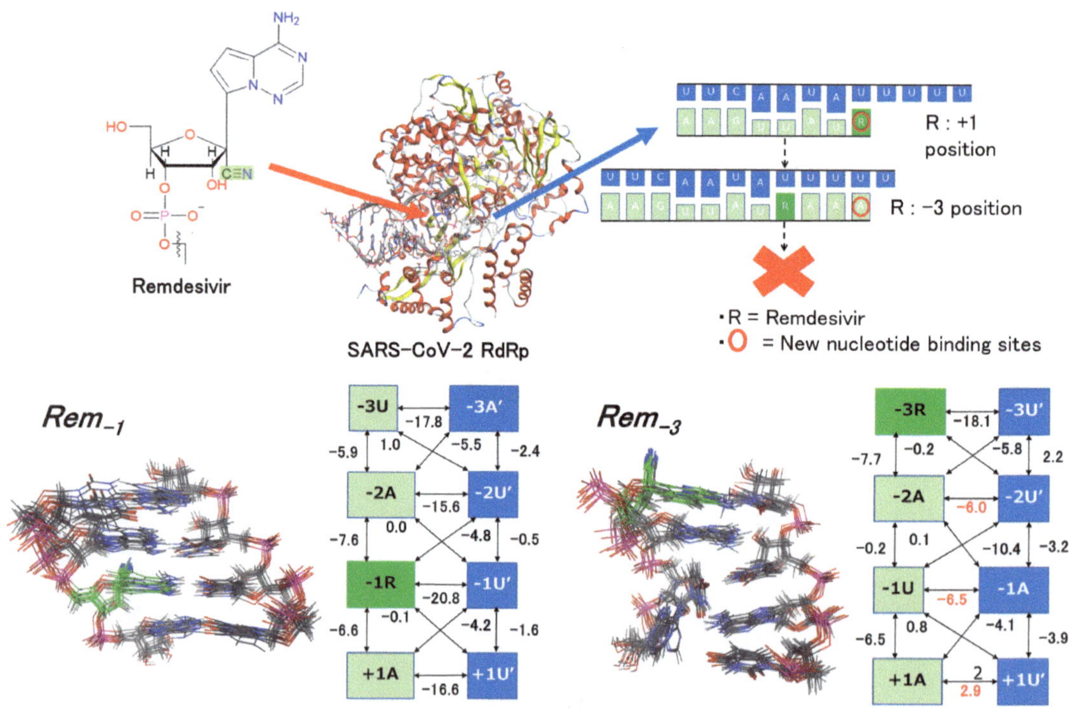

Fig. 2 Dynamical interaction of RNA-dependent RNA polymerase with Remdesivir

biologically relevant, dynamic conditions rather than relying solely on static representations.

3 Collaborative Research with Experimental Studies Using the MD+FMO Method

Using the predictive capabilities of this MD+FMO method has recently led to increased collaboration with experimental research [13–17].

In a cryo-EM structural study of the human herpesvirus helicase–primase complex, the ligand conformation was determined by evaluating interaction complementarity derived from FMO calculations, enabling refinement of the complex structure with an α-subfamily–selective ligand [13]. In cryo-EM analysis, oxygen and nitrogen atoms are often difficult to distinguish, whereas FMO calculations can discriminate these atom types based on their interaction characteristics. In addition, four structural clusters were identified from a total of 25 μs of MD trajectories, and FMO calculations were performed for a representative structure from each cluster. Cluster weighting enabled estimation of ensemble-averaged interaction energies comparable to those obtained by uniform sampling. The FMO results quantitatively showed how the two ligands interact with distinct amino acid residues (Fig. 3). These analyses indicated that the resistance of amenamevir to the Phe360 and Asn902 mutations,

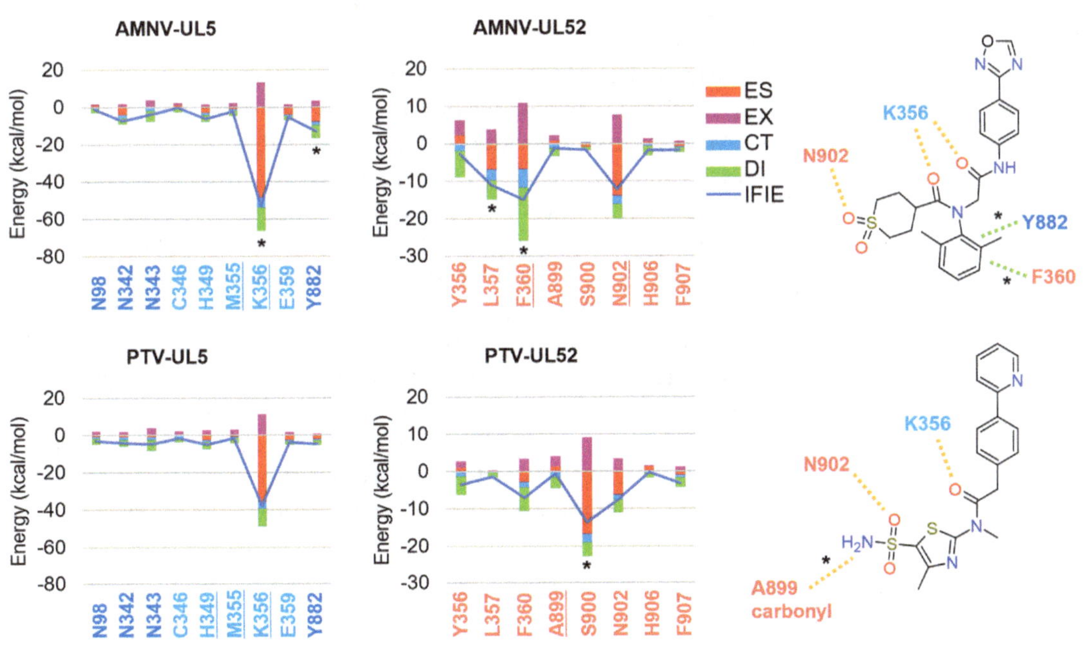

Fig. 3 Interactions between the human herpesvirus helicase–primase complex and two ligands, amenamevir (AMV) and pritelivir (PTV). Figure 4(f–k) in [13] by Sato et al. is licensed under CC BY 4.0

and of pritelivir to the Ala899 mutation, arises from strong hydrogen bonds and CH/π interactions with these residues.

MD+FMO combined with experimental approaches has also contributed to elucidating translation mechanisms [14]. MD+FMO calculations for the complex of desmethyl desamino pateamine A (DMDA-PatA), a small molecule that selectively inhibits mRNA translation, with the dead-box RNA-binding protein eIF4A and RNA characterized key interaction features. Although experiments showed preferential binding of DMDA-PatA to the RNA GNG motif, dynamical FMO interaction analysis, together with the design of an inactive DMDA-PatA derivative, indicated that the positive charge on the compound's tertiary amine drives G-selectivity. These findings provide guidance for designing eIF4A inhibitors.

Similarly, the plant-derived compound rocaglamide A (RocA) inhibits eIF4A and exhibits anticancer activity, but its mechanism remained unclear beyond evidence from sequence-binding experiments indicating a preference for purine-rich RNA and an X-ray structure of the quaternary complex containing an ATP-like compound. Before publication of the crystal structure, FMO calculations predicted hydrogen bonding and CH/π interactions between RocA and specific RNA sequences, strengthening the proposed structural basis [3]. Subsequent MD+FMO analysis showed that purine-sequence mutations weaken RNA binding by altering interactions between RNA bases and surrounding residues at the binding interface (Fig. 4). This analysis also identified two residues essential for RNA binding (Phe163 and Gln195) and three residues that modulate RNA binding (Phe192, Arg282, and Asp164). Selected functional roles were supported by mutagenesis experiments. Together, these studies demonstrate that FMO-based analysis provides a strong complement to experimental approaches in molecular biology.

FMO has also provided insights into the sequence-dependent RNA cleavage mechanism of 8–17 deoxyribozymes (DNAzymes) relevant to nucleic acid therapeutic development [15]. Available X-ray structures have largely involved complexes in which the RNA substrate was replaced by DNA and further stabilized by proteins. Using these structures as starting points, structural modeling and MD-based relaxation followed by dynamical FMO interaction analysis showed that the pattern, stability, and spatial distribution of hydrogen-bond bridges between base pairs in the active site vary with RNA sequence. These results support molecular design strategies grounded in the structural and mechanistic basis of sequence preference and complement conventional comprehensive experimental analyses.

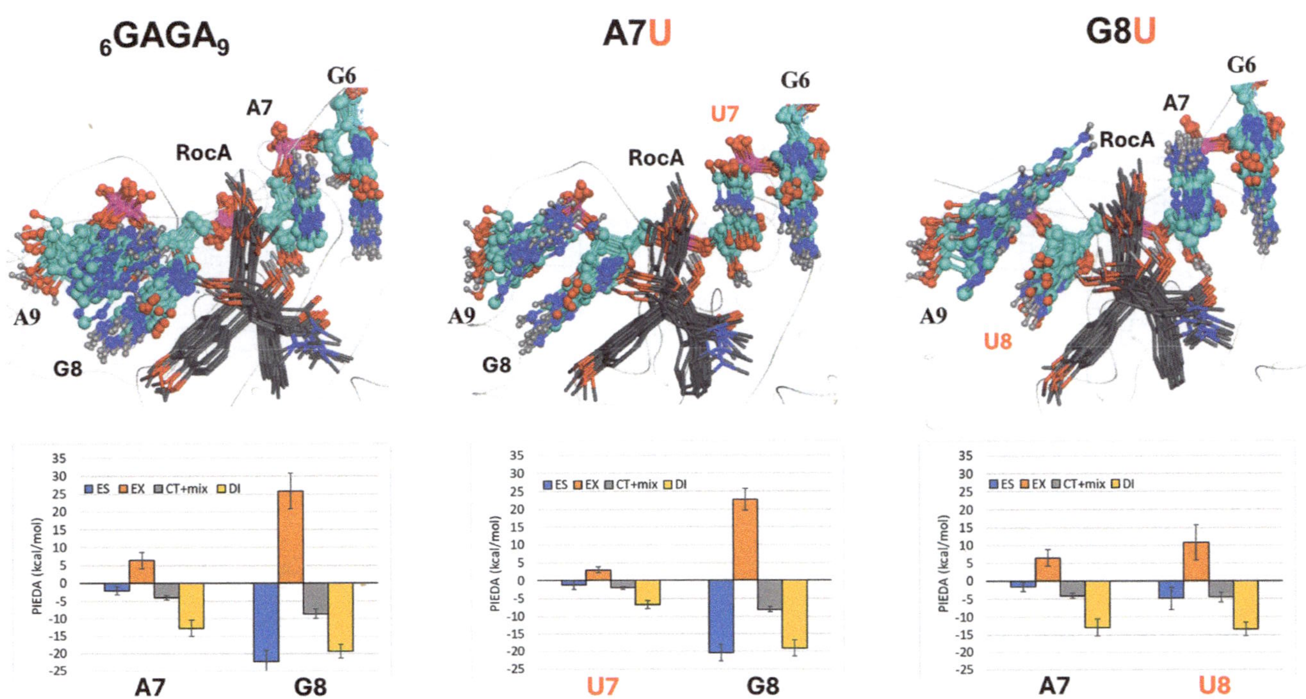

Fig. 4 Fluctuation structure and dynamic interaction analysis around RocA in each RNA sequence complex

4 Concluding Remarks

As described in this chapter, integrating classical MD simulations with FMO calculations provides deeper insight into structures examined in structural biology and can also accommodate structures that are modestly perturbed relative to known conformations. When used in a complementary manner with experimental research, this approach is expected to help elucidate diverse mechanisms in molecular biology. For example, defining how point mutations in viral antigens alter antigen recognition is critical for preparedness against emerging viruses and future pandemics [16]. In addition, this framework has been shown to support synergistic analyses with NMR for targets in which dynamic behavior is essential, including membrane proteins [17]. In drug discovery, it is applicable across diverse modalities, including small molecules, medium-sized molecules, protein-protein interactions, nucleic acids, glycans, and lipids. A major advantage of the FMO method, as a first-principles approach, is that it enables quantitative energy evaluations that are independent of force-field parameters across diverse molecular species. Methods for predicting the structures of proteins and their complexes using machine learning and artificial intelligence are advancing rapidly. For the FMO community, these developments offer an opportunity to greatly expand the number of structures amenable to computational analysis, and continued mutual progress is increasingly anticipated.

Acknowledgments

The author thanks Dr. Yuma Handa, Mr. Shuhei Miyakawa, Mr. Toma Miyagishi, Prof. Toru Sengoku, Dr. Shintaro Iwasaki, Dr. Takuhiro Ito, and Prof. Takao Yamaguchi for their collaborations and helpful discussions. The author also thanks Prof. Yuji Mochizuki and Dr. Tatsuya Nakano for general discussions related to FMO calculations, and Dr. Chiduru Watanabe, Dr. Teruki Honma, Prof. Koichiro Kato, Prof. Yoshiharu Mori, Prof. Shigenori Tanaka, and Dr. Daisuke Takaya for the FMO application for drug discovery research. This research was partially supported by the Platform Project for Supporting Drug Discovery and Life Science Research (Basis for Supporting Innovative Drug Discovery and Life Science Research (BINDS)) from the Japan Agency for Medical Research and Development (AMED) under Grant Number JP25ama121030. We would like to thank Editage (www.editage.jp) for English language editing.

References

1. Kitaura et al (1999) Fragment molecular orbital method: an approximate computational method for large molecules. Chem Phys Lett 313:701–706. https://doi.org/10.1016/S0009-2614(99)00874-X
2. Takeuchi K, Ikeda Y, Senda M, Harada A, Okuwaki K, Fukuzawa K, Nakagawa S, Yu HY, Nagase L, Imai M, Sasaki M, Lo Y-H, Ito D, Osaka N, Fujii Y, Sasaki AT, Senda T (2022) The GTP responsiveness of PI5P4Kβ evolved from a compromised trade-off between activity and specificity. Structure 30:886. https://doi.org/10.1016/j.str.2022.04.004
3. Iwasaki S, Iwasaki W, Takahashi M, Sakamoto A, Watanabe C, Shichino Y, Floor SN, Fujiwara K, Mito M, Dodo K, Sodeoka M, Imataka H, Honma T, Fukuzawa K, Ito T, Ingolia NT (2019) The translation inhibitor rocaglamide targets a bimolecular cavity between eIF4A and polypurine RNA. Mol Cell 73(4):738–748 e739. https://doi.org/10.1016/j.molcel.2018.11.026
4. Takebe K, Suzuki M, Kuwada-Kusunose T, Shirai S, Fukuzawa K, Takamiya T, Uzawa N, Iijima H (2023) Structural and computational analyses of the unique interactions of opicapone in the binding pocket of catechol o-methyltransferase: a crystallographic study and fragment molecular orbital analyses. J Chem Inf Model 63(14):4468–4476. https://doi.org/10.1021/acs.jcim.3c00331
5. Takaya D, Ohno S, Miyagishi T, Tanaka S, Okuwaki K, Watanabe C, Kato K, Tian Y-S, Fukuzawa K (2024) Quantum chemical calculation dataset for representative protein folds by the fragment molecular orbital method. Sci Data 11(1):1164. https://doi.org/10.1038/s41597-024-03999-2
6. Fukuzawa K, Kato K, Watanabe C, Kawashima Y, Handa Y, Yamamoto A, Watanabe K, Ohyama T, Kamisaka K, Takaya D, Honma T (2021) Special features of COVID-19 in the FMODB: fragment molecular orbital calculations and interaction energy analysis of SARS-CoV-2-related proteins. J Chem Inf Model 61:4594. https://doi.org/10.1021/acs.jcim.1c00694
7. Takaba K, Watanabe C, Tokuhisa A, Akinaga Y, Ma B, Kanada R, Araki M, Okuno Y, Kawashima Y, Moriwaki H, Kawashita N, Honma T, Fukuzawa K, Tanaka S (2022) Protein-ligand binding affinity prediction of cyclin-dependent kinase-2 inhibitors by dynamically averaged fragment molecular orbital-based interaction energy. J Comput Chem 43:1362. https://doi.org/10.1002/jcc.26940
8. Hatada R, Okuwaki K, Akisawa K, Mochizuki Y, Handa Y, Fukuzawa K, Komeiji Y, Okiyama Y, Tanaka S (2021) Statistical interaction analyses between SARS-CoV-2 main protease and inhibitor N3 by combining molecular dynamics simulation and fragment molecular orbital calculation. Appl Phys Express 14(2):027003–027007. https://doi.org/10.35848/1882-0786/abdac6
9. Handa Y, Okuwaki K, Kawashima Y, Hatada R, Mochizuki Y, Komeiji Y, Tanaka S, Furuishi T, Yonemochi E, Honma T, Fukuzawa K (2024) Prediction of binding pose and affinity of nelfinavir, a SARS-CoV-2 main protease repositioned drug, by combining docking, molecular dynamics, and fragment molecular orbital calculations. J Phys Chem B 128(10):2249–2265. https://doi.org/10.1021/acs.jpcb.3c05564
10. Fedorov DG, Kitaura K (2007) Pair interaction energy decomposition analysis. J Comput Chem 28(1):222–237. https://doi.org/10.1002/jcc.20496
11. Kato K, Honma T, Fukuzawa K (2020) Intermolecular interaction among Remdesivir, RNA and RNA-dependent RNA polymerase of SARS-CoV-2 analyzed by fragment molecular orbital calculation. J Mol Graph Model 100:107695–107699. https://doi.org/10.1016/j.jmgm.2020.107695
12. Miyakawa S, Okuwaki K, Kawashima Y, Handa Y, Furuishi T, Yonemochi E, Kato K, Fukuzawa K (2025) Dynamical fragment molecular orbital interaction analysis of SARS-CoV-2 RNA-dependent RNA polymerase and remdesivir. ACS Omega 10(49):60874–60885. https://doi.org/10.1021/acsomega.5c09502
13. Sato K, Ishida H, Miyagishi T, Kobayashi S, Kise Y, Hamada K, Okada C, Oguni A, Nureki O, Kono H, Fukuzawa K, Sengoku T (2026) Structure and inhibition mechanism of the helicase-primase complex from human herpesvirus 1. Cell Chem Biol 33:1–12. https://doi.org/10.1016/j.chembiol.2026.03.006
14. Saito H, Handa Y, Chen M, Schneider-Poetsch T, Shichino Y, Takahashi M, Romo D, Yoshida M, Furstner A, Ito T, Fukuzawa K, Iwasaki S (2024) DMDA-PatA mediates RNA sequence-selective translation repression by anchoring eIF4A and DDX3 to GNG motifs. Nat Commun 15(1):7418. https://doi.org/10.1038/s41467-024-51635-9
15. Nakajima S, Miyakawa S, Ohno S, Ikemoto H, Maruyama R, Chiba K, Suginaka R, Oda C, Kasahara Y, Fukuzawa K, Obika S, Yamaguchi T (2025) Sequence-dependent RNA cleavage by the 8-17 DNAzyme: identification of preferred tetranucleotide motifs. Bioorg Med Chem 134:118520. https://doi.org/10.1016/j.bmc.2025.118520

16. Hyodo A, Seki F, Fukuda K, Tashiro K, Kitai Y, Akahori Y, Watabe H, Katoh H, Osaki R, Takaya D, Kawashita N, Fukuhara H, Ikegame S, Yoshikawa T, Eunsil P, Morikawa S, Yamaguchi R, Lee B, Maenaka K, Shirai T, Fukuzawa K, Tanaka S, Takeda M (2025) Evolutionary and structural basis of SLAMF1 utilization in morbilliviruses-Implications for host range and cross-species transmission. PLoS Pathog 21(6):e1012990. https://doi.org/10.1371/journal.ppat.1012990
17. Ueda T, Tsuchida T, Kurita M, Mizumura T, Imai S, Shiraishi Y, Kofuku Y, Miyakawa S, Fukuzawa K, Takeuchi K, Shimada I (2025) Structural basis of the residence time of adenosine A(2A) receptor ligands revealed by NMR. Chem Sci 16(38):17948–17955. https://doi.org/10.1039/d5sc02398j

Chapter 4

Advancing Medicinal Chemistry Through FMO: Sygnature's Platform

Alexander Heifetz, Girinath G. Pillai, Anna Cavalleri, Maryamdokht Taimoory, Louise Birch and Colin Sambrook Smith

Abstract

The FMO Sygnature Platform (FMO-SP) transforms the fragment molecular orbital method (FMO) into a practical framework for drug discovery by combining quantum-level accuracy with workflow automation and intuitive visualization. FMO-SP quantifies residue-level stabilizing and destabilizing interactions, including solvation and polarisation effects, that are often missed by classical force fields. Three showcases illustrate its broad applicability. In kinase design (p38α MAPK), integration with nonequilibrium simulation/molecular dynamics (NES/MD) linked energetic magnitude with dynamic persistence, distinguishing consistent hot spots from transient interactions and informing strategies for potency. In safety pharmacology, application to the hERG potassium channel revealed how burial of a ligand's basic center within the channel's negative electrostatic field drives blockade, highlighting pK_a reduction as a rational approach to mitigate cardiotoxicity. In metabolic GPCRs (GLP-1R), analysis of Danuglipron demonstrated how electrostatic anchors and hydrophobic reinforcement explain potency trends, guiding next-generation oral agonist design. Beyond protein–ligand systems, FMO-SP extends to PPIs, ADCs, Molecular Glue Degraders (MGDs), PROTACs, and RNA–ligand complexes, and ongoing integration with AI pipelines promises scalable and interpretable hybrid approaches. Together, these advances establish FMO-SP as a practical quantum-based platform that delivers actionable insights to accelerate medicinal chemistry optimization.

Keywords Fragment molecular orbital (FMO), FMO-SP, Quantum mechanics, Protein–ligand interactions, Structure-based drug design (SBDD), Pair interaction energy (PIE), Pair interaction energy decomposition analysis (PIEDA), Desolvation, Polarization, Electrostatic complementarity, Lead optimization, Homology modelling, Docking poses, Molecular dynamics (MD), Antibody–drug conjugates (ADC), Molecular glues, PROTACs, Protein–protein interactions (PPIs), hERG, Automation, High-throughput screening, AI-driven drug design

1 Introduction

Understanding protein–ligand binding at the atomic level is essential for optimizing potency and selectivity in structure-based drug design (SBDD). Even with high-quality crystal structures, standard molecular mechanics (MM) or visual inspection often fail to capture

Alexander Heifetz (ed.), *Quantum Mechanics Applications in Drug Design*, Methods in Molecular Biology, vol. 3061,
https://doi.org/10.1007/978-1-0716-5420-0_4,

key interactions, such as CH/π, halogen/π, cation/π contacts, nonclassical hydrogen bonds, and hydrophobic effects, that are poorly parameterized in most MM models [1]. This limitation highlights the need for computational methods that can accurately quantify these interactions, which are critical for rationalizing structure–activity relationships (SAR) and guiding the design of next-generation compounds [2].

Quantum mechanical (QM) approaches are the gold standard for describing such interactions, but the steep computational cost of conventional QM calculations on protein-sized systems has historically made them impractical for the fast turnarounds required in drug discovery. This challenge was addressed in 1999, when Prof. Kitaura and co-workers [1] introduced the Fragment Molecular Orbital (FMO), method which divides a macromolecule, such as a protein–ligand complex, into fragments (e.g., residues or ligand subunits) and computes pair-interaction energies for each fragment and fragment pair. The method scales nearly linearly and is highly parallelizable. Pair interaction energies (PIEs) can be further decomposed into electrostatics, exchange-repulsion, charge transfer, dispersion, and, when extended, solvation terms. This pair interaction energy decomposition analysis (PIEDA) [3–10] reveals the physical origin of each contact, enabling precise differentiation between polar, hydrophobic, and steric contributions.

This innovation made QM analysis feasible for large biomolecules, uncovering interaction types previously overlooked in drug design and correcting misconceptions about protein–ligand binding. As Werner Heisenberg once remarked: *"The more accurate the calculations become, the more the concepts tend to vanish into thin air."* Despite this breakthrough, turning academic FMO into a practical industrial tool required extensive development and hands-on expertise. Today, FMO is well established in academic research and increasingly adopted in pharmaceutical drug discovery [11–24].

To support client projects, Sygnature Discovery has developed its own FMO-based platform, FMO-SP (Fig. 1). Sygnature is a leading integrated drug discovery CRO, delivering programs from target validation through to preclinical candidate selection, with multidisciplinary teams spanning medicinal chemistry, computational sciences, DMPK, in vitro/in vivo biology, and structural biology.

FMO-SP (Fig. 1) provides residue-resolved, quantitative insights into interaction strength (in kcal/mol), the chemical nature of interactions (e.g., electrostatic, hydrophobic), and key energy components such as desolvation and depolarization [26, 27]. Results are presented in intuitive formats—column plots, heatmaps, and 3D visualizations in MOE or PyMOL—allowing medicinal chemists to connect calculated interactions directly to specific ligand features. Because these data are quantitative, FMO-SP highlights opportunities to create new interactions or strengthen existing ones. Understanding the nature

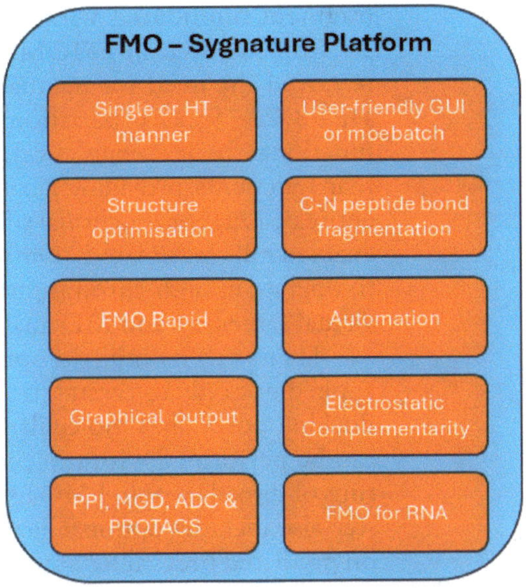

Fig. 1 Key features of the FMO–Sygnature Platform (FMO-SP). The platform supports high-throughput FMO with automation, structure optimization, and C–N fragmentation [25]. It provides graphical output and electrostatic complementarity

of each interaction guides whether a modification should enhance hydrophobic or hydrophilic character. These insights feed directly into the design cycle, guiding analog prioritization for synthesis and helping to save time and resources in medicinal chemistry.

It is important to note that FMO-SP primarily evaluates enthalpic contributions to binding. Binding free energy, however, also depends on entropy, solvation, ligand strain, and conformational changes in both ligand and receptor during binding. While these contributions are not explicitly captured, FMO-SP metrics often correlate well with experimental affinities, particularly within congeneric ligand series where entropic and conformational effects are broadly conserved. In this way, FMO-SP is most effective for guiding SAR analysis and the design of new analogs, rather than for absolute potency prediction.

FMO-SP also incorporates subsystem analysis (SA), developed by Dr. Dmitri Fedorov [28], which calculates additional energy contributions such as desolvation and polarization. These terms, added to total and pair interaction energies (TIE and PIE), provide a richer context for rationalizing SAR, supporting scoring, and guiding compound design.

Historically, FMO workflows required significant manual curation, since fragment definitions could sever peptide bonds in ways that obscured direct residue–ligand mapping. In 2024, a new C–N fragmentation scheme replaced the older CA–C approach, aligning fragment boundaries with conventional residue definitions (Fig. 2). Implemented in FMOMP2 (see Sect. 2.1), this improvement removed the need for manual energy reassignment. Alternatively, DFTB-PA [9, 27] (partition analysis with tight-binding density functional theory) enables calculations using larger segments rather than fragments, providing flexibility to define atom groupings such as entire residues or functional subunits. Together, these advances have enabled fully automated, user-friendly workflows, allowing FMO-SP to operate in high-throughput mode for applications ranging from virtual screening and MD analysis to AI-driven design.

Thanks to its built-in optimization module, FMO-SP can be applied not only to crystal structures but also to homology models, docking poses, and molecular dynamics trajectories. Optimization can be performed with molecular mechanics (MM) prior to FMO runs or directly within FMO using DFTB-PA [27]. The optimization step also includes optional ligand protonation, assigning states based on predicted pKa values at physiological pH (7.4) to ensure realistic modeling of binding interactions.

Another useful feature is the computation of polarization-dependent atomic charges, such as Mulliken charges, for electrostatic complementarity analysis [29, 30]. These maps reveal regions of attraction or repulsion between protein and ligand, adding another layer of design insight.

FMO-SP has been successfully applied to a wide range of targets, including protein–ligand complexes, GPCRs, and RNA–ligand systems, as well as emerging modalities such as ADCs and molecular glues. Ongoing development is expanding its scope to hERG 3D models, protein–protein interactions (PPIs) [31], and PROTACs, while integrating with virtual screening and AI-driven platforms. These efforts are supported by strong academic collaborations and active participation in the FMODD consortium (https://fmodd.jp/activities-en/) [32].

As described in Chap. 2, computational chemistry plays an important role in improving the efficiency of drug discovery. Drug design remains one of the most challenging stages of pharmaceutical R&D, and computational approaches help guide decision-making

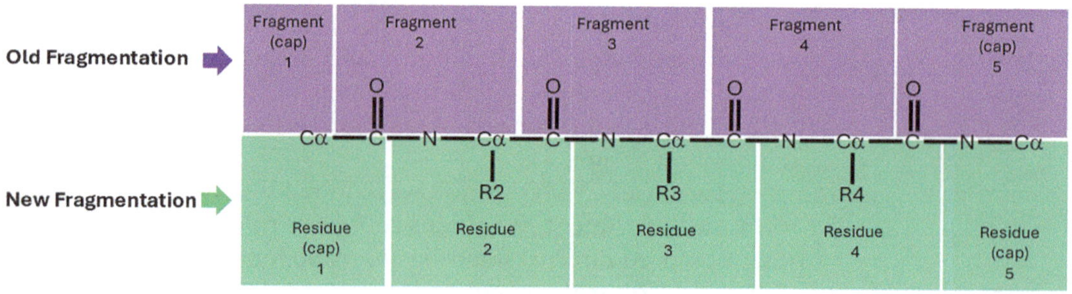

Fig. 2 Comparison of fragmentation schemes in FMO [25]

throughout hit identification, hit-to-lead, and lead optimization. By providing molecular-level insights into protein–ligand interactions, these methods reduce unnecessary experimental effort and enable resources to be focused more effectively. [33–39]

Within this broader context, SBDD stands out as one of the most widely applied computational methods [40, 41]. Yet its effectiveness is still constrained by oversimplified scoring functions that neglect subtle but critical enthalpic and entropic contributions. By providing quantum-level, residue-resolved interaction insights, FMO-SP overcomes these blind spots, transforming SBDD into a more predictive, quantitative, and mechanistically interpretable tool for modern drug discovery.

2 Sygnature's Fragment Molecular Orbital Platform (FMO-SP)

FMO-SP applies QM to reveal critical protein–ligand and protein–protein interactions that traditional force fields or visual inspection often miss—even when high-resolution structural data are available. FMO-SP (Fig. 3) provides quantitative insight into interaction strength (in kcal/mol), interaction type (electrostatic or hydrophobic), and key energy components such as desolvation and depolarization that drive molecular binding. The method computes interactions between fragment pairs, accounting for effects such as polarization and charge transfer. Interaction energies are analyzed using PIEDA [7], which separates contributions into electrostatics, dispersion, exchange-repulsion, and charge transfer, allowing medicinal chemists to interpret the results in chemically intuitive terms. The platform

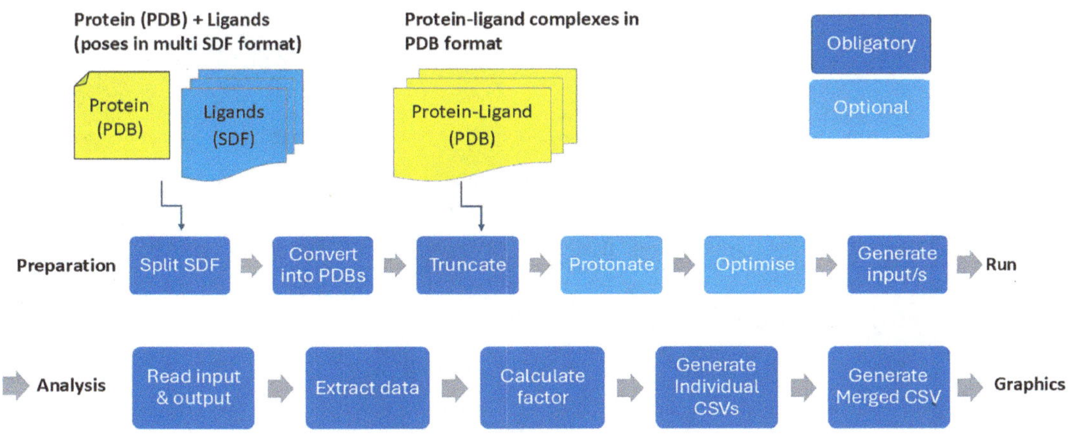

Fig. 3 Workflow of the FMO–Sygnature Platform (FMO-SP). The pipeline spans from protein–ligand preparation and optimization through GAMESS FMO calculations to automated extraction, analysis, and generation of CSV outputs for graphical visualization

enables rapid, flexible, and efficient analysis—typically completing an average protein–ligand calculation in just a few minutes.

The FMO-SP workflow consists of four sequential stages: **Input → Run → Analysis → Graphics** (Fig. 3).

- **Stage 1—Input:** Converts protein–water–ligand complexes into FMO-ready input files.
- **Stage 2—Run:** Executes the FMO calculations, integrated within GAMESS, using the generated input(s).
- **Stage 3—Analysis:** Extracts the pair and total interaction energy terms calculated by FMO-SP (Table 1) for individual systems in CSV format, along with merged and transposed datasets for cross-system comparisons such as SAR analysis.

Table 1
Energy terms and chemical factor reported by FMO-SP in Stage 3 (analysis)

Term	Description
PIEtotal	Total pair interaction energy between a residue and the ligand
PIEes	Electrostatic energy contribution. The electrostatic component arises from the Coulomb interaction between polarized charge distributions of fragments
PIEex	Exchange-repulsion energy. The exchange repulsion term is derived from the interaction between fragments situated in close proximity and is always repulsive; it is due to the Pauli repulsion and is related to the overlap of two occupied orbitals
PIEct	Charge-transfer energy. The charge transfer term arises from the interaction between occupied orbitals of a donor and unoccupied orbitals of an acceptor
PIEdisp	Dispersion energy. The dispersion arises as the interaction between instantaneous dipole moments of two fragments; it is hydrophobic (nonpolar) in nature and is obtained in PIEDA from the correlation energy of electrons
PIEsolv	Solvation energy – the extent to which the interaction is shielded from water
Factor	$f_{ij}=(\mathrm{PIE}_{ij}^{es}-\mathrm{PIE}_{ij}^{solv})/(\mathrm{PIE}_{ij}^{es}-\mathrm{PIE}_{ij}^{solv}+\mathrm{PIE}_{ij}^{disp})$ Chemical factor distinguishes between electrostatic (ratio above 0.5) and hydrophobic (ratio below 0.5)
TIE	Total interaction energy—sum of all PIEtotal values
TIE-SIG	Significant total interaction energy from PIEtotal values with an absolute magnitude ≥ 3 kcal/mol
Lig_pol	Ligand polarization energy
Lig_solv	Energy cost or gain of transferring the ligand from water into the binding site
Prot_pol	Protein polarization energy
Prot_solv	Energy cost of removing solvent from the protein surface to accommodate the ligand
Bind	Sum of *TIE*, *Lig_pol*, *Lig_solv*, *Prot_pol*, and *Prot_solv*

- **Stage 4—Graphics:** Produces visual outputs—such as column plots for individual ligands and heatmaps for ligand sets, to support interpretation and decision-making.

2.1 Step 1: Input Preparation

This stage converts protein–water–ligand complexes into inputs for FMO calculations. Two input scenarios are supported (Fig. 3):

- **Multiple protein–water–ligand complexes (PDB)**—a directory containing pre-built complexes, suited to cases where the protein conformation differs for each ligand (e.g., molecular dynamics trajectories or induced-fit docking results).
- **Single protein (PDB) with multiple ligands (SDF)**—a directory containing a single protein structure and multiple ligands, used when the protein conformation is the same for all ligands (e.g., rigid docking workflows).

The Input stage is fully automated via an SVL script in MOE (v. 2024.0601), with a Python wrapper enabling standalone execution via *moebatch*. The system is prepared by truncating residues and water molecules within 4.5 Å of ligand atoms. Peripheral atoms in truncated regions are capped to prevent artificial charges, and gaps in residues are similarly capped to preserve chemical integrity.

Before input generation, an **optional structure optimization step** can refine the geometry of the complex. Optimization may be performed either with molecular mechanics (MM) before the FMO run or directly within the run using DFTB-PA [27]. In the MM-based protocol, a constrained minimization with the semi-empirical Amber10:EHT force field (implemented in MOE) is applied, allowing each atom to deviate by 0.5–1.5 Å from its original position. Protein and water atoms are restrained, while the ligand is fixed, relieving peripheral strain without altering the binding mode. For efficiency, optimization is applied only to the truncated system.

This optimization stage was introduced for user convenience; however, FMO-SP itself does not alter or "optimize" structures as part of its core function. Instead, it takes the structure as provided and generates a detailed "snapshot" of the interactions—analogous to a camera capturing an image without changing the subject. Users are therefore expected to complete all structure preparation prior to running FMO-SP.

An **optional ligand protonation step** can also be performed before optimization, assigning the preferred protonation state based on predicted pKa values at physiological pH 7.4. This ensures realistic electrostatic and hydrogen-bonding patterns in subsequent FMO calculations.

FMO-SP employs a **C–N fragmentation scheme** aligned with conventional residue boundaries, replacing the older CA–C approach [25]. It also supports DFTB-PA segmentation [27], enabling calculations on user-defined segments such as entire residues or functional subunits (Fig. 2).

Subsystem analysis (SA), developed by Dr. Dmitri Fedorov [28, 42], embeds inter-fragment effects into fragment-level properties such as energy and charge, providing a chemically intuitive breakdown of protein–ligand interactions. SA quantifies each fragment's contribution to binding—including often-overlooked terms such as desolvation and polarization energies—and tracks their changes during the binding process ($P + L \rightarrow PL$)—Fig. 4. This yields a residue-level energy profile that integrates both intrinsic and environmental effects, offering a precise and interpretable map of how individual residues influence binding and guiding optimization strategies.

Finally, FMO-SP incorporates a polarizable continuum solvent model (PCM) to account for the influence of the aqueous environment on protein–ligand binding. PCM enables implicit treatment of bulk solvent dielectric effects, providing a more realistic estimate of interaction energies compared to gas-phase calculations. By combining PCM with PIEDA and SA, FMO-SP captures both intrinsic residue energetics and their modulation by solvation. This is particularly important for charged and polar interactions, where

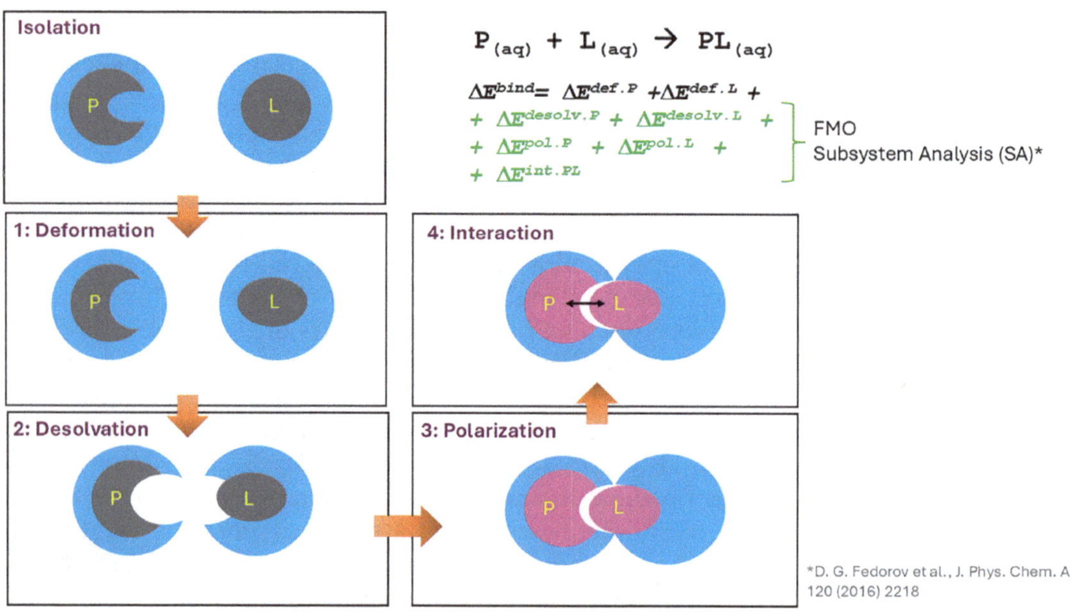

Fig. 4 Subsystem analysis (SA) of protein–ligand binding in FMO. Binding free energy is decomposed into four steps: (1) protein/ligand deformation, (2) desolvation, (3) mutual polarization, and (4) direct interaction. This framework provides a physically meaningful breakdown of binding energetics. The figure was originally designed by Dr. Dmitri Fedorov (Quantum Mechanics in Drug Discovery, Springer, 2020) and adapted for this chapter with his kind permission

shielding from the solvent can significantly attenuate or even reverse apparent binding strengths.

2.2 Step 2: Run

In this stage, the prepared systems (FMO inputs) are subjected to the actual FMO computations using the FMO module of GAMESS (US). As of July 15, 2024 (R2 Patch 1), this is the most recent publicly released version supporting FMO functionality.

2.2.1 Execution Environment

FMO-SP automates the setup and submission of GAMESS jobs, ensuring reproducibility and scalability. Calculations are parallelized to enable efficient use of multi-core processors and high-performance computing clusters.

2.2.2 FMO Computational Modes

FMO-SP supports two computational modes (Table 2):

- **MP2** (second-order Møller–Plesset perturbation theory) with the @c631@ basis set—highly accurate but slower. For example, a system of ~46 residues requires ~80 min on 14 CPU cores.
- **DFTB-PA** (density-functional tight-binding with partition analysis) using the 3ob-3-1 parameter set—significantly faster, with the same system completing in ~3 min, making it well suited for high-throughput screening or SAR analysis.

Table 2
Comparison of FMO computational modes [44]

Feature	FMO with MP2 [43]	FMO with DFTB-PA [27]
Method type	Ab initio (MP2 —second order perturbation)	Semi-empirical (DFTB-PA; DFTB with partition analysis)
Accuracy	High (explicit correlation)	Moderate (parameterized)
Fragmentation	Fragments (residues, waters, ligands)	Fragments + segments
Speed	Slow/high computational cost	Fast/low computational cost
System size	Medium to large	Very large (proteins, etc.)
Electron correlation	Explicitly treated	Approximate
Parameter use	No parameters	Needs parameter sets
Best for	Precise energy decomposition	SAR, scoring function, virtual screening
Energy decomposition	High-quality IFIE (Inter-Fragment Interaction Energy)	Trend-level IFIE accuracy
Geometry optimization	Feasible, slow	Fast and practical
Excited states	Possible	Limited
Common software	GAMESS, ABINIT-MP	GAMESS

Thus, while MP2 provides benchmark accuracy for detailed studies (e.g., crystal structures, critical design decisions), DFTB-PA offers a pragmatic balance between speed and accuracy for large-scale or iterative workflows.

2.2.3 Alternative FMO Platforms

FMO methodology is also implemented in other platforms:

- **ABINIT-MP**—a specialized FMO program with extensive QM/MM and solvent models.
- **PAICS**—a parallelized system for high-performance FMO computations.
- **OpenFMO**—an open-source, MIT-licensed FMO framework designed for distributed computing.

2.3 Step 3: Analysis

The analysis stage of FMO-SP is responsible for extracting and organizing the interaction energies generated during the FMO calculations. At the residue–ligand level, FMO decomposes binding into pair interaction energies (PIEs), which quantify the contribution of each protein fragment (typically a residue or water molecule) to ligand binding. These PIE values are further decomposed into chemically meaningful terms using PIEDA and SA. The full set of energy terms calculated in this stage is summarized in Table 1.

The outputs of this stage are delivered in CSV format, ensuring compatibility with downstream data analysis workflows. For each system, FMO-SP generates an individual CSV file containing residue-level and global energy terms. These files are then automatically merged and transposed into a single CSV matrix, enabling straightforward cross-comparison of multiple ligands against the same target. Such aggregated data are particularly valuable in SAR analysis, where medicinal chemists can directly correlate structural modifications with changes in residue-level energetics.

The residue-level resolution provided by this analysis enables the identification of both favorable interactions and unfavorable contributions. By quantifying these effects in a chemically intuitive format, Stage 3 forms the essential bridge between raw FMO outputs and actionable insights for drug design.

2.3.1 Chemical Factor

In this stage, the FMO-SP also computes a chemical factor f_{ij} defined as:

$$f_{ij} = \frac{PIE_{ij}^{esp} - PIE_{ij}^{solv}}{PIE_{ij}^{esp} - PIE_{ij}^{solv} + PIE_{ij}^{disp}}$$

This factor provides a dimensionless ratio that classifies the chemical nature of the interaction between fragment i and fragment j:

- $f_{ij} > 0.5$ interaction is predominantly **electrostatic/hydrophilic**

- $f_{ij} < 0.5$ interaction is predominantly **hydrophobic**

This classification helps in guiding design strategies—for example, deciding whether to introduce a polar group to strengthen an electrostatic contact, or to enhance hydrophobic complementarity in a nonpolar pocket

2.4 Step 4: Graphics

The Graphics stage transforms the numerical outputs generated in the Analysis stage of FMO-SP into clear, interpretable visualizations for medicinal chemistry and project teams. While the raw results are stored as CSV files, the graphical representation provides an immediate understanding of residue-level interactions and comparative trends across compound series. Following previous reports, we considered any interaction with an absolute PIE value of ≥3.0 kcal/mol to be significant, and used this threshold in both column plots and heatmaps.

2.4.1 Visual Outputs

- **Column plots** (Fig. 5b): Display the contribution of individual residues to ligand binding for a single complex, highlighting hotspots (favorable interactions) and unfavorable terms (repulsions, desolvation penalties). These plots make it straightforward to identify residues where optimization is most impactful.
- **Heatmaps** (Fig. 5d, e): Compare residue-level energetics across multiple ligands simultaneously. Heatmaps allow chemists to detect conserved interaction patterns, highlight residues driving selectivity, and directly link SAR changes to shifts in FMO-derived energies.
- **3D visualization** (Fig. 5c): Integration with MOE and PyMOL (ongoing development) overlays FMO interaction energies onto the 3D protein–ligand complex, allowing intuitive mapping of numerical results back onto structural features.

2.4.2 Workflow Integration

Visualization is supported through two complementary routes:

1. **Automated Python— post-processing pipeline** generates heatmaps column plots as part of the FMO-SP workflow, ensuring reproducibility and consistency across projects.
2. **KNIME Hub**—Users can generate the same plots interactively by uploading the merged CSV file (for heatmaps across multiple ligands) or individual CSV files (for column plots of single ligands, Fig. 5b). This option provides a flexible, user-driven interface for quickly exploring data without needing to run scripts locally.

By combining quantitative (kcal/mol) outputs with intuitive visual formats, Stage 4 serves as the decision-making bridge between computational chemistry and medicinal chemistry. Residues critical for binding can be prioritized for strengthening, unfavorable terms

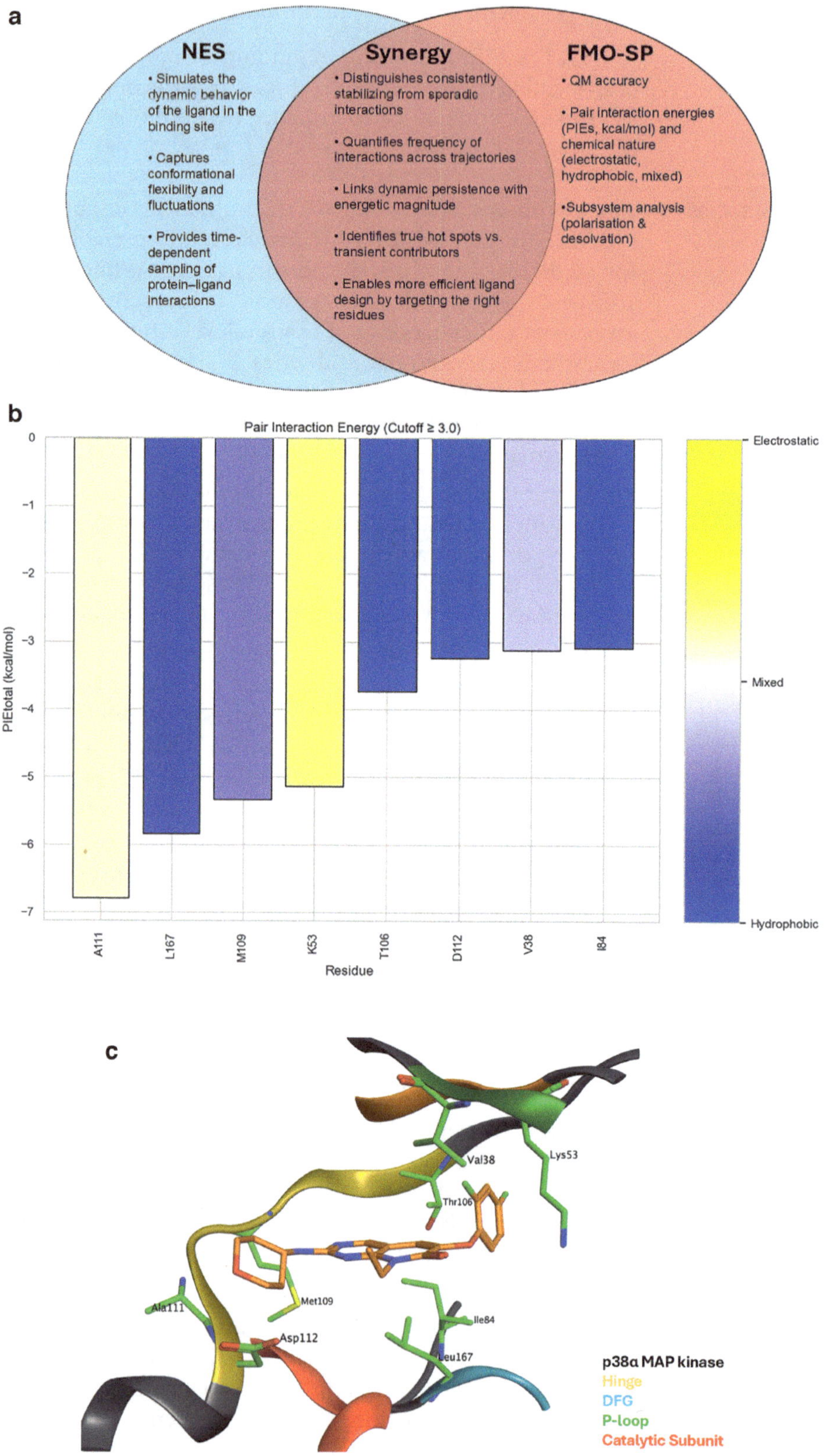

Fig. 5 (continued)

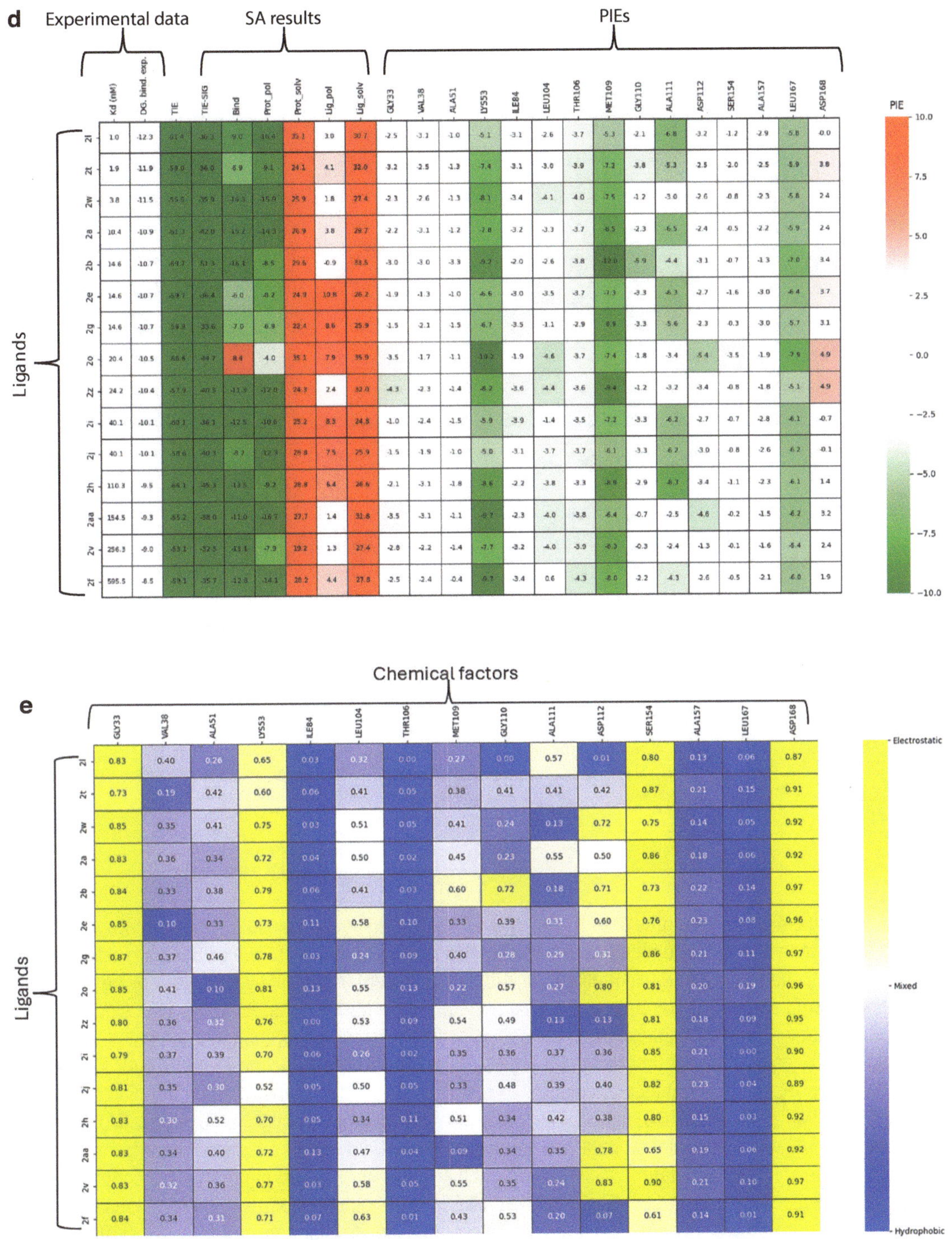

Fig. 5 (continued)

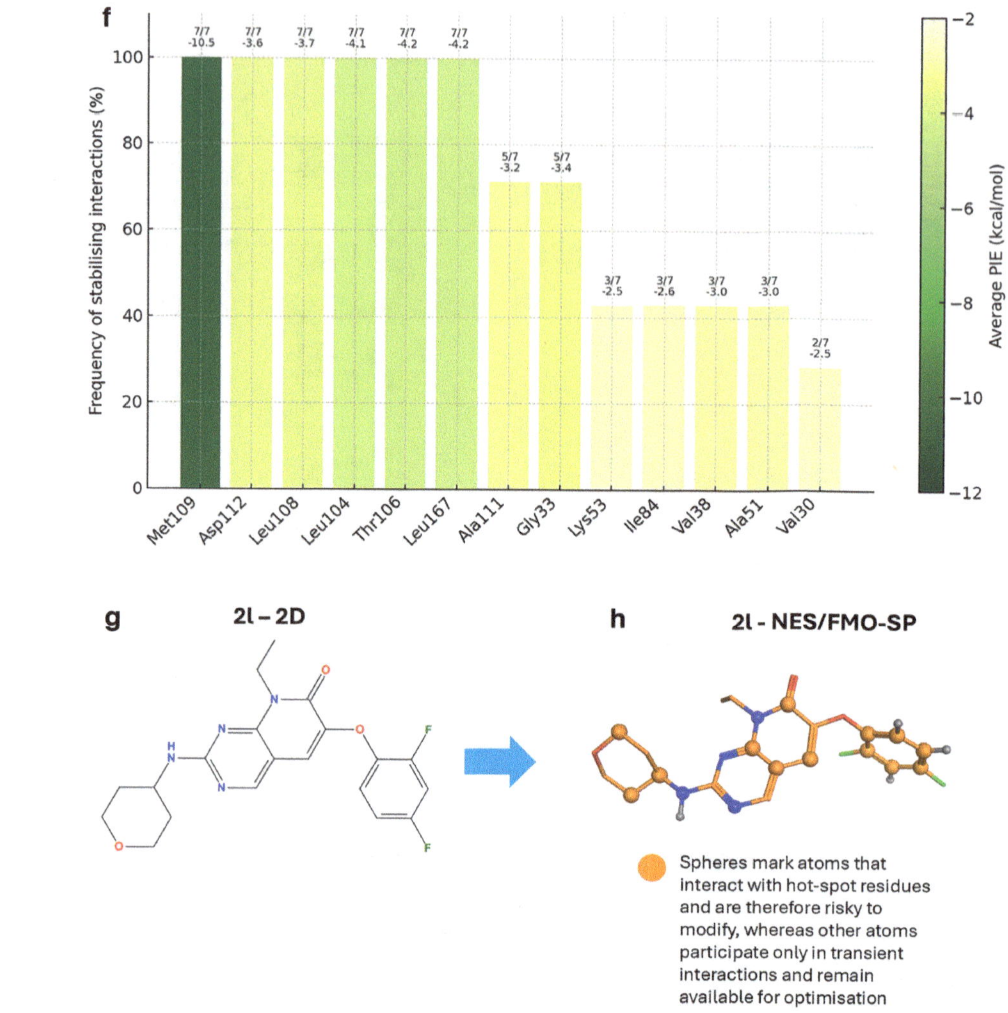

Fig. 5 (**a**) Complementary strengths of NES and FMO-SP. (**b**) Sorted column plot showing pair interaction energies (PIEs) between ligand 2l and residues of p38α MAP kinase (PDB: 3MPT [53]). Residues are ordered by PIE magnitude, with chemical nature (electrostatic, hydrophobic, mixed) indicated by color. (**c**) Key protein–ligand interactions identified by FMO-SP analysis for ligand 2l and p38α MAP kinase. Residues with significant PIEs (|PIE| ≥3.0 kcal/mol) are highlighted: green indicates favorable interactions, magenta indicates unfavorable interactions. (**d**) Heatmap of pair interaction energies (PIEs) for 15 ligands bound to p38α MAP kinase (PDB: 3MPT). Rows correspond to ligands and columns to residues, with color intensity indicating binding strength: green = favorable (≤−3.0 kcal/mol), red = unfavorable (≥ +3.0 kcal/mol). Subsystem analysis (SA), total interaction energy (TIE), and experimental data are also included. (**e**) Heatmap of chemical factor for the same 15 ligands, distinguishing electrostatic (yellow), hydrophobic (blue), and mixed (light shades) contributions. (**f**) Stabilization frequency and average PIEs across seven NES/MD trajectories for the p38α MAP kinase–ligand 2l complex. Bar height shows how often each residue passed the stabilization threshold (≤ −3.0 kcal/mol), while bar color indicates average PIE (green = strong, yellow = weak). (**g**) Example of how NES/FMO-SP analysis is converted into design recommendations for ligand 2l. Orange spheres mark atoms interacting with hot-spot residues (risky to modify), while other atoms represent transient interactions (available for modification)

flagged for redesign, and ligand modifications guided by a chemically interpretable picture of the binding environment.

2.5 Electrostatic Complementarity and FMO-SP (FMO-SP-ESC)

Electrostatic potential complementarity (ESC), originally developed by Katzir, Eisenstein, and Heifetz et al. [45]. and successfully applied to reduce hERG liability in Histamine H3 receptor binders [46], is one of the new features of FMO-SP. It maps regions on the ligand surface where electrostatics either attract or repel the protein, providing a direct visual and quantitative framework for SAR analysis and ligand design.

The electrostatic complementarity score (*ESCS*) between protein and ligand is defined as the overlap of their electrostatic potentials at each grid point (i, j, j):

$$ESCS = \sum_{i=1}^{N}\sum_{j=1}^{N}\sum_{k=1}^{N} P_{i,j,k}^{prot} \ X \ P_{i,j,k}^{lig}$$

where $(P_{i,j,k}^{prot})$ and $(P_{i,j,k}^{lig})$ are the protein and ligand electrostatic potentials at grid point (i, j, j). Electrostatic potentials are computed in MOE by solving the linearized Poisson–Boltzmann equation via a finite-difference method on a fine 0.5 Å grid, with grid dimensions adjusted to capture the full electrostatic field.

Regions of attraction or repulsion are visualized by mapping Lennard-Jones particles at grid points contributing significantly to the ESCS. Significance is defined by ranking all grid points by interaction strength and selecting the top 20%. Visualization in the MOE grid module highlights complementary and noncomplementary electrostatic zones on the ligand interface.

The accuracy of ESC depends on the predicted binding mode and the reliability of charges. MM-derived charges often miss self-polarization and local dielectric effects, while FMO-derived charges capture inter-fragment quantum interactions and provide a chemically realistic description of charge distribution. In FMO-SP, these polarization-dependent charges (e.g., Mulliken) are used to compute ESCS and generate complementarity maps.

We designate this implementation as FMO-SP-ESC, which integrates quantum-level polarization into electrostatic complementarity analysis, enhancing predictive power in SBDD and guiding ligand optimization with high precision.

2.6 Extended Capabilities of FMO-SP Beyond Protein–Ligand Complexes

While the FMO Sygnature Platform (FMO-SP) is highly effective for analyzing classical protein–ligand complexes, its scope is not limited to this domain. By leveraging its quantum-level energetics, subsystem analysis, and flexible fragmentation strategies, FMO-SP can be applied to a wide range of molecular systems where binding interfaces are complex or nontraditional.

Extended applications include:

- Protein–protein interactions (**PPIs**) [47]: Quantifying residue-level contributions at large, diffuse interfaces and identifying stabilizing versus destabilizing regions.
- Antibody–drug conjugates (**ADCs**): Disentangling energetic contributions of linker, payload, and antibody residues to optimize conjugation strategies.
- Molecular Glue Degraders (**MGDs**) and **PROTACs**: Mapping the cooperative energetics underlying ternary complex formation.
- **RNA–ligand** systems: Capturing electrostatic and stacking interactions critical for nucleic acid recognition.

These capabilities highlight that FMO-SP is not restricted to small-molecule binding but extends to emerging modalities where quantum-level precision is essential to disentangle complex interaction networks.

A detailed exploration of these applications, with case studies and retrospective analyses, is presented in Chap. 6. Furthermore, as described in Chap. 2, we are actively working on integrating FMO-SP with AI, creating hybrid pipelines that combine quantum-level interpretability with large-scale predictive power.

3 Showcases

3.1 Integrated FMO-SP and NES/MD Analysis of p38α MAP Kinase Binders

Protein–ligand binding is an inherently dynamic process, governed by a network of residue–water–ligand interactions. A deep understanding of these interactions cannot be obtained from a single static snapshot, since ligands continuously adjust within the protein pocket, forming and breaking interactions over time. To capture this behavior, it is essential to go beyond static analysis and employ an integrated framework that combines quantum-level energetics (FMO-SP) with dynamic simulations(NES/MD) [48–51].

NES/MD simulates the temporal evolution of ligand behavior within the binding site, while FMO-SP provides the enthalpic rationale for this behavior by identifying which residues contribute most strongly to stabilization. Together, the two methods link dynamic persistence with energetic magnitude, yielding a comprehensive and mechanistically interpretable picture of protein–ligand recognition. This integration supports the rational design of compounds with improved potencies and potentially longer residence times.

3.1.1 Demonstration on p38α MAP Kinase Binders

To demonstrate this integrated approach, we analyzed 15 known binders of p38α MAP kinase [52] (PDB: 3MPT). Instead of relying on static overlays, ligand–protein complexes were generated from the average structures of 6 ns NES/MD trajectories, thereby capturing conformational relaxation prior to quantum analysis.

The complementary strengths of NES and FMO-SP are illustrated in a Venn diagram (Fig. 5a), which highlights how their synergy enables the identification of consistently stabilizing interactions, quantifies their frequency across trajectories, and links persistence with energetic magnitude to inform design strategies.

3.1.2 Key Binding Determinants

Across the dataset, eight residues consistently emerged as key drivers of binding affinity (Fig. 5b,c):

- **Hinge region**: Thr106 (gatekeeper), Met109 (hydrophobic anchor), Ala111 and Asp112 (mixed contributions).
- **DFG motif**: Leu167 (hydrophobic).
- **P-loop/β-strand**: Val38.
- **Hydrophobic pocket**: Ile84.
- **Catalytic machinery**: Lys53 (electrostatic).

These findings are consistent with mutagenesis data: disruption of hinge hydrogen bonds abolishes potency, while mutations within the DFG motif alter selectivity.

3.1.3 Correlation with Experimental Data

Ligands forming strong hinge and hydrophobic interactions correlated with nanomolar Kd values, whereas those lacking hinge stabilization or incurring penalties from Lys53 or Asp112 typically exhibited micromolar affinities. Importantly, residues that consistently stabilized binding across NES/MD trajectories overlapped with those most predictive of experimental potency, underscoring the complementarity of the two approaches (Fig. 5d, e).

3.1.4 Dynamic Case Study: Ligand 21

We further applied the integrated framework to ligand **21**, analyzing seven NES/MD trajectories from the 6 ns simulation. For each trajectory, residue–ligand PIEs were computed using FMO-SP. An interaction was classified as "stabilizing" if the PIE was ≤ -3.0 kcal/mol.

To visualize these results, a combined frequency/PIE plot was generated (Fig. 5f). The frequency (*y*-axis) reflects how often each residue crossed the stabilization threshold across the seven trajectories (range: 0–7), while the color gradient encodes the average PIE strength (dark green = stronger, yellow = weaker).

This integrated view not only identifies residues providing the strongest energetic contributions but also highlights how consistently they do so across the dynamic ensemble. FMO-SP pinpointed hinge and hydrophobic residues as dominant energetic stabilizers, while NES/MD quantified their persistence over time. Together, the methods distinguish true hot spots from transient or fluctuating interactions.

For ligand **21**, the analysis revealed:

- **Hot spots**: Met109, Asp112, Thr106, Ala111, and Leu167—interactions with these residues must be preserved during optimization.
- **Transient interactions**: Val28, Val38, Ala49, Val35, Lys51, Ile84, Ile82, and Cys31—these residues explain SAR variability and ligand-dependent effects, suggesting that strengthening their interactions could enhance potency and selectivity.

The integration of these insights into design is illustrated in Fig. 5g, where hot-spot and transient interactions are mapped directly onto the ligand scaffold to distinguish "risky" and "design-accessible" regions.

3.1.5 Complementarity and Implications for Design

When analysis is limited to a single static snapshot, hot spots and transient interactions may appear indistinguishable, since both can contribute similar energies at one instant. By incorporating NES/MD trajectories into the FMO-SP framework, it becomes possible to resolve which interactions are consistently stabilizing and which are sporadic or ligand-dependent. This distinction provides medicinal chemists with a much more reliable basis for optimization.

Furthermore, the occurrence of interactions in some trajectories but not others highlights flexible or context-dependent regions of the binding pocket. Mapping this information onto the ligand structure reveals which regions are "nonnegotiable" (i.e., interacting with hot-spot residues and therefore risky to modify) and which are "design-accessible" (i.e., interacting primarily with transient residues).

Such knowledge is critical for guiding design strategies. Chemists can focus on reinforcing hot-spot interactions, or alternatively, on enhancing weaker transient interactions that contribute in a ligand-dependent manner. At a minimum, scaffold regions that already form strong hot-spot interactions should not be altered to avoid potency loss. Moreover, these strategies may contribute to longer residence times—a property increasingly recognized as important for efficacy and selectivity. Although residence time was not directly measured in this study, the integrated FMO-SP + NES/MD framework provides a strong foundation for its future exploration.

3.2 FMO-SP Analysis of hERG Liability

Unwanted binding to the human ether-à-go-go–related gene (hERG) potassium channel is a major safety concern in drug discovery [54]. Such binding can cause QT interval prolongation and life-threatening arrhythmias, making regulatory assessment of hERG liability mandatory for all small-molecule drug candidates. Addressing hERG risk early reduces late-stage attrition and ensures cardiovascular safety.

3.2.1 Structural Features of hERG Binding

Unlike many protein targets, hERG binding requires only simple pharmacophore elements—typically an aromatic group and a basic center [54–56].

- Many small modifications in ligand structure can dramatically alter hERG affinity.
- Medicinal chemists face the challenge of reducing hERG binding without losing potency at the primary target.
- Regions where such modifications can be applied are referred to as "hot spots."

3.2.2 Role of Electrostatics

As a potassium channel, hERG presents a predominantly negative electrostatic environment [46].

- Ligand modifications that increase negative charge may create repulsion and reduce binding.
- The challenge is to introduce changes in a spatially selective manner, preserving potency at the therapeutic target.
- Here, Electrostatic Complementarity (ESC) analysis becomes valuable: it maps areas of complementarity or clash between ligand and protein, guiding rational design [46].

3.2.3 Structural Dataset and Methodology

We applied FMO-SP and ESC analysis to the cryo-EM structure of hERG (KCNH2) bound to Pimozide (PDB: 8ZYQ [57]; resolution 3.18 Å).

- Pimozide is a potent hERG blocker (IC_{50} ~18 nM).
- The structure captures how a prototypical hERG inhibitor is anchored within the channel pore.

3.2.4 Key Binding Determinants Identified by FMO-SP

The FMO-SP results (Fig. 6a) highlight the dual anchoring mechanism:

- Electrostatic interactions: strong contributions from S624A, Y652A and F656A.
- Hydrophobic/π–π interactions: notable roles for Y652A and F656A.
- Secondary stabilization: additional contributions from T623A and S649A.

Together, these interactions explain the high affinity of Pimozide for hERG.

3.2.5 Insights from Electrostatic Complementarity Analysis (ESC)

ESC analysis (Fig. 6b) confirmed that the primary driver of hERG liability is the electrostatic attraction between:

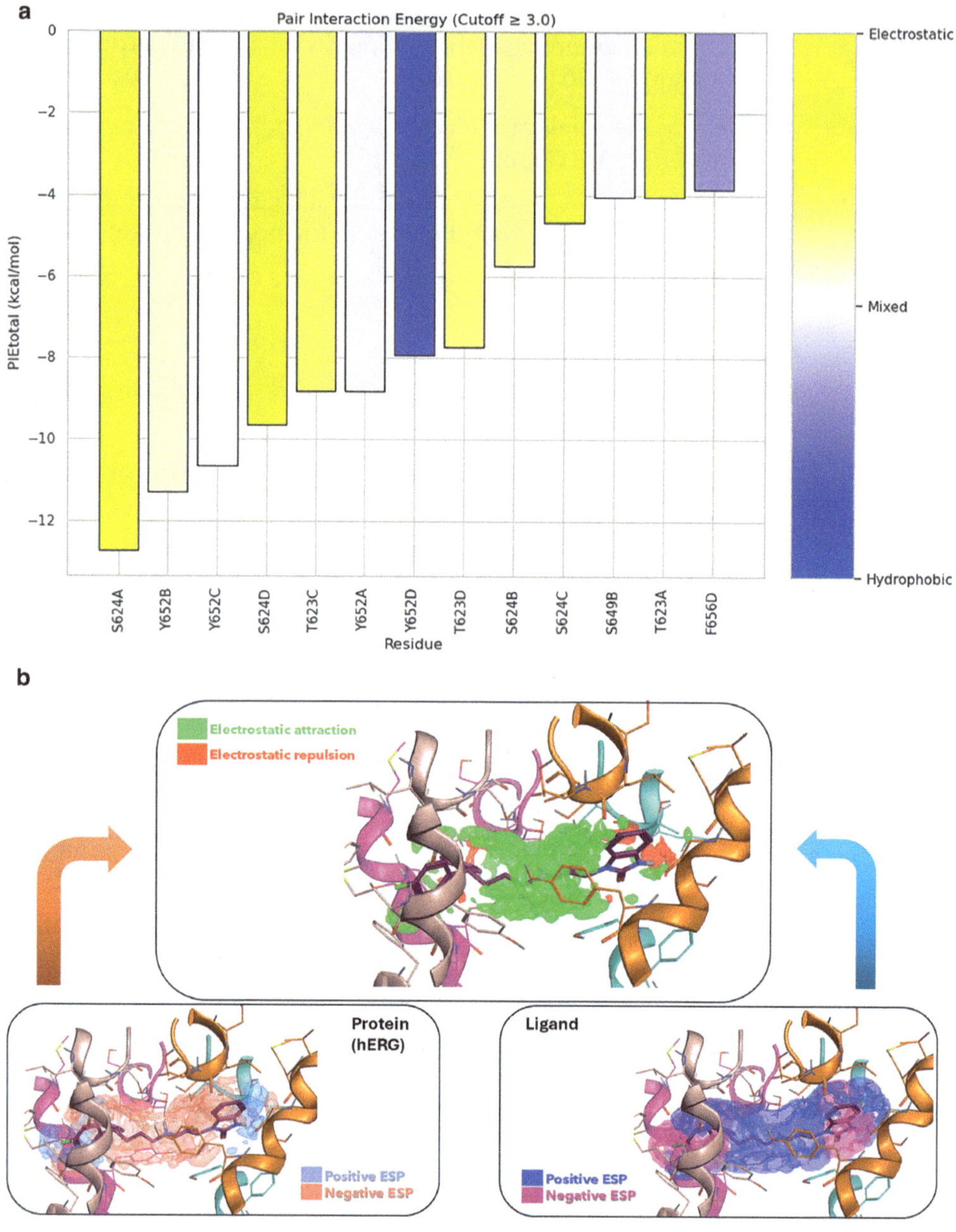

Fig. 6 (**a**) Column plot of pair interaction energies (PIEs) derived from the cryo-EM structure of the human hERG (KCNH2) potassium channel bound to Pimozide (PDB code: 8ZYQ). The *x*-axis shows interacting residues, while the *y*-axis represents PIE values (kcal/mol). Each column corresponds to an individual residue–ligand interaction, with the column length indicating interaction strength and the color denoting chemical nature: yellow = electrostatic, blue = hydrophobic, and light blue = mixed contributions. Only interactions above the threshold (|PIE| ≥3.0 kcal/mol) are displayed. (**b**) Electrostatic complementarity (ESC) analysis [46] of the cryo-EM structure of the human hERG (KCNH2) bound to Pimozide (PDB code: 8ZYQ) using the ESC/FMO-SP approach. The top panel shows regions of electrostatic attraction (green) and repulsion (red) between the ligand and protein. The bottom panels depict electrostatic surface potentials (ESP): protein (left) with positive ESP in blue and negative ESP in red, and ligand (right) with positive ESP in blue and negative ESP in red. Together, these views illustrate the strong electrostatic complementarity that stabilizes Pimozide binding within the hERG channel cavity

- the positively charged basic center of the ligand, which is positioned directly in the center of the channel's negative electrostatic field, and
- the residues lining the pore, notably Ser624A, Tyr652A, and Phe656A, which collectively contribute to the negative electrostatic environment surrounding the ligand

This spatial arrangement produces an exceptionally strong stabilizing force. It also explains why ligands containing a basic center so frequently exhibit hERG liability: the interaction is not a simple "salt bridge," as is sometimes assumed, but rather the consequence of the ligand's charge being buried inside a highly concentrated negative electrostatic environment.

From a medicinal chemistry perspective, one rational optimization strategy is to reduce the pK_a of the basic center, thereby lowering its protonation level and positive charge at physiological pH. This weakens the electrostatic attraction to hERG while preserving potency at the intended target if applied carefully.

3.2.6 Rational Optimization Strategies

Based on FMO-SP and ESC, several strategies emerge for reducing hERG liability:

- Lower the pK_a of the basic center to reduce its protonation and positive charge, weakening attraction to the negative electrostatic field of the channel.
- Reduce or shield the basic center through steric or electronic modifications.
- Introduce negative charge at carefully chosen positions to weaken electrostatic complementarity.

These strategies emphasize that spatial precision is critical: modifications must be targeted to the correct hot spots, balancing reduced hERG liability with retention of activity at the primary target.

3.2.7 Summary

The combined FMO-SP and ESC analyses uncover the precise electrostatic and structural anchors underlying hERG binding—particularly, the placement of the ligand's positively charged basic center into the core of the channel's negative electrostatic field. This arrangement creates a very strong stabilizing interaction, which explains why many drug-like molecules with basic centers exhibit hERG liability—not due to a simplistic "salt bridge," but because of their strategic charge placement.

From a design perspective, one powerful strategy is to lower the pK_a of the basic center, reducing its protonation—and thus its positive charge—under physiological conditions. This weakens the electrostatic attraction to the hERG channel, thereby reducing liability while aiming to preserve therapeutic potency.

Importantly, these modifications must be subtle and precisely targeted. General strategies such as lowering LogP or LogD may not be sufficient, since what matters is **where** the change is applied. Introducing polarity or charge at the wrong position can be ineffective—or even disastrous—by weakening binding to the primary target while leaving hERG affinity unchanged. Here, FMO-SP and ESC analyses provide unique guidance by identifying the exact hot spots on a ligand that drive hERG binding, enabling chemists to make rational, spatially targeted modifications that reduce hERG liability without compromising desired activity.

3.3 FMO-SP Analysis of Metabolic GPCRs

3.3.1 GPCRs as Drug Targets

G-protein coupled receptors (GPCRs) remain one of the most attractive drug development targets, with ~40% of all prescribed medicines acting on this family. Despite ~800 GPCRs encoded in the human genome, only ~50 have been successfully exploited to date, leaving significant untapped therapeutic potential. As will be described in Chap. 15, FMO methods have already been applied extensively to support GPCR drug design [58, 59].

3.3.2 Metabolic GPCRs in Drug Discovery

Metabolic GPCRs function as molecular switches, converting hormonal and nutrient-derived signals into cellular responses that regulate blood sugar, appetite, fat storage, and energy balance [60–62]. Their pivotal role in metabolism, combined with the emergence of blockbuster therapies, has made them some of the most exciting contemporary drug discovery targets [60–62].

3.3.3 Advances in Structural Biology

The structural biology of GPCRs has been transformed by cryo-electron microscopy (cryo-EM). Once elusive due to their dynamic and membrane-embedded nature, metabolic GPCRs can now be visualized at sub-2.5 Å resolution. At Sygnature Discovery, the Protein Science team has established strong expertise in preparing cryo-EM–ready samples of metabolic GPCRs, enabling structural insights that complement computational approaches such as FMO-SP.

3.3.4 Therapeutic Breakthroughs and Unmet Needs

Recent breakthrough medicines—including Ozempic, Wegovy, and Mounjaro—target the glucagon-like peptide-1 receptor (GLP-1R), which regulates satiety and glycaemic control [60–62]. However, their peptide-based nature requires subcutaneous injection, limiting patient convenience and adherence.

In 2022, Pfizer reported PF-06882961 (Danuglipron) [52], the first orally bioavailable small-molecule GLP-1R agonist [ref]. This milestone illustrated the opportunity to move beyond peptides. Here, we provide a retrospective analysis of how the design process for Danuglipron could have been accelerated and rationalized using FMO-SP.

However, despite this breakthrough, Danuglipron ultimately did not progress beyond clinical development. On April 14, 2025,

Pfizer officially discontinued all development of Danuglipron. The search for new small-molecule GLP-1R agonists therefore continues, driven by the substantial therapeutic potential of oral alternatives to peptide-based drugs. In this context, FMO-SP can play a critical role, not only in retrospectively rationalizing SAR but also in prospectively guiding optimization. As will be described here, FMO-SP is well positioned to accelerate the discovery of next-generation GLP-1R agonists by identifying key residue-level interactions, quantifying energetic trade-offs, and ranking optimization strategies.

3.3.5 FMO-SP Analysis of Danuglipron

FMO-SP analysis of Danuglipron and related GLP-1R agonists highlights the residue-level interactions driving potency and efficacy (Fig. 7a–d).

- Structural modeling: Danuglipron binds deep within the GLP-1R orthosteric site, engaging residues across TM2, TM3, ECL1, and ECL4.
- Key stabilizing residues: W33, D198, K197, Q221, F230, Q234, R380, F381, and the disulfide-linked C226–C296.
- These interactions form a diverse network of electrostatic, hydrophobic, and mixed interactions.

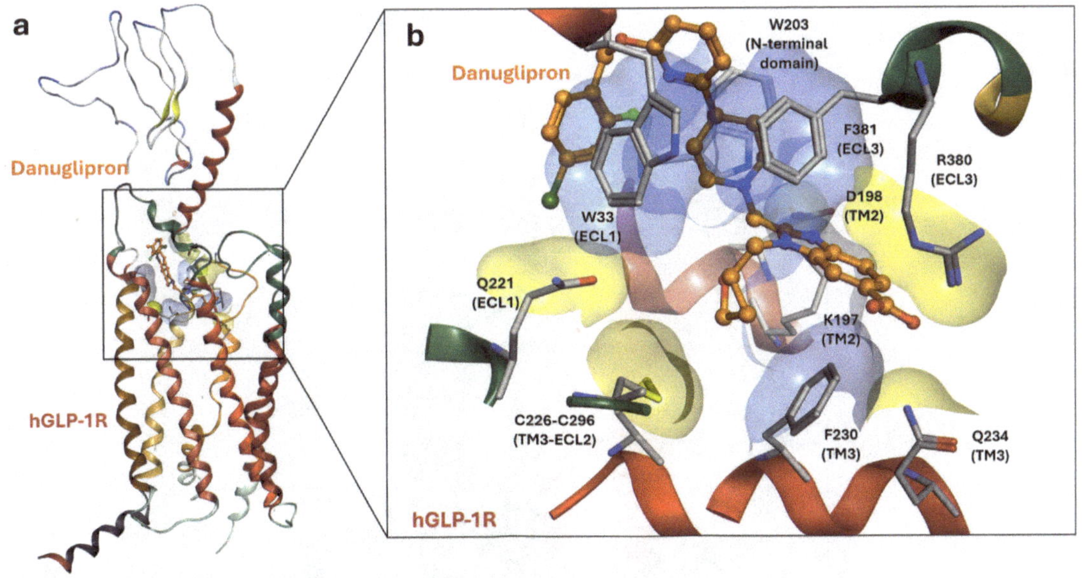

Fig. 7 (**a**) Predicted binding mode of Danuglipron (orange) within the cryo-EM structure of human GLP-1 receptor (GLP-1R, PDB code 7S15 [52]), shown in side view. The transmembrane helices are displayed as ribbons, with Danuglipron in stick representation. (**b**) Enlarged binding site view highlighting key GLP-1R residues responsible for interactions with Danuglipron. Residues are color-coded according to interaction type: electrostatic, hydrophobic, and mixed. Key contributors include W33 (N-terminal extracellular domain), D198 and K197 (TM2), Q221 (ECL1), F230 and Q234 (TM3), R380 and F381 (ECL3), and the disulfide-linked C226–C296 (TM3–ECL2).

c

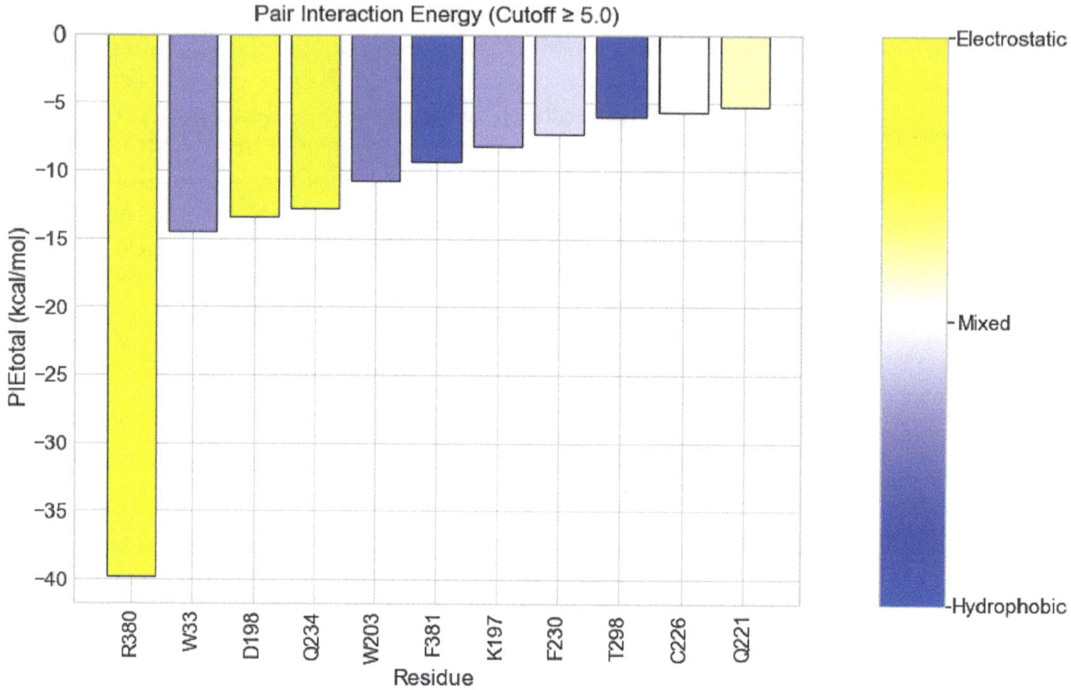

d

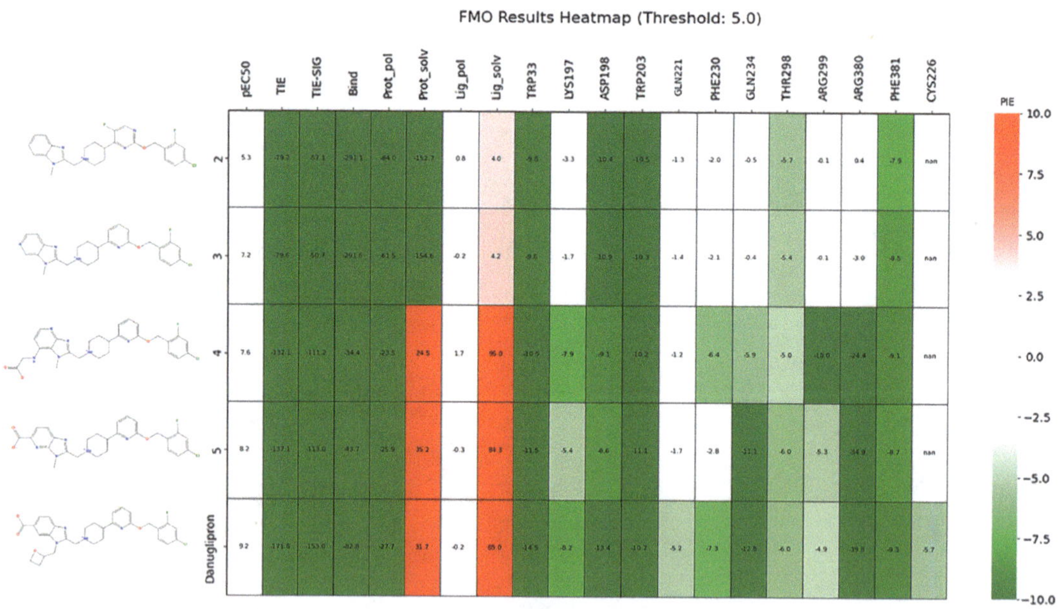

Fig. 7 (Continued) (**c**) Column plot of pair interaction energies (PIEs) for the predicted binding mode of Danuglipron (orange) in the crystal structure of human GLP-1R (PDB 7S15 [52]). FMO-SP analysis (cutoff ≤ −5 kcal/mol) identified 11 major interactions: four electrostatic, six hydrophobic, and one mixed. (**d**) Heatmap of FMO-SP results for five GLP-1R agonists (compounds 2–5 and Danuglipron). Each column shows PIE contributions (kcal/mol) with receptor residues. Experimental potency (pEC50) correlates strongly with FMO-SP-derived TIE and TIE-SIG values ($R^2 = 0.75$), supporting an association between these interactions and compound potency

3.3.6 Energetic Hot Spots and Correlation with Potency

- FMO-SP (Fig. 7c): identified 11 dominant interactions stronger than −5 kcal/mol. Electrostatics (D198, K197, and R380) were complemented by hydrophobic stabilization (F230, F381).
- Correlation with potency (Fig. 7d): Total interaction energy (TIE) correlated strongly ($R^2 \approx 0.75$) with experimental pEC50 values. This suggests that efficacy is determined by the additive effect of a small number of strong enthalpic interactions. FMO-SP not only reproduced mutagenesis data (e.g., the essential roles of W33 and R380) but also provided a predictive framework to rationalize SAR trends.

3.3.7 Insights for Future Design

Across all compounds, a consistent set of interaction hot spots (W33, D198, R380, and F381) emerges as essential anchors, while variations in secondary residues (Q221, Q234, and F230) explain SAR differences.

- Potency is tuned by the balance of electrostatic anchoring and hydrophobic reinforcement.
- Small-molecule agonists successfully mimic the electrostatic anchors of peptide ligands, underlining the principle of peptide-like mimicry in GPCR drug design.
- FMO-SP thus provides guidance for designing small molecules that replicate peptide interaction patterns, maximizing affinity and efficacy.

3.3.8 Beyond Small Molecules

The scope of FMO-SP extends beyond classical protein–ligand analysis. It can also be applied to peptide–protein complexes, such as Semaglutide bound to GLP-1R (PDB: 7KI0 [63]), which will be discussed further in Chap. 6.

4 Discussion and Conclusion

FMO-SP provides a practical way to integrate QM into modern drug discovery. By automating fragmentation, execution, analysis, and visualization, it delivers residue-level interaction energies with chemical interpretability at a speed compatible with medicinal chemistry timelines. This balance of quantum accuracy and workflow efficiency makes it well-suited for routine structure-based design, extending the reach of computational chemistry into areas where molecular mechanics alone is insufficient.

Our showcases highlight the versatility of FMO-SP across distinct challenges in drug discovery. In kinase design (p38α MAPK), integration with nonequilibrium simulation/molecular dynamics

(NES/MD) allowed us to connect energetic magnitude with dynamic persistence, distinguishing consistent hot spots from transient interactions. This analysis not only explained potency differences within a congeneric series but also suggested strategies for improving ligand residence time, a property increasingly recognized as critical for efficacy. By combining enthalpic and dynamic perspectives, the FMO-SP + NES/MD framework establishes a more reliable foundation for optimization than either method alone.

In safety pharmacology, application to the hERG potassium channel demonstrated how FMO-SP and electrostatic complementarity (ESC) together reveal the physicochemical origin of blockade. Rather than a simplistic salt-bridge model, the results showed that the burial of a protonated basic center within the highly negative electrostatic environment of the channel pore drives strong stabilization. This insight suggests a clear medicinal chemistry strategy: reducing the pK_a of the basic center to lower its protonation state at physiological pH, thereby decreasing affinity for hERG while maintaining potency at the therapeutic target. Such spatially precise design rules exemplify the kind of actionable guidance that FMO-SP can provide in addressing off-target liabilities.

In metabolic GPCRs(GLP-1R), retrospective analysis of Danuglipron revealed how electrostatic anchoring and hydrophobic reinforcement explain potency trends within the series. FMO-SP correctly reproduced the essential roles of residues such as W33 and R380 while identifying secondary contributions that account for SAR variability. These findings illustrate how quantum-level decomposition can rationalize why some modifications maintain efficacy while others do not, providing a roadmap for the discovery of next-generation oral agonists. Given the discontinuation of Danuglipron, such insights remain highly relevant as efforts to identify nonpeptidic GLP-1R modulators continue.

Together, these examples illustrate several broader points. First, FMO-SP excels at identifying interaction hot spots that must be preserved, while also flagging design-accessible regions where modifications are less risky. Second, by decomposing energetics into electrostatic, hydrophobic, and solvation-driven contributions, it provides medicinal chemists with chemically interpretable levers for optimization. Third, integration with dynamic simulations, such as NES/MD or complementary methods, such as ESC, broadens the scope of analysis, moving beyond static snapshots to address both stability and persistence.

Importantly, the platform is not restricted to classical protein–ligand complexes. As outlined in Sect 2.6 and described in detail in Chap. 6, FMO-SP is equally applicable to protein–protein interactions (PPIs), antibody–drug conjugates (ADCs), Molecular Glue Degraders (MGDs), PROTACs, and RNA–ligand systems. These modalities often involve complex binding interfaces that challenge conventional scoring functions, but where quantum-level residue resolution can disentangle stabilizing from destabilizing contributions.

Beyond these applications, FMO-SP is also well positioned to complement virtual screening (VS) workflows. While slower than docking- or AI-based scoring methods, its automation and workflow integration make it practical as a high-accuracy refinement filter. Applying FMO-SP to the top-ranked 1000–10,000 hits from VS campaigns can focus attention on compounds that establish the most critical residue-level interactions for potency and selectivity. This strategy combines the breadth of VS with the depth of quantum mechanics, ensuring that only the most promising candidates progress to experimental validation.

While still the most widely used and successful strategy, SBDD is constrained by the approximations of molecular mechanics and the difficulty of reliably capturing enthalpic contributions. By contrast, FMO-SP provides quantum-level energetics with practical usability, addressing a major blind spot in conventional workflows. Looking ahead, integration of FMO-SP with AI pipelines, as described in Chap. 2, aims to combine quantum-level interpretability with large-scale predictive power, enabling workflows that are both predictive and mechanistically transparent.

In conclusion, the FMO-SP represents a significant advance in making QM accessible and impactful for medicinal chemistry. By bridging static and dynamic perspectives, supporting both potency and safety optimization, and extending to emerging modalities, FMO-SP offers a versatile and practical solution to some of the most persistent challenges in drug discovery. Coupled with automation, virtual screening integration, and AI-driven scalability, it stands as a blueprint for the next generation of structure-based design pipelines—reducing design risk, accelerating optimization, and providing medicinal chemists with actionable insights across modalities.

5 Take-Home Key Messages on FMO-SP

- Quantum accuracy at practical speed: FMO-SP applies QM to large biomolecular systems with workflow automation, delivering chemically realistic interaction energies at residue resolution.
- Beyond static analysis: By integrating with NES/MD, FMO-SP links enthalpic magnitude with dynamic persistence, distinguishing consistent hot spots from transient interactions.
- Design guidance: FMO-SP identifies stabilizing versus destabilizing interactions, highlighting residues that must be preserved and regions of the ligand that are design-accessible.
- Broad applicability: While highly effective for protein–ligand complexes, FMO-SP extends to PPIs, ADCs, MGDs,

PROTACs, and RNA–ligand systems, with details explored in Chap. 6.

- AI integration: As discussed in Chap. 2, ongoing work is combining FMO-SP with AI pipelines to scale quantum-level interpretability to large chemical spaces.
- Practical drug discovery impact: FMO-SP reduces design risk, accelerates optimization, and supports potency, selectivity, and residence-time strategies, complementing virtual screening and experimental validation.

Acknowledgments

The authors would like to express their deepest gratitude to Dr. Dmitri Fedorov (AIST, Japan) for his invaluable support, guidance, and enduring friendship, which have been a constant source of inspiration in the development of the FMO-SP platform. His pioneering and fundamental contributions to the Fragment Molecular Orbital method have laid the foundation of our efforts and continue to inspire new directions. We also warmly thank Professor Kaori Fukuzawa (Osaka University, Japan; Chair of FMODD) and her students for their scientific inspiration, collaboration and stimulating discussions, which have profoundly shaped our understanding of FMO applications in drug design. Our sincere appreciation extends to Dr. Teruki Honma (RIKEN, Japan) and his team for their insightful perspectives, collaboration and encouragement, which have enriched both the scientific and collaborative spirit of this work. Finally, the authors gratefully acknowledge Barbara Sander and Andrew Henry (CCG Support) for their expertise and kind assistance with SVL scripting, as well as their patience in responding to our many requests. Together, these colleagues and friends have not only advanced the science but have also contributed to the spirit of collaboration, openness, and shared discovery that underpins this work.

References

1. Fedorov DG, Nagata T, Kitaura K (2012) Exploring chemiestry with the fragment molecular orbital method. Phys Chem Chem Phys 14:7562
2. Fukuzawa K, Tanaka S (2022) Fragment molecular orbital calculations for biomolecules. Curr Opin Struct Biol 72:127–134
3. Fedorov DG (2021) Electron density from the fragment molecular orbital method combined with density-functional tight-binding. Chem Phys Lett 780:138900
4. Fedorov DG, Inostroza D, Courbiere B et al (2025) Decomposition analysis for visualization of noncovalent interactions based on the fragment molecular orbital method. J Chem Theory Comput 21:4435–4446
5. Fedorov DG, Nakamura T (2022) Free energy decomposition analysis based on the fragment molecular orbital method. J Phys Chem Lett 13:1596–1601
6. Pruitt SR, Nakata H, Nagata T et al (2016) Importance of three-body interactions in

molecular dynamics simulations of water demonstrated with the fragment molecular orbital method. J Chem Theory Comput 12:1423–1435
7. Fedorov DG, Kitaura K (2007) Pair interaction energy decomposition analysis. J Comput Chem 28:222–237
8. Fedorov DG (2022) Polarization energies in the fragment molecular orbital method. J Comput Chem 43:1094–1103
9. Vuong VQ, Nishimoto Y, Fedorov DG et al (2019) The fragment molecular orbital method based on long-range corrected density-functional tight-binding. J Chem Theory Comput 15:3008–3020
10. Fedorov DG (2020) Three-body energy decomposition analysis based on the fragment molecular orbital method. J Phys Chem A 124:4956–4971
11. Akisawa K, Hatada R, Okuwaki K et al (2021) Interaction analyses of SARS-CoV-2 spike protein based on fragment molecular orbital calculations. RSC Adv 11:3272–3279
12. Watanabe C, Okiyama Y, Tanaka S et al (2021) Molecular recognition of SARS-CoV-2 spike glycoprotein: quantum chemical hot spot and epitope analyses. Chem Sci 12:4722–4739
13. Handa Y, Okuwaki K, Kawashima Y et al (2024) Prediction of binding pose and affinity of nelfinavir, a SARS-CoV-2 main protease repositioned drug, by combining docking, molecular dynamics, and fragment molecular orbital calculations. J Phys Chem B 128:2249–2265
14. Takebe K, Miyakawa S, Sangawa T et al (2025) Combined structural and FMO-based insights into shaft pilin polymerization mechanism in Streptococcus sanguinis. Int J Biol Macromol 332:148264
15. Fukuzawa K, Kato K, Watanabe C et al (2021) Special features of COVID-19 in the FMODB: fragment molecular orbital calculations and interaction energy analysis of SARS-CoV-2-related proteins. J Chem Inf Model 61:4594–4612
16. Watanabe C, Tanaka S, Okiyama Y et al (2023) Quantum chemical interaction analysis between SARS-CoV-2 main protease and Ensitrelvir compared with its initial screening hit. J Phys Chem Lett 14:3609–3620
17. Takaba K, Watanabe C, Tokuhisa A et al (2022) Protein-ligand binding affinity prediction of cyclin-dependent kinase-2 inhibitors by dynamically averaged fragment molecular orbital-based interaction energy. J Comput Chem 43:1362–1371
18. Shino A, Otsu M, Imai K et al (2023) Probing RNA–small molecule interactions using biophysical and computational approaches. ACS Chem Biol 18:2368–2376
19. Takebe K, Suzuki M, Kuwada-Kusunose T et al (2023) Structural and computational analyses of the unique interactions of opicapone in the binding pocket of catechol O-methyltransferase: a crystallographic study and fragment molecular orbital analyses. J Chem Inf Model 63:4468–4476
20. Hengphasatporn K, Harada R, Wilasluck P et al (2022) Promising SARS-CoV-2 main protease inhibitor ligand-binding modes evaluated using LB-PaCS-MD/FMO. Sci Rep 12: 17984
21. Morao I, Fedorov DG, Robinson R et al (2017) Rapid and accurate assessment of GPCR–ligand interactions Using the fragment molecular orbital-based density-functional tight-binding method. J Comput Chem 38:1987–1990
22. Heifetz A, Trani G, Aldeghi M et al (2016) Fragment molecular orbital method applied to lead optimization of novel interleukin-2 inducible T-cell kinase (ITK) inhibitors. J Med Chem 59:4352–4363
23. Hatada R, Okuwaki K, Mochizuki Y et al (2020) Fragment molecular orbital based interaction analyses on COVID-19 main protease–inhibitor N3 complex (PDB ID: 6LU7). J Chem Inf Model 60:3593–3602
24. Choi J, Kim H-J, Jin X et al (2018) Application of the fragment molecular orbital method to discover novel natural products for prion disease. Sci Rep 8: 13063
25. Fedorov DG (2024) The peptide bond: resonance increases bond order and complicates fragmentation. ChemPhysChem 25:e202400170
26. Nagata T, Fedorov DG, Sawada T et al (2012) Analysis of solute–solvent interactions in the fragment molecular orbital method interfaced with effective fragment potentials: theory and application to a solvated Griffithsin–carbohydrate complex. J Phys Chem A 116:9088–9099
27. Nishimoto Y, Fedorov DG (2016) The fragment molecular orbital method combined with density-functional tight-binding and the polarizable continuum model. Phys Chem Chem Phys 18:22047–22061
28. Fedorov DG, Kitaura K (2016) Subsystem analysis for the fragment molecular orbital method and its application to protein–ligand binding in solution. J Phys Chem A 120:2218–2231
29. Takaya D, Ohno S, Miyagishi T et al (2024) Quantum chemical calculation dataset for representative protein folds by the fragment molecular orbital method. Sci Data 11:1164
30. Watanabe C, Watanabe H, Okiyama Y et al (2019) Development of an automated fragment molecular orbital (FMO) calculation protocol toward construction of quantum mechanical calculation database for large biomolecules. Chem-Bio Inform J 19:5–18

31. Monteleone S, Fedorov DG, Townsend-Nicholson A et al (2022) Hotspot identification and drug design of protein–protein interaction modulators using the fragment molecular orbital method. J Chem Inf Model 62:3784–3799
32. Takaya D, Watanabe C, Nagase S et al (2021) FMODB: The World's First Database of quantum mechanical calculations for biomacromolecules based on the fragment molecular orbital method. J Chem Inf Model 61:777–794
33. Nippa DF, Boddy AJ, Atz K et al (2025) Accelerating compound synthesis in drug discovery: the role of digitalisation and automation. RSC Med Chem 16:5753. https://doi.org/10.1039/D5MD00672D
34. Schneider G (2018) Automating drug discovery. Nat Rev Drug Discov 17:97–113
35. Sadybekov AV, Katritch V (2023) Computational approaches streamlining drug discovery. Nature 616:673–685
36. Plowright AT, Johnstone C, Kihlberg J et al (2012) Hypothesis driven drug design: improving quality and effectiveness of the design-make-test-analyse cycle. Drug Discov Today 17:56–62
37. Walters WP, Wang R (2020) New trends in virtual screening. J Chem Inf Model 60:4109–4111
38. Baker LM, Aimon A, Murray JB et al (2020) Rapid optimisation of fragments and hits to lead compounds from screening of crude reaction mixtures. Commun Chem 3:122
39. Xia Y, Wang Y, Wang Z et al (2024) A comprehensive review of molecular optimization in artificial intelligence-based drug discovery. Quant Biol 12:15–29
40. Sadybekov AV, Katritch V (2023) Computational approaches streamlining drug discovery. Nature 616:673–685
41. Lewis R, Ertl P, Jacoby E et al (2005) Computational chemistry at Novartis. Chimia 59:545
42. Fedorov DG, Kitaura K (2004) The importance of three-body terms in the fragment molecular orbital method. J Chem Phys 120:6832–6840
43. Nagata T, Fedorov DG, Ishimura K et al (2011) Analytic energy gradient for second-order Møller-Plesset perturbation theory based on the fragment molecular orbital method. J Chem Phys 135:044110
44. Fedorov DG (2017) The fragment molecular orbital method: theoretical development, implementation in GAMESS, and applications. WIREs Comput Mol Sci 7:e1322
45. Heifetz A, Katchalski-Katzir E, Eisenstein M (2002) Electrostatics in protein–protein docking. Protein Sci 11:571–587
46. Davenport AJ, Möller C, Heifetz A et al (2010) Using electrophysiology and *in silico* three-dimensional modeling to reduce human *Ether-à-go-go* related gene K^+ channel inhibition in a histamine H3 receptor antagonist program. Assay Drug Dev Technol 8:781–789
47. Monteleone S, Fedorov DG, Townsend-Nicholson A et al (2022) Hotspot identification and drug design of protein–protein interaction modulators using the fragment molecular orbital method. J Chem Inf Model 62:3784–3799
48. Cossins BP, Foucher S, Edge CM et al (2009) Assessment of nonequilibrium free energy methods. J Phys Chem B 113:5508–5519
49. Poole WG, Samways ML, Branduardi D et al (2025) Accelerating fragment-based drug discovery using grand canonical nonequilibrium candidate Monte Carlo. Nat Commun 16: 6198
50. Melling OJ, Samways ML, Ge Y et al (2023) Enhanced grand canonical sampling of occluded water sites using nonequilibrium candidate Monte Carlo. J Chem Theory Comput 19:1050–1062
51. Cossins BP, Foucher S, Edge CM et al (2008) Protein–Ligand binding affinity by nonequilibrium free energy methods. J Phys Chem B 112:14985–14992
52. Griffith DA, Edmonds DJ, Fortin J-P et al (2022) A small-molecule oral agonist of the human glucagon-like peptide-1 receptor. J Med Chem 65:8208–8226
53. Down K, Bamborough P, Alder C et al (2010) The discovery and initial optimisation of pyrrole-2-carboxamides as inhibitors of p38alpha MAP kinase. Bioorg Med Chem Lett 20:3936–3940
54. Kalyaanamoorthy S, Barakat KH (2018) Development of safe drugs: the hERG challenge. Med Res Rev 38:525–555
55. Butler A, Helliwell MV, Zhang Y et al (2020) An update on the structure of hERG. Front Pharmacol 10:1572
56. Kalyaanamoorthy S, Lamothe SM, Hou X et al (2020) A structure-based computational workflow to predict liability and binding modes of small molecules to hERG. Sci Rep 10: 16262
57. Miyashita Y, Moriya T, Kato T et al (2024) Improved higher resolution cryo-EM structures reveal the binding modes of hERG channel inhibitors. Structure 32:1926–1935.e3
58. Lorente JS, Sokolov AV, Ferguson G et al (2025) GPCR drug discovery: new agents, targets and indications. Nat Rev Drug Discov 24:458–479
59. Hauser AS, Attwood MM, Rask-Andersen M et al (2017) Trends in GPCR drug discovery:

new agents, targets and indications. Nat Rev Drug Discov 16:829–842
60. Gogineni P, Melson E, Papamargaritis D et al (2024) Oral glucagon-like peptide-1 receptor agonists and combinations of entero-pancreatic hormones as treatments for adults with type 2 diabetes: where are we now?. Expert Opin Pharmacother 25:801–818
61. Zheng Z, Zong Y, Ma Y et al (2024) Glucagon-like peptide-1 receptor: mechanisms and advances in therapy. Signal Transduct Target Ther 9:234
62. Drucker DJ (2022) GLP-1 physiology informs the pharmacotherapy of obesity. Mol Metab 57:101351
63. Zhang X, Belousoff MJ, Liang Y-L et al (2021) Structure and dynamics of semaglutide- and taspoglutide-bound GLP-1R-Gs complexes. Cell Rep 36:109374

Chapter 5

Protein–Protein Interactions with the Fragment Molecular Orbital Method

Dustin C. Woods, Stefania Monteleone and Stéphanie Labouille

Abstract

Protein–protein interactions (PPIs) are essential for the function of many proteins, and their dysregulation contributes to a wide range of diseases. As such, PPIs have emerged as challenging yet promising targets for drug discovery. However, the structural and energetic complexity of PPIs has limited the success of traditional small-molecule strategies, underscoring the need for accurate and interpretable computational tools to characterize interfacial energetics. The fragment molecular orbital (FMO) quantum mechanics method offers a comprehensive and computationally inexpensive means of identifying the strength and the chemical nature of the molecular interactions taking place at the protein–protein interface of large biomolecular systems. The application of FMO to protein–protein interfaces enables residue-level identification of binding *hotspots* and classification of key energetic contributions, including direct, intramolecular, and water-mediated interactions.

Keywords Protein–protein interactions, FMO, Hotspot residues, Structure-based drug design, PROTAC, PPI inhibitors, Molecular glues, Ternary complexes

1 Introduction

Protein–protein interactions (PPIs) play a defining role in nearly every biological process [1], mediating signaling cascades, structural assemblies, the regulation of enzymatic activity, cell growth, apoptosis, DNA replication, and transcription. These interactions generate a complex network, called the *interactome*, in which disruptions can propagate through signaling pathways and contribute to diverse pathologies [2–5]. The interfaces that drive PPIs are typically large and topographically complex, but binding is often dominated by a limited set of energetically critical *hotspot* residues that account for a disproportionate fraction of the binding free energy [6]. The ability to quantify the energetic contributions of individual residues across an interface, whether for the purpose of designing stabilizers, inhibitors, or degraders, has therefore become a central goal of both computational chemistry and structure-based drug design.

Alexander Heifetz (ed.), *Quantum Mechanics Applications in Drug Design*, Methods in Molecular Biology, vol. 3061,
https://doi.org/10.1007/978-1-0716-5420-0_5,

The growing understanding of PPI energetics has driven the emergence of PPI-focused drug discovery strategies [3]. PPI inhibitors aim to disrupt native protein interfaces, an approach that has yielded clinical successes against targets such as MDM2–p53 [7, 8], the menin-KMT2A PPI in acute myeloid leukemia [9], and others [10–12]. Conversely, molecular glues act by inducing or stabilizing novel PPIs, promoting functional associations that enhance degradation or signaling [13]. These mechanisms have recently converged with the advent of proteolysis-targeting chimeras (PROTACs), bifunctional molecules that recruit an E3 ligase to a protein of interest, forming a ternary complex that induces ubiquitination and degradation [14]. Computational modeling of PROTAC ternary complexes has underscored the importance of accurately describing cooperative PPI energetics, a challenge that purely empirical methods often fail to capture [14].

Early efforts to identify *hotspots* relied primarily on site-directed mutagenesis (SDM), most notably alanine scanning [15]. While effective, SDM is time-consuming [16], expensive, and limited by its inability to probe large numbers of residues or transient complexes. To mitigate these bottlenecks, computational methods emerged, such as molecular mechanics (MM) based scanning and free-energy decomposition [17]. However, these methods often depend on parameterized force fields that struggle to describe non-covalent forces abundant on protein interfaces, such as CH/π [18], cation/π [19], halogen/π [20], and nonclassical hydrogen bonds [21–23]. Conventional MM force fields tend to approximate these terms with simplified parameter sets, which can obscure subtle, but energetically significant interactions [24, 25]. These limitations motivated the development of the Fragment Molecular Orbital (FMO) method, which combines the rigor of quantum mechanics (QM) with practical computational scaling to accurately model these effects and account for polarization, charge redistribution, and electron delocalization [26]. Applied to protein–protein interfaces, FMO extends quantum mechanical methods to quantify residue–residue interactions, accurately identifying *hotspots* and stabilizing contributions often missed by classical approaches. By decomposing these interactions into intuitive maps, FMO offers a structure-based alternative to experimental *hotspot* screening.

2 FMO Applied to PPIs

FMO has been extensively applied to study protein-ligand interactions [27–29]. In these calculations, one typically fragments the protein by cutting backbone Cα–C bonds and treats the ligand as a single fragment. For protein–protein complexes, or to some extent for any complex involving a larger ligand, this strategy becomes computationally intractable. Instead, one needs to fragment both partners. While this is not technically difficult to do, visualization and interpretation of the results require careful consideration. Multiple protocols have been reported in the literature [30–34], which mainly differ by how they display the results (Fig. 1).

In FMO-PPI [34] (Fig. 1A), for instance, pair interaction energies are represented as a 2D heat map and may include intra- and intermolecular interactions, as well as water bridges and ligands. Alternatively, the heatmap may be colored according to chemical factor, i.e., the ratio between electrostatics and dispersion terms. Chemical factors provide an intuitive metric to describe the energetic composition, electrostatic vs. hydrophobic, of each fragment–fragment interaction. FMO-PPI excludes backbone–backbone interactions from the analysis because these interactions cannot be probed experimentally and make the results more difficult to interpret. This is done by subtracting pair interaction energies (PIEs) computed

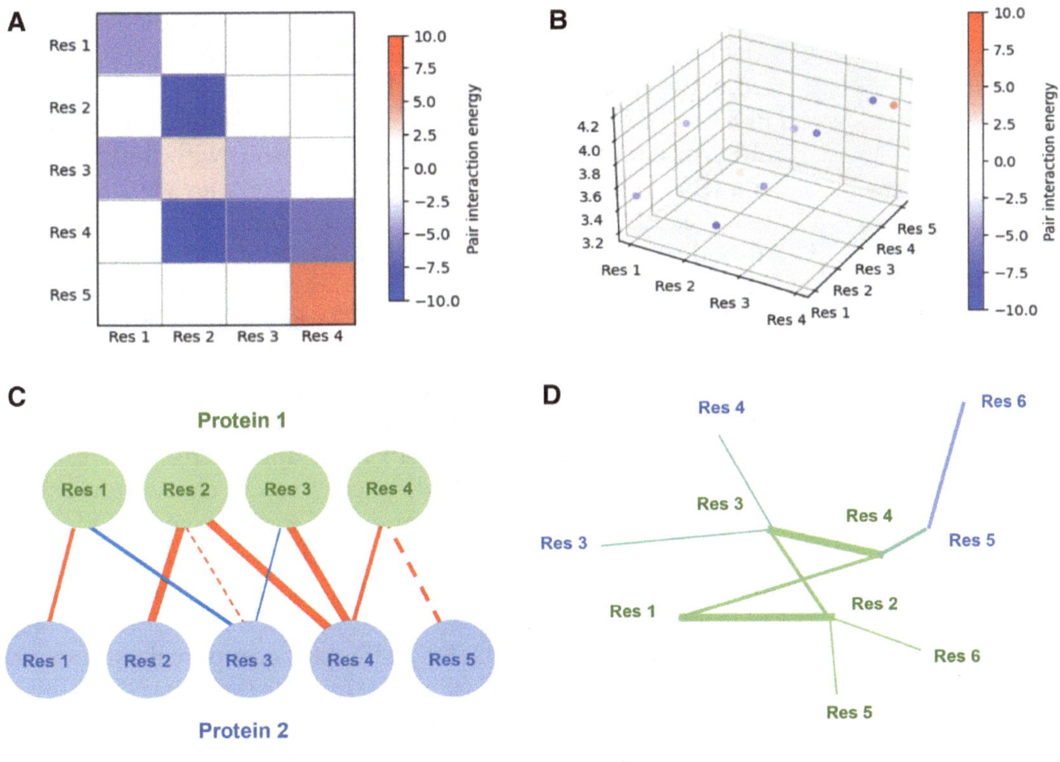

Fig. 1 Different representations of FMO-PPI results. (**A**) 2D heat map, colored by strength of pair interaction energies (PIEs). Residues from protein 1 and protein 2 are plotted along the x- and y-axis, respectively. (**B**) 3D-scattered pair plot (3D-SPIEs). Distance between residues is shown along the z-axis. Dots are colored by the strength of pair interaction energies. (**C**) PIE map. Line thickness is proportional to the strength of pair interaction energies. Lines are colored by their chemical character: dark blue for hydrophobic interactions (driven by dispersion) and red for electrostatics interactions. Dashed lines indicate repulsion, and non-dashed lines represent attraction. (**D**) PIE–PRN plot. Line thickness is proportional to pair interaction energies, while neither the sign of the interactions nor their chemical character is represented.

for an artificial backbone-only complex (all side chains replaced by hydrogen atoms) from the PIEs computed using the wild-type all-atom complex. This reduces the number of interactions detected, enhancing interpretability of the results.

In 3D-SPIEs [33] (Fig. 1B), pair interaction energies between fragments are represented as a 3D scatter plot, where x- and y-axes indicate fragments (one protein per axis), z-axis corresponds to the distance between the fragments, and points are colored according to their PIE values. Other representations include PIE-maps and Protein Residue Networks (PIE–PRN) [30] (Fig. 1C, D).

3 Hotspots Identification

Because they often have large and flat interfaces, PPIs are known to be difficult targets for small-molecule inhibitors. However, only a few residues typically contribute to most of the binding affinity. Understanding these *hotspots* can help identify novel potent inhibitors.

The 3D-SPIEs method was used to analyze *hotspots* at the interface between cell death 1 (PD-1) and programed death ligand-1 (PD-L1) [33]. The authors identified two *hotspot* regions in the wild-type complex. By comparing these results with those obtained for a series of related complexes (PD-1 in complex with antibodies, as well as PD-L1 in complex with small molecules, peptides, and antibodies), the authors concluded that only one *hotspot* region is critical for inhibition.

Another application of 3D-SPIEs was the analysis of the YAP–TEAD interface [35]. From the FMO results, the authors derived a pharmacophore model and proceeded with a virtual screening campaign that led to the identification of a novel micromolar inhibitor.

In a separate work, FMO-PPI was applied to study protein–protein complexes spanning different protein classes, including β -lactamase, interferon, kinase, and GPCR. This method identified *hotspots* consistent with SDM experiments but also revealed additional features of the interactions driving protein–protein binding [34] (Fig. 2).

The analysis of the INFα2–IFNAR2 complex illustrates how FMO–PPI can better rationalize SDM. While the mutation of residue $L30^{IFN\alpha2}$ into alanine has a dramatic effect on the formation of the INFα2–IFNAR2 complex, FMO-PPI found that this residue is not directly interacting with IFNAR2 but instead is forming a strong interaction with $R33^{IFN\alpha2}$, which in turn is involved in many intermolecular IFNAR2-interactions. These results suggest that $L30^{IFN\alpha2}$ plays a critical role in stabilizing the bioactive conformation of $R33^{IFN\alpha2}$, thereby indirectly affecting complex stability.

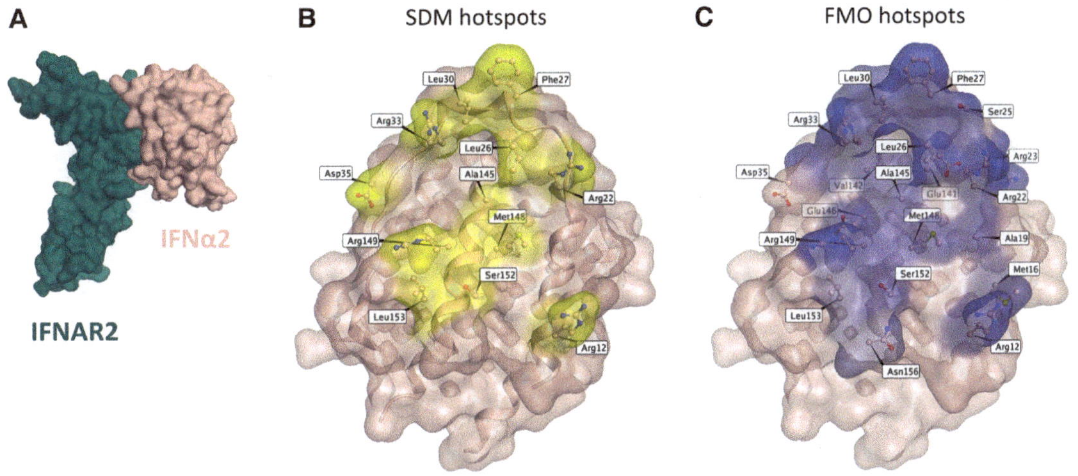

Fig. 2 Hotspot identification using FMO. (**A**) INFα2-IFNAR2 complex (PDB ID: 3S9D), where IFNAR2 and IFNα 2 are colored dark teal and pale pink, respectively. (**B**) INFα2 *hotspots* identified by Site Directed Mutagenesis (SDM) experiments are labeled and highlighted in yellow. (**C**) INFα2 *hotspots* predicted by FMO calculations are labeled and colored according to their relative PIE values. FMO results show consistency with SDM experiments and provide additional information on the strength of the interactions (darker blue indicates lower PIEs and thus stronger interactions)

4 Structure-Based Drug Design of PPI Modulators Using FMO

The next step towards the design of a novel PPI modulator is targeting the identified *hotspots* in a virtual screening campaign or in more advanced phases, such as hit-to-lead and lead optimization. The strategy may be slightly different depending on the mode of action of the ligand, for instance: (a) PPI inhibitor, (b) molecular glue, or (c) PROTAC.

Previous publications report on FMO can be applied to guide the SBDD towards potent PPI inhibitors [35, 36]. In this case, as the goal is to inhibit protein–protein interactions, it is recommended to start with the FMO analysis of the protein–protein interface to identify key *hotspots*. After mapping them on the surface of the two (or more) proteins, a potential ligand binding site can be selected considering that it should include at least one *hotspot* and be druggable to enable ligand binding. Recently, this methodology was retrospectively applied to the design of a potent inhibitor of KRAS and SOS1 interaction reported in literature [37], demonstrating how FMO can identify and target *hotspots* toward the design of a PPI inhibitor [34] (Fig. 3). The hit-to-lead and lead optimization phases will then follow traditional structure-based drug design approaches while being facilitated using FMO [35, 36].

For molecular glues, the aim is to stabilize the interface between two proteins. In the absence of a trimeric structure, ligand design

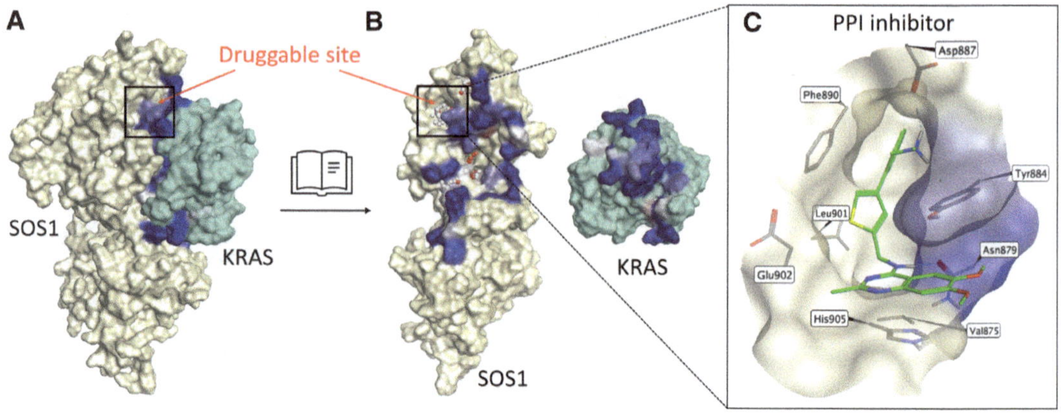

Fig. 3 Structure-based design of a PPI inhibitor using FMO. (**A**) SOS1-KRAS complex (PDB ID: 6EPL), where SOS1 and KRAS are pale yellow and mint green, respectively. (**B**) Mapping of the *hotspots* on the surfaces of the disassembled proteins. SOS1 and KRAS were rotated in opposite directions by 90° on the vertical axis and translated to expose the interfaces of the two proteins. Each interface residue is colored by its highest Δ PIE value, where the shade of blue darkens with the strength of the PIEs. Potential ligand binding pockets on SOS1 were identified by Site Finder [38, 39] in the Molecular Operating Environment (MOE) [40] and marked by spheres (white spheres represent the hydrophobic area and red spheres the polar one). (**C**) BAY-293–SOS1 complex (PDB ID: 5OVI), where the carbons are colored green and dark grey for the ligand and receptor sidechains, respectively. *Hotspot* residues are labeled and highlighted in blue, and the ligand is represented in green sticks for carbon atoms

becomes challenging, as the protein–protein complex forms only in the presence of the glue. However, given an experimental structure, FMO can be applied to identify key interactions between the components and to design an optimized lead compound with higher activity and potentially stronger target degradation. Predicting the effect of ligand modifications is even more challenging for molecular glues than for PPI inhibitors, largely because these systems are highly flexible and accommodate substantial protein-induced fit upon ligand binding. As a result, activity cliffs are common. However, given that interactions between molecular glues and their target proteins are predominantly hydrophobic, FMO enables quantification of subtle variations in protein–ligand and protein–protein interactions, informing the design of more potent lead compounds. For instance, the FMO analysis of the CRBN-CK1α-lenalidomide complex [34] showed that the molecular glue forms a key CH–π interaction with G40 in the degron loop of CK1α. Not only is this interaction difficult to quantify if not using QM calculations, but also this is the only interaction between the ligand and the target protein, highlighting the crucial role of this residue in the formation of the trimeric complex.

Finally, PROTACs constitute another class of PPI modulators, which are heterobifunctional small molecules comprising two warheads and a linker. Typically, the warheads are known ligands: one for the E3 ubiquitin ligase and the other for the protein of interest.

Here, the main challenge is the formation of the trimeric complex upon coupling of the warheads using a generally long and flexible linker. Again, if the experimental structure of the complex is known, FMO can be applied to analyze its PPIs and protein–PROTAC interactions to better guide the design of improved compounds. In a recent study, FMO was used to rank predicted PROTAC ternary complexes, whereby both the protein–ligand and protein–protein interactions were computed to select near-native structural models [41]. In one of the systems under consideration (VHL–BRD4), the PPIs constituted about 44% of the total interaction energy computed by FMO on the crystal structure, demonstrating their crucial role in the achievement of high cooperativity of that specific PROTAC. Furthermore, QM calculations have been used to predict the degradation potential of a PROTAC, as suggested by Nandy et al. [42].

5 Challenges and Future Perspectives

Despite its quantum-level accuracy, a central challenge for FMO is that its application relies on static structures, whereas realistic energetics emerge from inherently dynamic ensembles that govern biomolecular interactions. To advance beyond static models, integrating molecular dynamics (MD) with FMO would capture heterogeneity, flexibility, and entropic effects critical for accurate energetics [43]. MD reveals transient contacts and water-mediated interactions that shape residue-pair QM energies central to FMO [44, 45]. While free-energy perturbation (FEP) [46] and MM/GB(PB)SA [47, 48] incorporate dynamics, they face cost–speed and interpretability trade-offs, respectively. MD–FMO workflows remain challenged by sampling adequacy, frame selection, and solvent complexity [49]. Enhanced sampling mitigates these challenges [50] yet raises computational demands. Ultimately, balancing ensemble realism with quantum-level granularity will define next-generation FMO [51].

Recent developments in FMO, such as automated integration within molecular modeling environments [52] and density-functional tight-binding (DFTB) acceleration [53], continue to expand the scope and practicality of the method. FMO–DFTB has recently been applied successfully to score and rank putative ternary complexes in targeted protein degradation [41]. It was shown to outperform classical energy functions in ranking near-native PROTAC poses and provided insights into the cooperative residue networks underlying ternary assembly.

Taken together, FMO and its derivatives represent a modern framework for understanding and designing PPIs. While faster heuristic or machine-learning approaches provide valuable triage, FMO uniquely combines accuracy, interpretability, and mechanistic granularity. It supports not only the identification of binding *hotspots* but also the discovery of druggable subpockets in multibody systems, and the rational design of both inhibitors and stabilizers.

As computational efficiencies and software integrations continue to improve, quantum-mechanical residue-pair analysis through FMO is expected to become an increasingly practical and informative tool for studying and designing protein–protein interactions.

References

1. Snider J, Kotlyar M, Saraon P, Yao Z, Jurisica I, Stagljar I (2015) Fundamentals of protein interaction network mapping. Mol Syst Biol 11(12):848. https://doi.org/10.15252/msb.20156351
2. Koh GCKW, Porras P, Aranda B, Hermjakob H, Orchard SE (2012) Analyzing protein–protein interaction networks. J Proteome Res 11(4):2014–2031. https://doi.org/10.1021/pr201211w
3. Lu H, Zhou Q, He J, Jiang Z, Peng C, Tong R, Shi J (2020) Recent advances in the development of protein–protein interactions modulators: mechanisms and clinical trials. Signal Transduct Target Ther 5(1):213. https://doi.org/10.1038/s41392-020-00315-3
4. White AW, Westwell AD, Brahemi G (2008) Protein–protein interactions as targets for small-molecule therapeutics in cancer. Expert Rev Mol Med 10:e8. https://doi.org/10.1017/S1462399408000641
5. Blazer LL, Neubig RR (2009) Small molecule protein–protein interaction inhibitors as CNS therapeutic agents: current progress and future hurdles. Neuropsychopharmacology 34(1):126–141. https://doi.org/10.1038/npp.2008.151
6. Keskin O, Gursoy A, Ma B, Nussinov R (2008) Principles of protein–protein interactions: what are the preferred ways for proteins to interact?. Chem Rev 108(4):1225–1244. https://doi.org/10.1021/cr040409x
7. Sun D, Li Z, Rew Y, Gribble M, Bartberger MD, Beck HP, Canon J, Chen A, Chen X, Chow D, Deignan J, Duquette J, Eksterowicz J, Fisher B, Fox BM, Fu J, Gonzalez AZ, Gonzalez-Lopez De Turiso F, Houze JB, Huang X, Jiang M, Jin L, Kayser F, Liu J, Lo M-C, Long AM, Lucas B, McGee LR, McIntosh J, Mihalic J, Oliner JD, Osgood T, Peterson ML, Roveto P, Saiki AY, Shaffer P, Toteva M, Wang Y, Wang YC, Wortman S, Yakowec P, Yan X, Ye Q, Yu D, Yu M, Zhao X, Zhou J, Zhu J, Olson SH, Medina JC (2014) Discovery of AMG 232, a potent, selective, and orally bioavailable MDM2–p53 inhibitor in clinical development. J Med Chem 57(4):1454–1472. https://doi.org/10.1021/jm401753e
8. Siriki A, Ragland G, Vasireddy J, Chitneni SK (2025) A concise review on MDM2 inhibitors and recent progress in radiopharmaceutical development for imaging MDM2 expression in tumors with PET or SPECT. Bioorg. Med. Chem. Lett. 129:130394. https://doi.org/10.1016/j.bmcl.2025.130394
9. Candoni A, Coppola G (2024) A 2024 update on menin inhibitors. a new class of target agents against KMT2A-rearranged and NPM1-mutated acute myeloid leukemia. Hematol Rep 16(2):244–254. https://doi.org/10.3390/hematolrep16020024
10. Deng J, Paulus A, Fang DD, Manna A, Wang G, Wang H, Zhu S, Chen J, Min P, Yin Y, Dutta N, Halder N, Ciccio G, Copland IIIJA, Miller J, Han B, Bai L, Liu L, Wang M, McEachern D, Przybranowski S, Yang C-Y, Stuckey JA, Wu D, Li C, Ryan J, Letai A, Ailawadhi S, Yang D, Wang S, Chanan-Khan A, Zhai Y (2022) Lisaftoclax (APG-2575) Is a novel BCL-2 inhibitor with robust antitumor activity in preclinical models of hematologic malignancy. Clin Cancer Res 28(24):5455–5468. https://doi.org/10.1158/1078-0432.CCR-21-4037
11. Blum KA, Supko JG, Maris MB, Flinn IW, Goy A, Younes A, Bobba S, Senderowicz AM, Efuni S, Rippley R, Colak G, Trojer P, Abramson JS (2022) A phase I study of pelabresib (CPI-0610), a small-molecule inhibitor of BET Proteins, in Patients with Relapsed or Refractory Lymphoma. Cancer Res Commun 2(8):795–805. https://doi.org/10.1158/2767-9764.CRC-22-0060
12. Vugmeyster Y, Ravula A, Rouits E, Diderichsen PM, Kleijn HJ, Koenig A, Wang X, Schroeder A, Goteti K, Venkatakrishnan K (2024) Model-informed selection of the recommended phase III dose of the inhibitor of apoptosis protein inhibitor, xevinapant, in combination with cisplatin and concurrent radiotherapy in patients with locally advanced squamous cell carcinoma of the head and neck. Clin Pharmacol & Ther 115(1):52–61. https://doi.org/10.1002/cpt.3065
13. Konstantinidou M, Arkin MR (2024) Molecular glues for protein-protein interactions: progressing toward a new dream. Cell

Chem Biol 31(6):1064–1088. https://doi.org/10.1016/j.chembiol.2024.04.002

14. Zattoni J, Vottero P, Carena G, Uliveto C, Pozzati G, Morabito B, Gitari E, Tuszynski J, Aminpour M (2025) A comprehensive primer and review of PROTACs and their In silico design. Comput Methods Programs Biomed 264:108687. https://doi.org/10.1016/j.cmpb.2025.108687
15. Clackson T, Wells JA (1995) A hot spot of binding energy in a hormone-receptor interface. Science 267(5196):383–386. https://doi.org/10.1126/science.7529940
16. Wells JA (1991) Systematic mutational analysis of protein-protein interfaces. In: Methods in enzymology, vol 202. Academic Press, pp 390–411. https://doi.org/10.1016/0076-6879(91)02020-A
17. Zoete V, Michielin O (2007) Comparison between computational alanine scanning and per-residue binding free energy decomposition for protein–protein association using MM-GBSA: application to the TCR-p-MHC complex. Proteins: Struct Funct Bioinf 67(4):1026–1047. https://doi.org/10.1002/prot.21395
18. Ozawa T, Okazaki K, Kitaura K (2011) CH/π hydrogen bonds play a role in ligand recognition and equilibrium between active and inactive states of the β2 adrenergic receptor: an ab initio fragment molecular orbital (FMO) study. Bioorg Med Chem 19(17):5231–5237. https://doi.org/10.1016/j.bmc.2011.07.004
19. Gallivan JP, Dougherty DA (1999) Cation-π interactions in structural biology. Proc Natl Acad Sci 96(17):9459–9464. https://doi.org/10.1073/pnas.96.17.9459
20. Lu Y-X, Zou J-W, Wang Y-H, Yu Q-S (2007) Substituent effects on noncovalent halogen/π interactions: theoretical study. Int J Quantum Chem 107(6):1479–1486. https://doi.org/10.1002/qua.21279
21. Johnston RC, Cheong PH-Y (2013) C–H···O non-classical hydrogen bonding in the stereomechanics of organic transformations: theory and recognition. Org Biomol Chem 11(31):5057–5064. https://doi.org/10.1039/C3OB40828K
22. Bissantz C, Kuhn B, Stahl M (2010) A medicinal chemist's guide to molecular interactions. J Med Chem 53(14):5061–5084. https://doi.org/10.1021/jm100112j
23. Lopes PEM, Guvench O, MacKerell AD (2015) Current status of protein force fields for molecular dynamics simulations. In: Kukol A (Ed.) Eds. Molecular modeling of proteins. Springer New York, New York, NY, 47–71 https://doi.org/10.1007/978-1-4939-1465-4_3
24. Raha K, Peters MB, Wang B, Yu N, Wollacott AM, Westerhoff LM, Merz KM (2007) The role of quantum mechanics in structure-based drug design. Drug Discov Today 12(17):725–731. https://doi.org/10.1016/j.drudis.2007.07.006
25. Tong Y, Mei Y, Li YL, Ji CG, Zhang JZH (2010) Electrostatic polarization makes a substantial contribution to the free energy of avidin–biotin binding. J Am Chem Soc 132(14):5137–5142. https://doi.org/10.1021/ja909575j
26. Fedorov DG, Kitaura K (2007) Extending the power of quantum chemistry to large systems with the fragment molecular orbital method. J Phys Chem A 111(30):6904–6914. https://doi.org/10.1021/jp0716740
27. Fukuzawa K, Tanaka S (2022) Fragment molecular orbital calculations for biomolecules. Curr Opin Struct Biol 72:127–134. https://doi.org/10.1016/j.sbi.2021.08.010
28. Heifetz A, Chudyk EI, Gleave L, Aldeghi M, Cherezov V, Fedorov DG, Biggin PC, Bodkin MJ (2016) The fragment molecular orbital method reveals new insight into the chemical nature of GPCR–ligand interactions. J Chem Inf Model 56(1):159–172. https://doi.org/10.1021/acs.jcim.5b00644
29. Alexeev Y, P. Mazanetz M, Ichihara O, G. Fedorov D (2012) GAMESS as a free quantum-mechanical platform for drug research. Curr Top Med Chem 12(18):2013–2033. https://doi.org/10.2174/1568026611212180008
30. Heifetz A, Sladek V, Townsend-Nicholson A, Fedorov DG (2020) Characterizing protein-protein interactions with the fragment molecular orbital method. In: Heifetz A (Ed.) Eds. Quantum mechanics in drug discovery. Springer US, New York, NY, 187–205 https://doi.org/10.1007/978-3-031-75845-4_1
31. Tanaka S (2023) Protein–protein interaction modelling with the fragment molecular orbital method. In: Tsumoto K, Kuroda D (Eds.)Eds. Computer-aided antibody design. Springer US, New York, NY, 295–305 https://doi.org/10.1007/978-1-0716-2609-2_16
32. Paciotti R, Agamennone M, Coletti C, Storchi L (2020) Characterization of PD-L1 binding sites by a combined FMO/GRID-DRY approach. J Comput Aided Mol Des 34(8):897–914. https://doi.org/10.1007/s10822-020-00306-0
33. Lim H, Chun J, Jin X, Kim J, Yoon J, No KT (2019) Investigation of protein-protein interactions and hot spot region between PD-1 and PD-L1 by fragment molecular orbital method. Sci Rep 9(1):16727. https://doi.org/10.1038/s41598-019-53216-z
34. Monteleone S, Fedorov DG, Townsend-Nicholson A, Southey M, Bodkin M, Heifetz A (2022) Hotspot identification and drug design of protein–protein interaction modulators using the fragment molecular orbital method. J Chem Inf Model 62(16):3784–3799. https://doi.org/10.1021/acs.jcim.2c00457

35. Kim J, Lim H, Moon S, Cho SY, Kim M, Park JH, Park HW, No KT (2021) Hot spot analysis of YAP-TEAD Protein-protein interaction using the fragment molecular orbital method and its application for inhibitor discovery. Cancers 13(16):4246. https://doi.org/10.3390/cancers13164246
36. Chuntakaruk H, Hengphasatporn K, Shigeta Y, Aonbangkhen C, Lee VS, Khotavivattana T, Rungrotmongkol T, Hannongbua S (2024) FMO-guided design of darunavir analogs as HIV-1 protease inhibitors. Sci Rep 14(1):3639. https://doi.org/10.1038/s41598-024-53940-1
37. Hillig RC, Sautier B, Schroeder J, Moosmayer D, Hilpmann A, Stegmann CM, Werbeck ND, Briem H, Boemer U, Weiske J, Badock V, Mastouri J, Petersen K, Siemeister G, Kahmann JD, Wegener D, Böhnke N, Eis K, Graham K, Wortmann L, von Nussbaum F, Bader B (2019) Discovery of potent SOS1 inhibitors that block RAS activation via disruption of the RAS–SOS1 interaction. Proc Natl Acad Sci 116(7):2551–2560. https://doi.org/10.1073/pnas.1812963116
38. Edelsbrunner H, Mücke EP (1994) Three-dimensional alpha shapes. ACM Trans Graph 13(1):43–72. https://doi.org/10.1145/174462.156635
39. Liang J, Woodward C, Edelsbrunner H (1998) Anatomy of protein pockets and cavities: measurement of binding site geometry and implications for ligand design. Prot Sci 7(9):1884–1897. https://doi.org/10.1002/pro.5560070905
40. Chemical Computing Group ULC SSW (2024) Molecular Operating Environment (MOE), Suite #910, Montreal, QC H3A 2R7, Canada H3A 2R72022.02. 2022.02 edn
41. Monteleone S, Morao I, Fedorov DG, Kellici TF (2025) Quantum mechanics-based ranking of predicted proteolysis targeting chimeras-mediated ternary complexes. ACS Med Chem Lett 16(3):420–427. https://doi.org/10.1021/acsmedchemlett.4c00534
42. Nandy A, Boppana K, Phukan S (2025) Mechanistic insights into PROTAC-mediated degradation through an integrated framework of molecular dynamics, free energy landscapes, and quantum mechanics: a case study on kinase degraders. J Comput Aided Mol Des 39(1):51. https://doi.org/10.1007/s10822-025-00630-3
43. Karplus M, McCammon JA (2002) Molecular dynamics simulations of biomolecules. Nat Struct Biol 9(9):646–652. https://doi.org/10.1038/nsb0902-646
44. Fedorov DG, Kromann JC, Jensen JH (2018) Empirical corrections and pair interaction energies in the fragment molecular orbital method. Chem Phys Lett 706:328–333. https://doi.org/10.1016/j.cplett.2018.06.025
45. Koyama Y, Ueno-Noto K, Takano K (2014) Affinity of HIV-1 antibody 2G12 with monosaccharides: A theoretical study based on explicit and implicit water models. Comput Biol and Chem 49:36-44. https://doi.org/10.1016/j.compbiolchem.2014.01.013
46. Zwanzig RW (1954) High-temperature equation of state by a perturbation method. I. Nonpolar gases. J. Chem Phys 22(8):1420–1426. https://doi.org/10.1063/1.1740409
47. Kollman PA, Massova I, Reyes C, Kuhn B, Huo S, Chong L, Lee M, Lee T, Duan Y, Wang W, Donini O (2000) Calculating structures and free energies of complex molecules: combining molecular mechanics and continuum models. Acc Chem Res 33(12):889–897. https://doi.org/10.1021/ar000033j
48. Homeyer N, Gohlke H 2012 Free energy calculations by the molecular mechanics Poisson–Boltzmann surface area method. Mol Inform 31(2):114–22. https://doi.org/10.1002/minf.201100135
49. Grossfield A, Zuckerman DM (2009) Quantifying uncertainty and sampling quality in biomolecular simulations. In: Ralph AW (eds) Annual reports in computational chemistry. Elsevier 5:23–48. https://doi.org/10.1016/S1574-1400(09)00502-7 ISSN 1574-1400, ISBN 9780444533593
50. Wang D, Wang Y, Chang J, Zhang L, Wang HEW (2022). Efficient sampling of high-dimensional free energy landscapes using adaptive reinforced dynamics. Nat Comput Sci. 2(1):20–29. https://doi.org/10.1038/s43588-021-00173-1 PMID: 38177702
51. Ibrahim PE, Zuccotto F, Zachariae U, Gilbert I, Bodkin M. Accurate prediction of dynamic protein–ligand binding using P-score ranking. J Comput Chem. 2024 45(20):1762. https://doi.org/10.1002/jcc.27370
52. Moriwaki H, Kawashima Y, Watanabe C, Kamisaka K, Okiyama Y, Fukuzawa K, Honma T (2024) FMOe: preprocessing and visualizing package of the fragment molecular orbital method for molecular operating environment and its applications in covalent ligand and metalloprotein analysis. J Chem Inf Model 64(18):6927–6937. https://doi.org/10.1021/acs.jcim.4c01169
53. Morao I, Heifetz A, Fedorov DG (2020) Accurate scoring in seconds with the fragment molecular orbital and density-functional tight-binding methods. In: Heifetz A (Ed.) Eds. Quantum mechanics in drug discovery. Springer US, New York, NY, 143–148 https://doi.org/10.1007/978-1-0716-0282-9_9

Chapter 6

Beyond Small Molecules: Applying Fragment Molecular Orbital Sygnature Platform (FMO-SP) to Emerging Modalities

Alexander Heifetz, Girinath G. Pillai, Maryamdokht Taimoory, Louise Birch and Colin Sambrook Smith

Abstract

Drug discovery is rapidly expanding beyond traditional small molecules to encompass new therapeutic modalities capable of addressing previously "undruggable" targets. Between 2015 and mid-2025, FDA approvals increasingly included biologics and emerging therapeutic modalities such as antibody–drug conjugates (ADCs), nucleic acid therapeutics, and other nontraditional approaches. In parallel, PROTACs and molecular glue degraders (MGDs) have expanded rapidly in clinical development. These innovations are reshaping therapeutic strategies in oncology, metabolic disease, and rare disorders, but they introduce major design challenges: large and shallow protein–protein interfaces, competition with extensive native interaction networks, ternary complex cooperativity, and the intricacies of nucleic-acid recognition. Addressing these problems requires accurate quantum-mechanical methods capable of resolving the underlying molecular interactions. The Fragment Molecular Orbital Sygnature Platform (FMO-SP) integrates quantum mechanics with automated workflows and intuitive visual analytics to provide residue-level, quantified energetic insights. These capabilities enable rational optimization of ADCs, PROTAC, and MGD ternary complexes, GPCR-peptide, PPi interfaces, and RNA/DNA modulators. Building on Chap. 4 (small molecules), this chapter demonstrates how FMO-SP elucidates mechanisms of action and guides design across emerging therapeutic modalities, supporting more predictive and resource-efficient drug discovery.

Keywords Fragment molecular orbital sygnature platform, Fragment molecular orbital, Quantum mechanics, Drug discovery, Protein–protein interactions, Therapeutic modalities, Biologics, Antibody–drug conjugates, Peptides, Peptidomimetics, Proteolysis-targeting chimeras, Molecular glues, Molecular glue degraders, Targeted protein degradation, RNA modulators, DNA modulators, Nucleic acid therapeutics, G-protein coupled receptors, Metabolic GPCRs, Semaglutide, GLP-1 receptor, Cereblon, GSPT1, Residue-level energetics, Structure-based drug design, Computational chemistry, Quantum-based workflows

1 Introduction

Drug discovery over the last decade has been transformed by the rise of new therapeutic modalities that extend beyond the traditional small-molecule paradigm [1–7]. Although small molecules remained the largest individual class of US FDA approvals between

Alexander Heifetz (ed.), *Quantum Mechanics Applications in Drug Design*, Methods in Molecular Biology, vol. 3061,
https://doi.org/10.1007/978-1-0716-5420-0_6,

2015 and the first half of 2025, biologics and emerging therapeutic modalities accounted for an increasing proportion of newly approved drugs (Table 1; Fig. 1). Collectively, these approvals demonstrate

Table 1
FDA approvals per year (2015–mid 2025) [8–12]

Year	Small molecules	Biologics	ADCs	PROTACs	Molecular glues	DNA/RNA modulators	Total
2015	34	10	0	0	0	1	45
2016	32	12	0	0	0	0	44
2017	29	13	1	0	0	1	44
2018	35	15	1	0	0	0	51
2019	31	14	2	0	0	1	48
2020	27	18	2	0	0	2	49
2021	28	20	2	0	0	1	51
2022	29	22	2	0	0	1	54
2023	30	23	1	0	0	2	56
2024	27	20	1	0	1	2	51
2025 (H1)	26	18	0	0	0	1	45

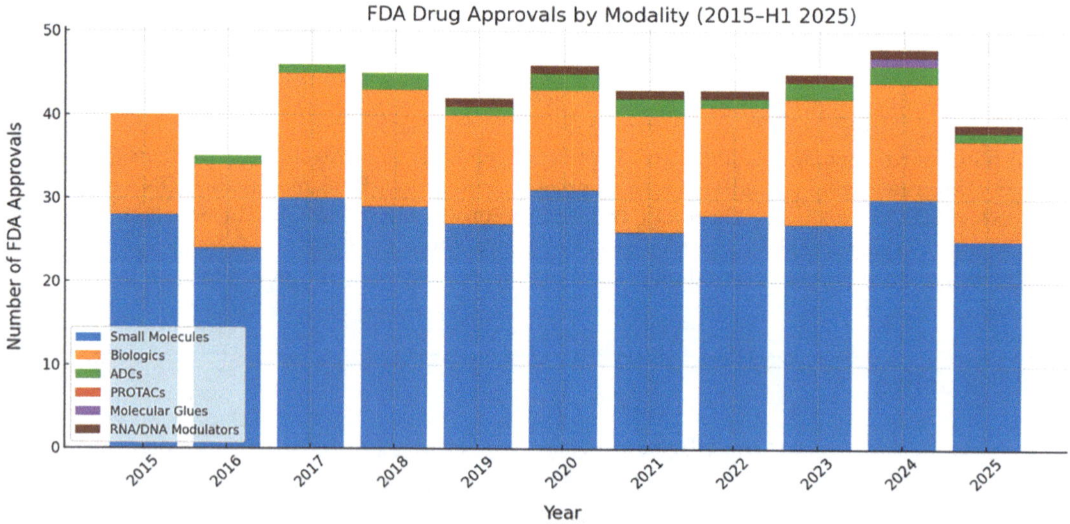

Fig. 1 Trends in FDA drug approvals by modality (2015-H1 2025) [8]. Stacked bar chart showing the annual number of FDA approvals for different therapeutic modalities. Small molecules remain the dominant class (≈ 65%–75% annually), while biologics account for ~25%–35%. Emerging modalities, including antibody–drug conjugates (ADCs), molecular glues, PROTACs, and RNA/DNA modulators, have begun to enter the market, with steady ADC approvals since 2017 while molecular glues and PROTACs represent rapidly developing areas of the therapeutic pipeline

a substantial diversification of the pharmaceutical pipeline between 2015 and mid-2025 [8–12].

Small molecules are low-molecular-weight compounds that typically act through binary binding to well-defined protein active or allosteric sites. Their strengths include oral bioavailability, ease of manufacture, and relatively low cost. However, they face persistent limitations [13–25]:

- Many disease-relevant proteins (e.g., transcription factors, scaffolding proteins, intrinsically disordered proteins) lack well-defined binding pockets and are considered **"undruggable"** [20].
- **ADMET challenges** [14]—poor solubility, limited bioavailability, metabolic instability, and rapid clearance—remain leading causes of attrition.
- **Off-target toxicity** [18] from diffuse distribution and unintended protein interactions often restricts dosing. In oncology, this is particularly problematic, as healthy proliferating tissues (e.g., bone marrow and gut epithelium) are damaged, leading to narrow therapeutic windows.
- **Drug resistance** [21], arising from mutations or adaptive signaling pathways, reduces the durability of therapeutic responses.
- **Drug–drug interactions** [21], especially via cytochrome P450 metabolism, complicate safety and clinical management.
- **hERG channel binding** [17] is a key liability, driving cardiotoxicity.
- **Blood–brain barrier (BBB) penetration** [22, 23] limits the effectiveness of small molecules for CNS indications.

To overcome these limitations, a new generation of modalities has emerged:

- **Biologics and Protein–Protein Interaction (PPI) Modulators** [20, 26–32]: Biologics (e.g., monoclonal antibodies, recombinant proteins, bispecifics, and peptides) are large, complex molecules derived from living systems. In this context, the biologic is the modality, and the PPI is the target. Unlike small molecules, biologics can engage broad, shallow, or flexible protein–protein interaction (PPI) surfaces that are typically considered **"undruggable."**
  - Biologics have transformed modern medicine, revolutionizing the treatment of cancers, autoimmune disorders, and inflammatory diseases. Their ability to bind with high specificity to extended protein surfaces makes them particularly effective where small molecules often fall short.
  - Protein–protein interaction (PPI) modulators aim to stabilize or disrupt multiprotein complexes that drive disease processes. Classic examples include the p53-MDM2

interaction (PDB: 1YCR) [33], critical in cancer biology; the BCL2-BAX interaction (PDB: 4LXD) [34], central to apoptosis regulation; and the immune checkpoint complex PD-1-PD-L1 (PDB: 4ZQK) [35], a major target in immuno-oncology. Unlike enzyme inhibition, these interactions typically involve large, flat, or dynamic interfaces—posing challenges for small molecules but offering opportunities for biologics, peptides, and glue-based strategies.

- **Molecular glues and molecular glue degraders (MGDs)** [36–42] are typically small molecules that stabilize novel protein–protein interactions between a target and an E3 ligase without the need for a linker. This minimalist approach enables degradation of proteins once considered intractable, including transcription factors. Recent clinical progress across emerging therapeutic modalities, including the approval of revumenib in 2024 [43], has highlighted the growing potential of innovative approaches to targeted protein degradation.
- **Proteolysis-targeting chimeras (PROTACs)** [44] are heterobifunctional small molecules that simultaneously bind a target protein and an E3 ligase, bringing them into proximity to trigger ubiquitination and proteasomal degradation. Unlike conventional inhibitors that block activity, PROTACs eliminate the entire protein, offering potential advantages in overcoming resistance and targeting non-enzymatic proteins. Several PROTACs are in advanced Phase II/III trials, but as of mid-2025 none have yet been approved (Fig. 3b).
- **Antibody–drug conjugates (ADCs)** [45, 46] and **Peptide--rug conjugates (PDCs)** [47] combine the targeting precision of monoclonal antibodies with the potency of cytotoxic small-molecule payloads. By selectively delivering the drug to diseased cells, ADCs minimize systemic toxicity and improve therapeutic windows. Since 2017, the FDA has approved 12 ADCs, nearly all in oncology, where selective tumor targeting is critical (Fig. 3b). Challenges remain in optimizing linker chemistry, payload stability, and controlled release, but ADCs represent one of the most successful non-traditional modalities.
- DNA/RNA Modulators **(Nucleic Acid Therapeutics)** [48–50] represent another fast-growing modality. Rather than targeting proteins, these drugs act upstream at the genetic level, altering transcription, splicing, or translation. Classes include antisense oligonucleotides (ASOs), siRNAs, mRNAs, CRISPR-based editors, and RNA-targeting small molecules. Key approvals include: Nusinersen (Spinraza, 2016) [51, 52]— antisense oligonucleotide for spinal muscular atrophy, Patisiran (Onpattro, 2018) [53]—siRNA for hereditary transthyretin amyloidosis, Inclisiran (Leqvio, 2020) [54]— siRNA for hypercholesterolemia,

mRNA vaccines (Pfizer-BioNTech, Moderna, 2020) [55]—for COVID-19 and Risdiplam (Evrysdi, 2020) [56]—small molecule that modulates SMN2 mRNA splicing. Together, these advances highlight nucleic acid drugs as a disruptive modality addressing diseases beyond the reach of classical small molecules.

As shown in Fig. 2, oncology accounts for nearly 40% of FDA approvals (2015-mid-2025), with modalities heavily concentrated in this area (63.6% of all modality approvals, Fig. 3a). Within oncology, ADCs dominate (≈60%), followed by PROTACs (20%), biologics (11.4%), and MGDs (8.6%) (Fig. 3b). By contrast, GLP-1 receptor (GLP-1R) peptide agonists, which drive ~14.5% of modality approvals (Fig. 2c) are not classical PPI modulators but exploit extended protein–ligand interactions reminiscent of PPIs, demonstrating the diversity of emerging therapeutic classes [57–60].

While these modalities open new therapeutic frontiers, their design is far more complex than that of traditional small molecules [1–4, 6]. Researchers must account for multi-domain architectures, linker flexibility, cooperative ternary binding, and non-canonical interactions, which are often beyond the reach of molecular mechanics (MM) and docking. Quantum mechanics (QM), particularly

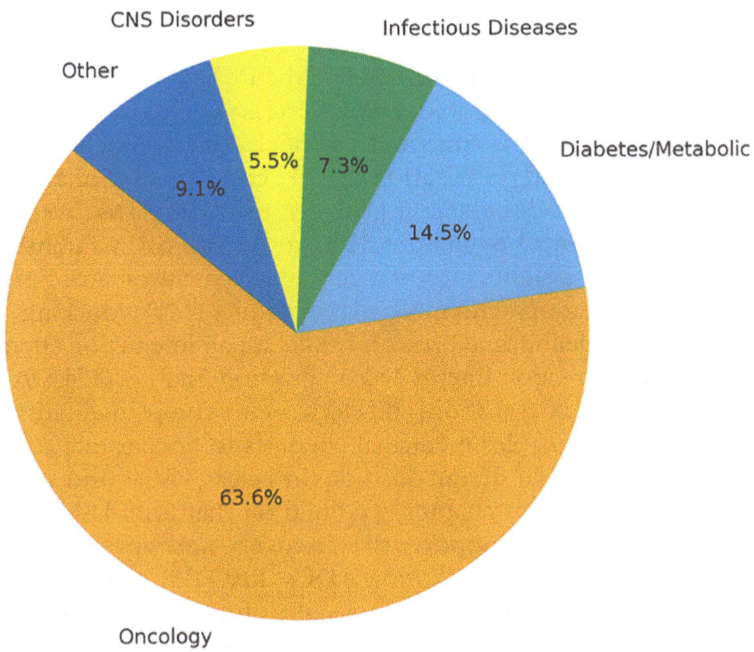

Fig. 2 Distribution of FDA-approved non-traditional modalities by therapeutic area (2015-H1 2025) [8]. Pie chart illustrating therapeutic focus of modality-based approvals. Oncology dominates with 63.6% of all approvals, followed by diabetes/metabolic diseases (14.5%), infectious diseases (7.3%), CNS disorders (5.5%), and other areas including autoimmune, cardiovascular, and respiratory indications (9.1%). These data highlight oncology as the leading field for new modalities.

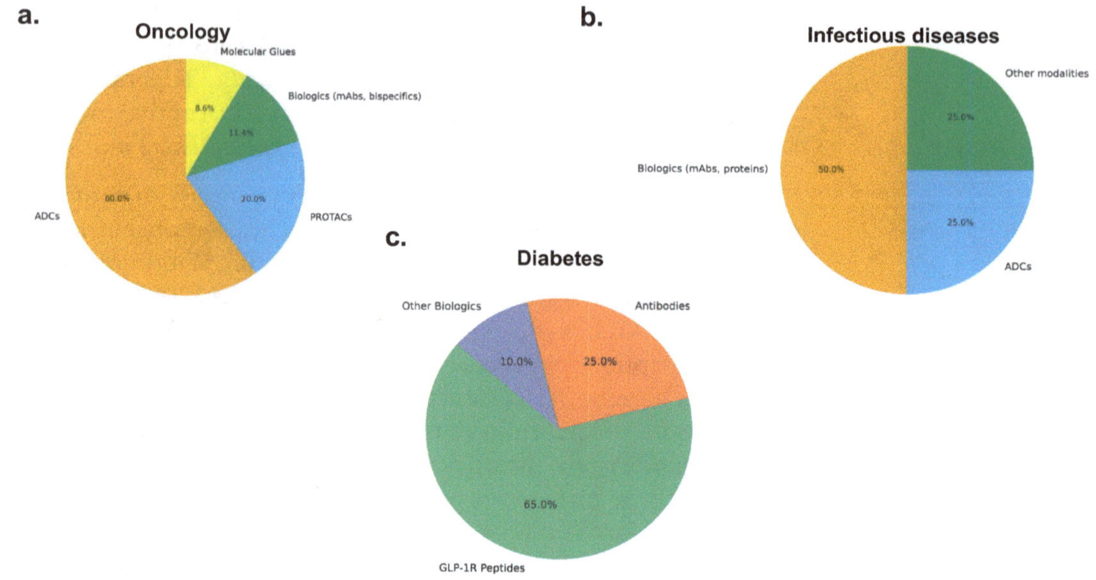

Fig. 3 Distribution of therapeutic modalities across disease areas [1–8]. (**a**) Oncology, (**b**) Metabolic disease, and (**c**) Infectious disease. The percentages reflect the relative contribution of emerging modalities in both FDA-approved drugs (2015-mid-2025) and active clinical trials, rather than approvals alone. Antibody–drug conjugates (ADCs) dominate oncology, while other modalities such as PROTACs and molecular glues are still in development but contribute significantly to the clinical pipeline

the Fragment Molecular Orbital (FMO) method, provides experiment-comparable residue-level accuracy, revealing the energetic and mechanistic contributions of electrostatics, charge transfer, dispersion, and polarization that are critical for rational design [61–65].

As discussed in Chap. 4, Sygnature Discovery has developed the FMO-Sygnature Platform (FMO-SP) to translate these quantum insights into practice. FMO-SP automates FMO calculations, integrates with molecular dynamics (MD), docking, and homology modeling, and presents results as pair interaction energies (PIE), Pair Interaction Energy Decomposition Analysis (PIEDA) [66], desolvation, polarization, and electrostatic complementarity maps [67, 68]. This enables medicinal chemists to link atomic-scale energetics to molecular design decisions in a clear, visual, and actionable manner.

In this chapter, we build on that foundation to illustrate how FMO-SP supports the discovery and optimization of emerging modalities—including ADCs, PROTACs, MGDs, and PPI modulators—where small molecules often fall short. By providing quantitative, residue-level insights into complex binding modes, FMO-SP helps to address the unique challenges of modality drug design, accelerating optimization and improving the likelihood of clinical success.

2 Extended Capabilities of FMO-SP Beyond Protein–Ligand Complexes: Methods and Showcases

While the FMO Sygnature Platform (FMO-SP) is highly effective for analyzing classical protein–ligand complexes, its scope extends well beyond this domain. By leveraging quantum-level energetics, subsystem analysis [69], and flexible fragmentation strategies [70]. FMO-SP can be applied to a wide range of molecular systems where binding interfaces are complex or non-traditional.

A detailed description of the FMO-SP methodology is provided in Chap. 4 of this book. In essence, FMO-SP applies QM to reveal critical protein–ligand and protein–protein interactions that are often missed by traditional force fields or visual inspection, even when high-resolution structural data are available. It provides quantitative insight into interaction strength (kcal/mol), interaction type (electrostatic or hydrophobic), and key energy components such as desolvation and depolarization that govern molecular binding.

The method computes interactions between fragment pairs while explicitly accounting for effects such as polarization and charge transfer. Interaction energies are further analyzed through PIEDA, which separates contributions into electrostatics, dispersion, exchange-repulsion, and charge transfer—allowing medicinal chemists to interpret results in chemically intuitive terms. Designed for speed and flexibility, the platform typically completes an average protein–ligand or protein–protein calculation within minutes.

As described in Chap. 4, the FMO-SP workflow consists of four sequential stages: Input → Run → Analysis → Graphics.

- **Stage 1:** Input: Converts protein–ligand-water-protein complexes into FMO-ready input files.
- **Stage 2**: Run: Executes FMO calculations within GAMESS using the generated input(s).
- **Stage 3:** Analysis: Extracts pairwise interaction energies for individual systems in CSV format, and generates merged and transposed datasets to enable cross-system comparisons such as structure–activity relationship (SAR) analysis.
- **Stage 4:** Graphics: Produces visual outputs—including column plots for individual ligands and heatmaps for ligand sets—to support interpretation and decision-making.

2.1 PPI and Molecular Glues—Methodology

FMO-SP provides four major categories of information about the protein–protein interactions (PPIs) formed between any two interacting proteins or protein-ligand–water–protein systems:

a) the list of PPIs that each interface residue forms with its neighboring amino acids,

b) the strength of these PPIs, reported in kcal/mol,
c) their chemical nature (electrostatic or hydrophobic), and
d) their interaction category, e.g., intermolecular, intramolecular, or water-bridged.

As previously reported [71], the key residues involved in protein–protein interactions are referred to as *hotspots*. Importantly, not only intermolecular PPIs contribute to the stability of protein–protein complexes, but also intramolecular interactions. Some hotspots play a dual role, engaging in both intra- and intermolecular interactions, while others have a sole role, participating exclusively in one or the other. In both cases, they stabilize the bioactive conformation of neighboring hotspots by forming intramolecular interactions, thereby reinforcing the overall interface. This interplay between intra- and intermolecular PPIs may also explain why proteins tend to associate: as complex formation progresses, both intra- and intermolecular PPIs increase, leading to enhanced stability of the secondary structure of each protein.

As described in Chap. 2, we performed FMO-SP calculations on truncated systems and, in the case of PPIs and molecular glues, specifically on the protein-(ligand)-protein interface. For defining the interface between protein A and protein B, we included every residue in which at least one atom lies within a distance of ≤4.5 Å from the partner protein, and vice versa. For molecular glues, the ligand was considered part of the interface if any of its atoms were within ≤ 4.5 Å of both protein A and protein B. In addition, to capture the full binding environment, we also included all protein atoms located within ≤4.5 Å of the ligand itself. This definition of the interface is critical, as it ensures a complete representation of how molecular glues function by stabilizing the protein–protein complex through simultaneous interactions with both proteins and their surrounding residues.

We adapted literature recommended protocol [71] by introducing ΔPIEij in place of PIEij (Eq. 1) to exclude backbone–backbone intramolecular interactions. We assumed that backbone–backbone interactions contribute predominantly to intramolecular stabilization and only minimally to intermolecular interactions. By applying this correction, we were able to reduce much of the "noise" in the data. To compute the backbone–backbone contribution, we deleted the side chains from the wild-type full-atom complex, leaving a system composed of Gly residues for the proteins, water molecules, and ligands (if present). The pairwise interaction energy between residues i and j (ΔPIE_{ij}) was then calculated by subtracting the backbone-only PIE (PIE_{ij}^{bb}) from the full-atom PIE (PIE_{ij}^{fa}), as shown below:

$$\Delta PIE_{ij} = PIE_{ij}^{fa} - PIE_{ij}^{bb} \tag{1}$$

2.2 PPI—FMO-SPAnalysis of the GLP-1R/Semaglutide Complex

2.2.1 Introduction

G-protein coupled receptors (GPCRs) remain among the most attractive classes of drug targets, with ~40% of all prescribed medicines acting on this family. Although the human genome encodes ~800 GPCRs, only ~50 have been successfully exploited, leaving substantial untapped therapeutic potential [72–74]. As outlined in Chap. 15, FMO methods are increasingly being applied to support GPCR drug design.

Within this family, metabolic GPCRs function as molecular switches that convert hormonal and nutrient signals into cellular responses regulating blood sugar, appetite, fat storage, and energy balance [73]. Their central role in metabolic homeostasis, combined with the success of blockbuster therapies, has made them some of the most clinically validated drug-discovery targets of the past decade. Notable examples include GLP-1 receptor (GLP-1R) agonists for type 2 diabetes and obesity, as well as GIP and dual GLP-1R/GIPR agonists, which are reshaping the therapeutic landscape [75–77] (Fig. 3c).

2.2.2 Case Study: GLP-1R/Semaglutide Complex

Semaglutide is a long-acting GLP-1R agonist peptide, clinically approved and marketed for the treatment of type 2 diabetes (Ozempic®, Rybelsus®*)* [75–77] and obesity (Wegovy®) [78]. Derived from the native GLP-1 peptide, it has been rationally modified to improve half-life and potency, exemplifying how peptide hormone engineering can produce blockbuster therapies.

2.2.3 Results from FMO-SP Analysis

FMO-SP has been applied to the analysis of the GLP-1R/Semaglutide complex (PDB: 7KI0) [79] (Fig. 4). The heatmap (Fig. 4a) reveals residue-level binding energetics and hotspot stabilization, while the SVL script interface (Fig. 4b) demonstrates how FMO-SP results can be transformed from numerical CSV data into an interactive 3D environment. The structural view (Fig. 4c) then highlights the spatial location of dominant interactions, directly linking energetic maps with receptor–peptide architecture.

The strongest stabilizing interactions are largely electrostatic, complemented by localized hydrophobic clamps (chain A corresponds to GLP-1R residues; chain C to Semaglutide):

- Electrostatic interactions
  - Asp215A–Arg36C (salt bridge; among the most stabilizing pairs)
  - Arg380A–Asp15C (robust electrostatic attraction at the lower interface)
  - Tyr205A–Ser17C (mixed, but electrostatics dominate)
- Hydrophobic interactions
  - Trp91A–Leu20C (N-terminal clamp on the peptide)
  - Trp211A–Phe28C (aromatic packing stabilizing the mid-peptide segment)

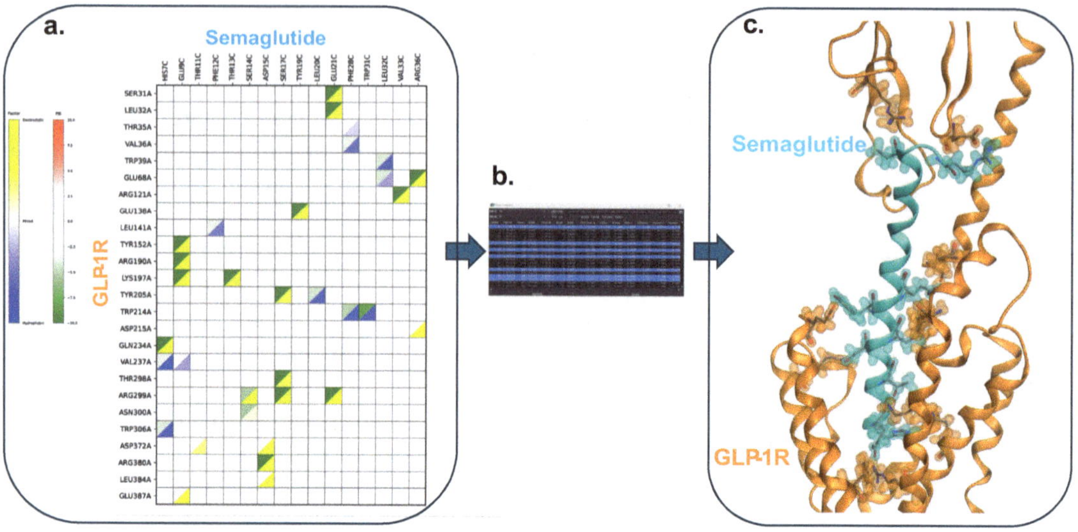

Fig. 4 FMO-SP analysis of the GLP-1R/Semaglutide complex (PDB: 7KI0). (**a**) Heatmap of significant intermolecular interactions between GLP-1R and Semaglutide ($|\Delta PIE| \geq 3.0$ kcal/mol). Each square represents a residue pair, with interaction energies (PIE: red = repulsion, green = attraction, kcal/mol) and chemical factor (blue = hydrophobic/dispersion, yellow = electrostatic, white = mixed). (**b**) Menu of the SVL script within FMO-SP that converts text-based CSV output into a 3D representation, enabling users to browse and visually inspect key hotspots (as demonstrated in panel c). (**c**) 3D structural view of GLP-1R (orange) bound to Semaglutide (cyan), showing the locations of the dominant $\Delta PIE \leq -10.0$ kcal/mol FMO-SP-identified interactions, with hotspots displayed as surfaces and sticks

- Val237A–His7C/Glu19C (mixed hydrophobic/electrostatic interactions that seat the peptide on the receptor shelf)

Together, these patterns show that Semaglutide's binding is driven by electrostatic anchoring at discrete hotspots—especially Asp/Arg salt bridges—while aromatic and aliphatic packing from residues such as Trp91A and Trp211A provide secondary consolidation.

2.2.4 Mechanistic Interpretation and Design Implications

From a drug design perspective, FMO-SP provides a ranked, residue-level roadmap:

- For peptide analogs: preserve or strengthen the top ΔPIE electrostatics (e.g., Asp215A/Arg36C; Arg380A/Asp15C), while tuning hydrophobic clamps to improve receptor residence time.
- For peptidomimetics and small molecules: mimic key electrostatic interactions through charged or polar groups, and reproduce hydrophobic interactions with constrained scaffolds projecting substituents into the same receptor pockets.

This ability to translate QM-level interaction maps into chemically intuitive design principles makes FMO-SP a valuable guide for developing next-generation GLP-1R modulators. A notable demonstration of such capabilities is provided by the retrospective analysis of PF-06882961 (Danuglipron) [80], the first orally bioavailable small-molecule GLP-1R agonist developed by Pfizer (see Chap. 4 for further details).

2.3 Molecular Glues—FMO-SP Analysis of the CRBN-CC-885-GSPT1 Ternary Complex

2.3.1 Introduction

Molecular glues (MGDs) are a rapidly emerging therapeutic modality that exploit protein–protein interactions (PPIs) to enable targeted protein degradation [36–42]. Unlike conventional inhibitors, MGDs promote the formation of ternary complexes between two proteins most often an E3 ubiquitin ligase and a disease-relevant target. By stabilizing interactions that do not naturally occur, MGDs trigger ubiquitination and proteasomal degradation of otherwise intractable proteins, providing a powerful strategy for drugging "undruggable" targets such as transcription factors and scaffolding proteins. In contrast to PROTACs, which are large bifunctional molecules with two linked ligands, MGDs are small, single-ligand compounds that subtly reshape surface complementarity at protein interfaces. This minimalist mode of action has positioned MGDs as a highly promising class of degraders, combining drug-like properties with the ability to rewire cellular protein networks.

As a representative example, we analyzed the CRBN-CC-885-GSPT1 ternary complex [81] (PDB: 5HXB, Fig. 5), which assembles only in the presence of the molecular glue CC-885. This complex has strong medical relevance. CRBN (cereblon) is the substrate receptor of the CRL4^{CRBN} E3 ubiquitin ligase complex, best known as the target of thalidomide and its analogs (IMiDs). CC-885 reprograms

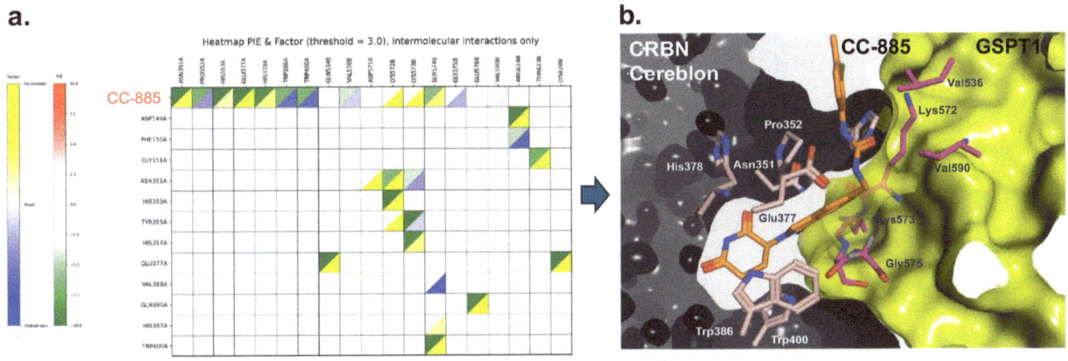

Fig. 5 FMO-SPanalysis of the CRBN–CC-885–GSPT1 ternary complex (PDB: 5HXB). (**a**) Heatmap of significant intermolecular interactions (|ΔPIE| ≥ 3.0 kcal/mol), with top stabilizing residue pairs highlighted. (**b**) 3D view of the ternary interface showing CC-885 (magenta) anchored to CRBN and making secondary contacts to GSPT1. Together, the PIE and structural views reveal how CC-885 reshapes and stabilizes the CRBN–GSPT1 interface.

CRBN to recruit the translation termination factor GSPT1, leading to its ubiquitination and proteasomal degradation. Degradation of GSPT1 selectively kills acute myeloid leukemia (AML) cells and has been proposed as a therapeutic strategy for hematological malignancies. More broadly, this case demonstrates how MGDs can redirect E3 ligases toward neo-substrates, extending targeted protein degradation beyond classical IMiD-sensitive proteins. FMO-SP provides residue-level mechanistic insight into how CC-885 stabilizes the CRBN–GSPT1 interface, clarifying why this ternary complex forms only in the ligand's presence and which interactions are most critical for its stability.

2.3.2 Results

FMO-SP identified a clear three-way stabilization network:

- CRBN–GSPT1 interface (A-B): 16 contacts, TIE ≈ −166 kcal/mol, best ≈ −34.4 kcal/mol (ASP149A–ARG614B salt bridge). Strongly electrostatic, forming the backbone of ternary assembly.
- CC-885 → CRBN (LIG-A): 7 contacts, TIE ≈ −94 kcal/mol, best ≈ −34.7 kcal/mol. Dominated by salt-bridge and H-bond-like interactions, establishing CRBN as the ligand's primary anchor.
- CC-885 → GSPT1 (LIG-B): 6 contacts, TIE ≈ −21 kcal/mol, best ≈ −6.4 kcal/mol. A weaker set of mixed electrostatic and hydrophobic contacts that reinforce recruitment of GSPT1.

Intramolecular stabilizers within CRBN (TIE ≈ −69 kcal/mol) and GSPT1 (TIE ≈ −145 kcal/mol) suggest that local preorganization further supports interface presentation. The breakdown of these interaction categories is summarized in Table 2, which highlights the number of contacts, total stabilization energy, strongest residue pairs, and dominant interaction character. Importantly, FMO-SP reproduced all experimentally validated critical interactions previously reported for this ternary system [81].

2.3.3 Mechanistic Interpretation

Without CC-885, the CRBN–GSPT1 (A–B) interface already shows some favorable electrostatics (TIE ≈ −166 kcal/mol), but the geometry is not sufficient to stably assemble into a productive ternary complex. FMO-SP highlights *why*:

- CRBN alone does not strongly recruit GSPT1. The direct A–B contacts, while electrostatic, are insufficiently stabilized and lack the "bridging" orientation.
- CC-885 provides a strong energetic anchor to CRBN (TIE ≈ −94 kcal/mol). This anchor repositions CRBN's binding surface, creating a complementary pocket that can now engage GSPT1.
- Secondary CC-885–GSPT1 contacts (TIE ≈ −21 kcal/mol) provide the final "lock," rigidifying the ternary assembly.

Table 2
Summary of stabilizing interactions in the CRBN-CC-885-GSPT1 ternary complex

Interaction category	Count	TIE (kcal/mol)	Strongest PIE	Character (dominant)
CRBN–GSPT1 (A–B)	16	−166.1	Asp149A/Arg614B (−34.4)	Predominantly electrostatic/salt bridges
CC-885 → CRBN (LIG-A)	7	−94.4	Strongest ≈ −34.7	Salt-bridge/H-bond dominated
CC-885 → GSPT1 (LIG-B)	6	−20.7	Strongest ≈ −6.4	Mixed: modest electrostatic + hydrophobic/π
CRBN–CRBN (A–A)	11	−68.6	Strongest ≈ −10.2	Intramolecular electrostatics/fold clamps
GSPT1–GSPT1 (B–B)	15	−145.4	Strongest ≈ −33.8	Intramolecular electrostatics/fold clamps

- In other words: without the glue, the interface is not geometrically preorganized, and the modest A–B interactions are too weak or unstable to form the complex. With CC-885, the system gains ~115 kcal/mol of extra stabilization (LIG-A + LIG-B), which shifts the equilibrium decisively toward ternary assembly.

This is consistent with Matyskiela et al. (*Nature* 2016), who showed experimentally that CRBN and GSPT1 do not associate detectably in the absence of CC-885.

2.3.4 Implications for Design

Residue-level quantification from FMO-SP offers direct guidance for next-generation molecular glue development:

- Strengthen essential hotspots: optimize ligands to stabilize the ASP149A–ARG614B salt bridge and other electrostatic clamps.
- Tune CRBN anchoring: preserve strong CC-885–CRBN H-bond/salt-bridge contacts while improving geometry for tighter fit.
- Enhance GSPT1 recruitment: design analogs that extend stabilizing interactions on the GSPT1 side.
- Engineer selectivity: exploit electrostatics for potency or hydrophobicity for drug-like properties, guided by factor values.

By translating quantum-level interaction maps into chemically intuitive design rules, FMO-SP accelerates the discovery of novel MGDs with improved stability, potency, and selectivity.

2.4 DNA/RNA Modulators (Nucleic Acid Therapeutics)—FMO-SP Analysis of the RNA-SSO Complex

2.4.1 Introduction

Nucleic acid therapeutics [48–50] are rapidly emerging as a new class of modalities, with applications spanning antisense oligonucleotides, siRNA, RNA aptamers, and small molecules that modulate RNA or DNA structures. Unlike proteins, nucleic acids present highly charged, flexible scaffolds with distinct secondary and tertiary architectures. Their recognition depends on a combination of hydrogen bonding, electrostatics, and base-stacking interactions, making them challenging to model with classical force fields.

FMO-SP provides a unique advantage in this domain by quantifying base–ligand interactions at the QM level. This allows precise dissection of the energetic balance driving nucleic acid recognition, including contributions from solvation and polarization that are often overlooked in force-field approaches.

2.4.2 Case Study: RNA-SSO Complex from the hepatitis C virus IRES (PDB: 3TZR)

As a representative system, we analyzed the RNA-SSO complex (PDB: 3TZR). This structure corresponds to a riboswitch-like RNA element from the hepatitis C virus (HCV) internal ribosome entry site (IRES) bound to a small-molecule ligand, SSO.

Medical relevance: The HCV IRES is a critical regulatory RNA element that enables cap-independent viral translation. Ligands such as SSO that bind to the IRES can disrupt ribosome recruitment, representing promising antiviral strategies. Understanding these interactions at the residue/base level is therefore important for rational drug design in the RNA therapeutic space.

2.4.3 Results from FMO-SP Analysis

The FMO-SP analysis of 3TZR highlights nine strong interactions between SSO and the RNA interface (Fig. 6a), involving six key bases (A109B, G110B, G52A, A53A, C111B, U56A) and three water molecules (Fig. 6b).

- Interaction character: All nine interactions are predominantly electrostatic in nature, consistent with the charged RNA backbone and polar ligand groups (Fig. 6b).
- Ranking of strongest contacts: A109B > G110B > G52A > A53A > C111B > U56A.
- Water-mediated interactions: Three water molecules act as bridges, further stabilizing ligand binding and expanding the interaction network.
- Visualization: The 3D mapping shows that SSO engages bases from both RNA chains (A and B), effectively clamping into the binding pocket.

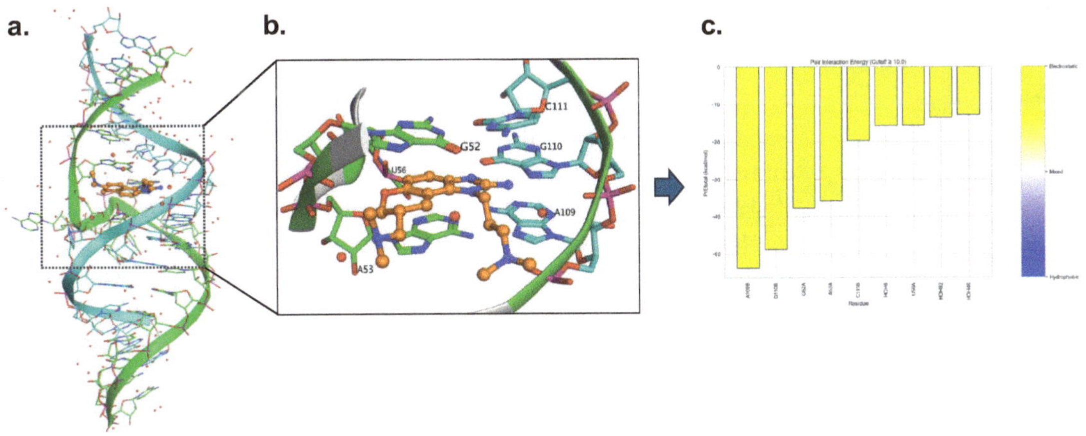

Fig. 6 FMO-SP analysis of the RNA-SSO complex (PDB: 3TZR). (**a**) Structural overview of the small-molecule ligand SSO (orange) bound to the riboswitch-like RNA from the Hepatitis C Virus IRES. (**b**) FMO-SP results mapped in 3D, showing nine strong interactions involving six key bases (colored green and cyan for chains A and B, respectively) and three water molecules that contribute to SSO binding. (**c**) Column plot of significant RNA base-SSO interactions ($|\Delta PIE| \geq 10.0$ kcal/mol). Each bar shows a base–ligand pair interaction energy (PIE: red = repulsion, green = attraction, kcal/mol) alongside its chemical factor (blue = hydrophobic/dispersion, yellow = electrostatic, white = mixed). All nine interactions are predominantly electrostatic in nature, with the most stabilizing contacts ranked as A109B > G110B > G52A > A53A > C111B > U56A, along with contributions from three water molecules. Together, these panels illustrate how FMO-SP partitions the RNA into fragments, quantifies individual base–ligand interactions, and highlights the complementary roles of electrostatics, hydrogen bonding, and water-mediated interactions in stabilizing RNA–ligand complexes

These results emphasize that ligand binding to RNA often combines direct base recognition with water-mediated electrostatic stabilization, reinforcing the need for quantum-level resolution.

2.4.4 Mechanistic Interpretation and Design Implications

The 3D interaction map (Fig. 6b) illustrates how SSO exploits phosphate backbone electrostatics, base-edge hydrogen bonding, and water-mediated contacts to stabilize its binding to the HCV IRES.

Design implications for nucleic acid modulators include:

- Optimizing charged or polar groups to maximize phosphate electrostatics.
- Incorporating H-bond donor/acceptor motifs that target base edges in loops or bulges.
- Exploiting structured water molecules as integral components of the binding interface.
- Designing planar or aromatic moieties that can enhance stacking with exposed bases.

These principles extend beyond viral RNAs to riboswitches, noncoding RNAs, and DNA secondary structures such as G-quadruplexes, highlighting the versatility of FMO-SP in nucleic acid drug design.

2.5 FMO-SP Analysis for PROTACs

2.5.1 Introduction

Proteolysis-targeting chimeras (PROTACs) are heterobifunctional small molecules that hijack the ubiquitin–proteasome system to eliminate disease-relevant proteins. Unlike occupancy-based inhibitors, PROTACs act catalytically: they induce the formation of a ternary complex between an E3 ubiquitin ligase and the target protein, leading to polyubiquitination and subsequent proteasomal degradation. A key pharmacological advantage is their ability to drug "undruggable" targets, including transcription factors, scaffolding proteins, and nuclear proteins, by exploiting induced PPIs rather than relying on active-site binding [44, 82].

Crucially, PROTAC potency and selectivity often depend on positive cooperativity within the ternary complex, arising from de novo PPIs formed between the ligase and target only in the presence of the PROTAC. Cooperativity can be quantified using biophysical assays such as surface plasmon resonance (SPR), where the cooperativity factor α compares binding affinities with and without the partner protein; $\alpha > 1$ indicates beneficial, synergy-driven cooperativity [83, 84].

Because PROTACs are large, highly flexible molecules—comprising two ligands connected by a linker—their physicochemical properties frequently fall beyond Lipinski's Rule of Five (bRo5). Rational design, therefore, requires understanding how linker length, composition, stereochemistry, and conformational preferences influence ternary-complex formation and energetics. Quantum-chemical methods capable of resolving residue-level interaction patterns offer a principled way to address these challenges and guide design [82, 85].

2.5.2 Case Study: Structure of the BRD4–MZ1 (PROTAC Ligand)–VHL Ternary Complex

We applied FMO-SP to the BRD4–MZ1–VHL ternary complex (PDB ID: 5T35) [86]. MZ1 links a BRD4-binding JQ1 derivative to a VHL ligand (VH032-derived hydroxyproline) via a PEG-like linker, bridging BRD4–BD2 and the VHL–ElonginC–ElonginB (VCB) complex. This structure provided the first evidence for cooperative recognition that underlies MZ1's selective degradation of BRD4 over BRD2/3 [85].

Our FMO-SP analysis focused on three mechanistic questions:

1. Which BRD4 and VHL residues most strongly stabilize the ternary complex?
2. How do linker orientation and ligand positioning influence cooperativity?
3. Why is the ternary complex more favorable than the binary interactions?

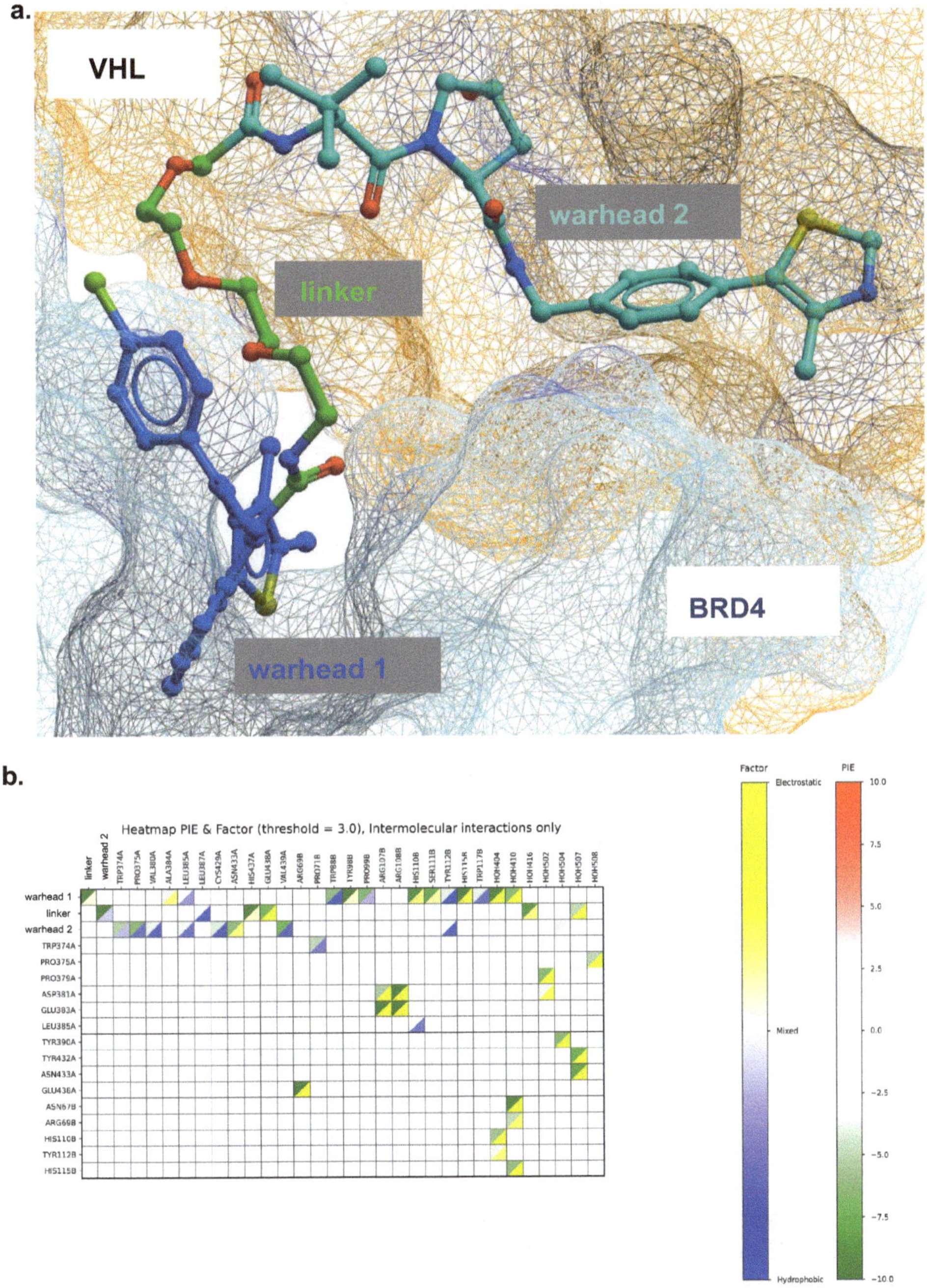

Fig. 7 (**a**) Structure of the BRD4–MZ1 (PROTAC ligand)– VHL ternary complex (PDB ID: 5T35) [86]. The MZ1 is fragmented into three fragments for FMO analysis: warhead 1 (BRD4-binding moiety, blue), linker (green), and warhead 2 (VHL-binding moiety, cyan). The protein partners BRD4 (light blue) and VHL (orange) are shown as surface meshes. This representation highlights how linker flexibility and spatial positioning enable cooperative ternary-complex formation by promoting de novo protein–protein interactions. (**b**) FMO heatmap summarizing pair interaction energies (PIEs) and factor contributions exceeding the 3.0 kcal/mol threshold, showing intermolecular interactions only. Each cell represents the intensity and type of interaction between individual residues and PROTAC fragments. Color coding reflects interaction components (electrostatics, exchange–repulsion, dispersion), enabling quantitative identification of stabilizing contacts within the complex. The color bar indicates the PIE scale in kcal/mol

To probe intramolecular interactions within MZ1, we fragmented the ligand into three components (Fig. 7a):

- warhead 1 (BRD4-binding fragment)
- linker (central flexible segment)
- warhead 2 (VHL-binding fragment)

FMO-SP enabled calculation of pair interaction energies (PIEs) between fragments, revealing how ligand architecture shapes ternary-complex energetics and degradation efficiency [86].

2.5.3 Results

FMO-SP identifies a strongly stabilizing BRD4–VHL (A–B) interface (TIE = −137.6 kcal/mol), dominated by a key salt bridge ASP381(A)/ARG108(B) (PIE = −34.1 kcal/mol)— See Table 3 and Fig. 7b for details.

Across ligand–protein contacts, MZ1→VHL (−73.9 kcal/mol) exceeds MZ1→BRD4 (−65.5 kcal/mol), reflecting the rich H-bond/salt-bridge network formed by the hydroxyproline-based VHL ligand.

Strong A–A and B–B intramolecular clamps (−79.0 and −84.9 kcal/mol) stabilize local folds and reduce entropic penalties, supporting productive cooperative assembly.

2.5.4 Mechanistic Interpretation

Previous structural studies [86] showed MZ1 adopting a compact, self-folded pose that precisely positions its warheads to create new BRD4–VCB PPIs absent in either binary complex. FMO-SP quantifies these induced interactions: the A–B TIE is the dominant stabilizing term, with the ASP381/ARG108 salt bridge and an ensemble of electrostatic contacts rationalizing the $\alpha > 1$ cooperativity observed biophysically.

The stronger VHL-side interactions (MZ1→B) arise from the hydroxyproline ligand's dense polar signature and fixed stereochemistry, while the BRD4-binding warhead engages a mixed ES/DI pattern within the acetyl-lysine pocket. Linker geometry that

Table 3

Summary of interaction categories and their energetic contributions in the BRD4-MZ1-VHL ternary complex (PDB ID: 5T35) from FMO-SP analysis

Interaction category	Count	TIE	Strongest PIE
A–B (A–B)	7	−137.6	ASP381A/ARG108B (−34.1)
MZ1 → A (LIG-A)	12	−65.5	Strongest ≈ −10.2
MZ1 → B (LIG-B)	9	−73.9	Strongest ≈ −18.07
A–A (A–A)	11	−79.0	Strongest ≈ −16.9
B–B (B–B)	17	−84.9	Strongest ≈ −13.7

pre-organises VHL-side contacts while allowing BRD4-side π-stacking maximizes cooperativity.

Large A–A and B–B TIEs reflect conformational preorganization—reducing entropy costs and improving interface complementarity—consistent with cooperative ternary complex formation.

2.5.5 Implications for SBDD

Optimizing VHL ligands: VH032-derived chemotypes with favorable exit vectors and stereochemistry can strengthen MZ1→B interactions and enhance degradation.

- Linker engineering: Short, directional linkers can boost cooperativity by aligning warheads and reducing conformational entropy—while overly rigid linkers risk misalignment.
- Beyond Ro5: Although PROTACs often violate Ro5, degradation efficiency depends more on ternary cooperativity than passive permeability. FMO-SP "hot-spot" mapping can guide modifications (ES/DI reinforcement, intramolecular H-bonds, conformational bias) to achieve cellular potency.

2.5.6 Intramolecular MZ1 Interactions and Linker Design Insights

FMO-SP analysis also reveals that MZ1 forms strong electrostatic intramolecular interactions between the linker and warhead 1 (BRD4-binding fragment), as well as pronounced hydrophobic interactions between the linker and warhead 2 (VHL-binding fragment) (Fig. 7b). These interactions help stabilize bioactive conformation of MZ1, positioning the two warheads for productive ternary-complex engagement.

A key advantage of FMO-SP is its ability to compute PIEs between individual PROTAC fragments. This enables systematic assessment of how linker chemistry length, polarity, rigidity, stereochemistry, and exit vector affects intramolecular stabilization. Identifying linkers that form favorable interactions with both warheads is particularly valuable for:

- pre-organizing the bioactive conformation,
- reducing conformational entropy penalties, and
- improving productive ternary-complex assembly,

which are critical challenges for long, flexible PROTAC molecules.

Thus, FMO-SP provides a rational, QM-based route to linker design, helping guide modifications that reinforce warhead–linker contacts and enhance overall degradation efficiency.

2.5.7 Summary

FMO-SP dissects the energetic drivers of MZ1's cooperative assembly in the BRD4–VCB ternary complex:

- The BRD4–VHL interface is the principal source of stabilization, with a defining salt bridge (ASP381/ARG108) underpinning positive cooperativity.

- VHL-side interactions dominate, highlighting the key role of hydroxyproline-based ligand design and informing VHL warhead optimization.
- Combining FMO-derived hot spots with biophysical cooperativity metrics offers a powerful framework for rational linker and warhead engineering.

FMO-SP shifts PROTAC design from empirical scanning to mechanistically informed optimization, directly linking molecular interactions to cooperativity and degradation outcomes. The MZ1 ternary complex remains a benchmark example of how PROTACs exploit induced PPIs to achieve selective protein degradation.

3 Conclusions

In this chapter, we moved beyond the comfort zone of classical small-molecule design and into the more intricate world of emerging therapeutic modalities. Biologics, peptideGPCR peptides, molecular glues, PROTACs, and nucleic acid modulators all expand what can be drugged, but they also stretch traditional modeling methods to their limits. Interfaces become larger and shallower, binding events become cooperative and multicomponent, and nucleic acids bring dense electrostatics and unconventional recognition motifs. In this setting, qualitative pictures and force-field scores are no longer enough.

The FMO-SP was developed precisely to meet this challenge. By embedding robust QM calculations in an automated, SAR-ready workflow, FMO-SP delivers residue-level energetics that can be read and acted on by medicinal chemists. Across the case studies in this chapter, the same pattern emerges: FMO-SP converts complex binding modes into quantitative maps of "who stabilizes what, and by how much," providing a direct line of sight from structure to design hypotheses.

For the GLP-1R/Semaglutide complex, FMO-SP disentangled electrostatic anchors, hydrophobic clamps, and water-mediated contacts that collectively explain high-affinity binding and receptor selectivity offering a blueprint for next-generation peptide analogues and peptidomimetics. In the CRBN–CC-885–GSPT1 ternary complex, FMO-SP quantified how a molecular glue reshapes and stabilizes a previously non-existent interface, pinpointing the hotspots and electrostatic clamps that make ternary assembly possible. For PROTACs, analysis of the BRD4–MZ1–VHL system showed how interfacial salt bridges, asymmetric warhead contributions, and intramolecular linker–warhead interactions cooperate to drive positive ternary cooperativity, turning "linker optimization" from trial-and-error into a guided exercise. Finally, in the RNA–SSO case, FMO-SP exposed the hierarchy of base contacts and water bridges that stabilize small-molecule binding to an HCV IRES element, illustrating its reach into nucleic acid therapeutics.

Taken together, these examples demonstrate that FMO-SP is not just a diagnostic tool; it is a design engine for modality-focused SBDD. It can rank and rationalize hotspots, compare alternative warheads and linkers, highlight opportunities for reinforcing key PPIs, and reveal where electrostatics, dispersion or charge transfer truly matter. Crucially, it does so in a way that integrates smoothly with existing workflows docking, MD, homology modeling and presents results as intuitive heatmaps and interaction tables that support rapid team decisions.

As drug discovery continues to pivot toward complex modalities, the central question is no longer "can we model this?" but "can we understand it well enough to design it rationally?". This chapter shows that FMO-SP helps answer that question in the affirmative. By bringing quantum-level insight into everyday project work, it enables more predictive, mechanism-aware optimization across modalities—and moves structure-based drug design for biologics, degraders, and nucleic acid modulators from aspirational to practical reality.

Acknowledgments

The authors would like to express their deepest gratitude to Dr. Dmitri Fedorov (AIST, Japan) for his invaluable support, guidance, and enduring friendship, which have been a constant source of inspiration in the development of the FMO-SP platform. His pioneering and fundamental contributions to the Fragment Molecular Orbital method have laid the foundation of our efforts and continue to inspire new directions.

References

1. Blanco M-J, Gardinier KM, Namchuk MN (2022) Advancing new chemical modalities into clinical studies. ACS Med Chem Lett 13:1691–1698
2. Blanco M-J, Gardinier KM (2020) New chemical modalities and strategic thinking in early drug discovery. ACS Med Chem Lett 11:228–231
3. Müller CE, Hansen FK, Gütschow M et al (2021) New drug modalities in medicinal chemistry, pharmacology, and translational science: joint virtual special issue by *Journal of Medicinal Chemistry*, *ACS Medicinal Chemistry Letters*, and *ACS Pharmacology & Translational Science*. ACS Med Chem Lett 12: 1508–1509
4. Valeur E, Guéret SM, Adihou H et al (2017) New modalities for challenging targets in drug discovery. Angew Chem Int Ed 56:10294–10323
5. Kansy M, Caron G (2021) New therapeutic modalities in drug discovery and development: insights & opportunities. ADMET DMPK 9:227–230
6. Choi Y, Vinks AA, Van Der Graaf PH (2023) Novel therapeutic modalities: the future is now. Clin Pharmacol Ther 114:493–496
7. Lai Y, Chu X, Di L et al (2022) Recent advances in the translation of drug metabolism and pharmacokinetics science for drug discovery and development. Acta Pharm Sin B 12:2751–2777
8. De La Torre BG, Albericio F (2025) The pharmaceutical industry in 2024: an analysis of FDA drug approvals from the perspective of molecules. https://www.preprints.org/manuscript/202501.0253/v1

9. Kayki-Mutlu G, Aksoyalp ZS, Wojnowski L et al (2024) A year in pharmacology: new drugs approved by the US Food and Drug Administration in 2023. Naunyn Schmiedebergs Arch Pharmacol 397:2949–2970
10. Yoon J, Lee S, Kim MJ et al (2025) Brief summary of the regulatory frameworks of regenerative medicine therapies. Front Pharmacol 15:1486812
11. Martins AC, Oshiro MY, Albericio F et al (2024) Food and Drug Administration (FDA) approvals of biological drugs in 2023. Biomedicines 12:1992
12. Strohl WR (2025) Structure and function of therapeutic antibodies approved by the US FDA in 2024. Antib Ther 8:197–237
13. Lee J, Beers JL, Geffert RM et al (2024) A review of CYP-mediated drug interactions: mechanisms and in vitro drug-drug interaction assessment. Biomolecules 14:99
14. Ferreira LLG, Andricopulo AD (2019) ADMET modeling approaches in drug discovery. Drug Discov Today 24:1157–1165
15. Henley MJ, Koehler AN (2021) Advances in targeting 'undruggable' transcription factors with small molecules. Nat Rev Drug Discov 20:669–688
16. Qin C, Wang Y-L, Zheng J et al (2025) Current perspectives in drug targeting intrinsically disordered proteins and biomolecular condensates. BMC Biol 23: 118
17. Kalyaanamoorthy S, Barakat KH (2018) Development of safe drugs: the hERG challenge. Med Res Rev 38:525–555
18. Lin A, Giuliano CJ, Palladino A et al (2019) Off-target toxicity is a common mechanism of action of cancer drugs undergoing clinical trials. Sci Transl Med 11:eaaw8412
19. Wu K, Kwon SH, Zhou X et al (2024) Overcoming challenges in small-molecule drug bioavailability: a review of key factors and approaches. Int J Mol Sci 25:13121
20. Xie X, Yu T, Li X et al (2023) Recent advances in targeting the "undruggable" proteins: from drug discovery to clinical trials. Signal Transduct Target Ther 8:335
21. Zhong L, Li Y, Xiong L et al (2021) Small molecules in targeted cancer therapy: advances, challenges, and future perspectives. Signal Transduct Target Ther 6:201
22. Pardridge WM (2005) The blood–brain barrier: bottleneck in brain drug development. NeuroRx 2:3–14
23. Wu D, Chen Q, Chen X et al (2023) The blood–brain barrier: structure, regulation and drug delivery. Signal Transduct Target Ther 8:217
24. Lei Z, Tian Q, Teng Q et al (2023) Understanding and targeting resistance mechanisms in cancer. MedComm 4:e265
25. Sun D, Gao W, Hu H et al (2022) Why 90% of clinical drug development fails and how to improve it?. Acta Pharm Sin B 12:3049–3062
26. Lu R-M, Hwang Y-C, Liu I-J et al (2020) Development of therapeutic antibodies for the treatment of diseases. J Biomed Sci 27:1
27. Shepard HM, Phillips GL, Thanos CD et al (2017) Developments in therapy with monoclonal antibodies and related proteins. Clin Med 17:220–232
28. Belvisi L, D'Andrea LD, Jiménez MA (2021) Editorial: peptides targeting protein-protein interactions: methods and applications. Front Mol Biosci 8:780106
29. Wang X, Ni D, Liu Y et al (2021) Rational design of peptide-based inhibitors disrupting protein-protein interactions. Front Chem 9:682675
30. Camps-Fajol C, Cavero D, Minguillón J et al (2025) Targeting protein-protein interactions in drug discovery: modulators approved or in clinical trials for cancer treatment. Pharmacol Res 211:107544
31. Scott DE, Bayly AR, Abell C et al (2016) Small molecules, big targets: drug discovery faces the protein–protein interaction challenge. Nat Rev Drug Discov 15:533–550
32. Wang L, Wang N, Zhang W et al (2022) Therapeutic peptides: current applications and future directions. Signal Transduct Target Ther 7:48
33. Kussie PH, Gorina S, Marechal V et al (1996) Structure of the MDM2 oncoprotein bound to the p53 tumor suppressor transactivation domain. Science 274:948–953
34. Souers AJ, Leverson JD, Boghaert ER et al (2013) ABT-199, a potent and selective BCL-2 inhibitor, achieves antitumor activity while sparing platelets. Nat Med 19:202–208
35. Zak KM, Kitel R, Przetocka S et al (2015) Structure of the complex of human programmed death 1, PD-1, and its ligand PD-L1. Structure 23:2341–2348
36. Tan X, Huang Z, Pei H et al (2024) Molecular glue-mediated targeted protein degradation: a novel strategy in small-molecule drug development. iScience 27:110712
37. Konstantinidou M, Arkin MR (2024) Molecular glues for protein-protein interactions: progressing toward a new dream. Cell Chem Biol 31:1064–1088
38. Dong G, Ding Y, He S et al (2021) Molecular glues for targeted protein degradation: from serendipity to rational discovery. J Med Chem 64:10606–10620
39. Sasso JM, Tenchov R, Wang D et al (2023) Molecular glues: the adhesive connecting targeted protein degradation to the clinic. Biochemistry 62:601–623

40. Rui H, Ashton KS, Min J et al (2023) Protein–protein interfaces in molecular glue-induced ternary complexes: classification, characterization, and prediction. RSC Chem Biol 4:192–215
41. Zhao L, Zhao J, Zhong K et al (2022) Targeted protein degradation: mechanisms, strategies and application. Signal Transduct Target Ther 7:113
42. Weagel EG, Foulks JM, Siddiqui A et al (2022) Molecular glues: enhanced protein–protein interactions and cell proteome editing. Med Chem Res 31:1068–1087
43. Abraham I, Martin P, Vaghela S et al (2025) Budget impact analysis of revumenib for the treatment of relapsed or refractory acute leukemias with a *KMT2A* translocation in the United States. J Manag Care Spec Pharm 31:680–693
44. Liu Z, Hu M, Yang Y et al (2022) An overview of PROTACs: a promising drug discovery paradigm. Mol Biomed 3:46
45. Riccardi F, Dal Bo M, Macor P et al (2023) A comprehensive overview on antibody-drug conjugates: from the conceptualization to cancer therapy. Front Pharmacol 14:1274088
46. Beck A, Goetsch L, Dumontet C et al (2017) Strategies and challenges for the next generation of antibody–drug conjugates. Nat Rev Drug Discov 16:315–337
47. Jadhav K, Abhang A, Kole EB et al (2025) Peptide-drug conjugates as next-generation therapeutics: exploring the potential and clinical progress. https://www.preprints.org/manuscript/202503.1521/v1
48. Egli M, Manoharan M (2023) Chemistry, structure and function of approved oligonucleotide therapeutics. Nucleic Acids Res 51:2529–2573
49. Cai Z, Ma H, Ye F et al (2025) Discovery of RNA-targeting small molecules: challenges and future directions. MedComm 6:e70342
50. Naeem S, Zhang J, Zhang Y et al (2025) Nucleic acid therapeutics: past, present, and future. Mol Ther Nucleic Acids 36:102440
51. Finkel RS, Mercuri E, Darras BT et al (2017) Nusinersen versus sham control in infantile-onset spinal muscular atrophy. N Engl J Med 377:1723–1732
52. Chiriboga CA (2017) Nusinersen for the treatment of spinal muscular atrophy. Expert Rev Neurother 17:955–962
53. Zahedipour F, Gol TM, Wisser S et al (2025) Revolutionizing pediatric gene and cell therapy: the hope for lipid-based nanoparticles in blood disorders. Curr Res Transl Med 73:103545
54. Li B, Chen Y, Zhang Y et al (2025) Adverse events associated with inclisiran: a real-world pharmacovigilance study of FDA adverse event reporting system (FAERS). Expert Opin Drug Saf 25(4):657–668
55. Szebeni J, Koller A (2025) Multisystem endothelial inflammation: a key driver of adverse events following mRNA-containing COVID-19 vaccines. Vaccines 13:855
56. Keritam O, Erdler M, Fasching B et al (2025) Efficacy and safety of risdiplam in adults with 5q-associated spinal muscular atrophy: a nationwide observational cohort study in Austria. eClinicalMedicine 88:103536
57. Ouma RBO, Ngari SM, Kibet JK (2024) A review of the current trends in computational approaches in drug design and metabolism. Discov Public Health 21:108
58. Drucker DJ (2022) GLP-1 physiology informs the pharmacotherapy of obesity. Mol Metab 57:101351
59. Son JW, Lim S (2024) Glucagon-like peptide-1 based therapies: a new horizon in obesity management. Endocrinol Metab 39:206–221
60. Zheng Z, Zong Y, Ma Y et al (2024) Glucagon-like peptide-1 receptor: mechanisms and advances in therapy. Signal Transduct Target Ther 9:234
61. Heifetz A, Trani G, Aldeghi M et al (2016) Fragment molecular orbital method applied to lead optimization of novel interleukin-2 inducible T-cell kinase (ITK) inhibitors. J Med Chem 59:4352–4363
62. Heifetz A, James T, Southey M et al (2020) Guiding medicinal chemistry with fragment molecular orbital (FMO) Method. In: Heifetz A (Ed.)Eds. Quantum mechanics in drug discovery. Springer US, New York, NY, 37–48
63. Handa C, Yamazaki Y, Yonekubo S et al (2022) Evaluating the correlation of binding affinities between isothermal titration calorimetry and fragment molecular orbital method of estrogen receptor beta with diarylpropionitrile (DPN) or DPN derivatives. J Steroid Biochem Mol Biol 222:106152
64. Fukuzawa K, Tanaka S (2022) Fragment molecular orbital calculations for biomolecules. Curr Opin Struct Biol 72:127–134
65. Fedorov DG, Nagata T, Kitaura K (2012) Exploring chemistry with the fragment molecular orbital method. Phys Chem Chem Phys 14:7562
66. Fedorov DG, Kitaura K (2007) Pair interaction energy decomposition analysis. J Comput Chem 28:222–237
67. Heifetz A, Katchalski-Katzir E, Eisenstein M (2002) Electrostatics in protein–protein docking. Protein Sci 11:571–587
68. Davenport AJ, Möller C, Heifetz A et al (2010) Using electrophysiology and *in silico* three-dimensional modeling to reduce human *Ether-à-go-go* related gene K^+ channel inhibition in a histamine H3 receptor antagonist program. Assay Drug Dev Technol 8:781–789

69. Fedorov DG, Kitaura K (2016) Subsystem analysis for the fragment molecular orbital method and its application to protein–ligand binding in solution. J Phys Chem A 120: 2218–2231
70. Fedorov DG (2024) The peptide bond: resonance increases bond order and complicates fragmentation. ChemPhysChem 25:e202400170
71. Monteleone S, Fedorov DG, Townsend-Nicholson A et al (2022) Hotspot identification and drug design of protein–protein interaction modulators using the fragment molecular orbital method. J Chem Inf Model 62:3784–3799
72. Hauser AS, Attwood MM, Rask-Andersen M et al (2017) Trends in GPCR drug discovery: new agents, targets and indications. Nat Rev Drug Discov 16:829–842
73. Recio C, Lucy D, Iveson P et al (2018) The role of metabolite-sensing G protein-coupled receptors in inflammation and metabolic disease. Antioxid Redox Signal 29:237–256
74. Congreve M, De Graaf C, Swain NA et al (2020) Impact of GPCR structures on drug discovery. Cell 181:81–91
75. Wilding JPH, Batterham RL, Calanna S et al (2021) Once-weekly semaglutide in adults with overweight or obesity. N Engl J Med 384:989–1002
76. Husain M, Birkenfeld AL, Donsmark M et al (2019) Oral semaglutide and cardiovascular outcomes in patients with type 2 diabetes. N Engl J Med 381:841–851
77. Marso SP, Bain SC, Consoli A et al (2016) Semaglutide and cardiovascular outcomes in patients with type 2 diabetes. N Engl J Med 375:1834–1844
78. Gomez-Peralta F, Choudhary P, Cosson E et al (2022) Understanding the clinical implications of differences between glucose management indicator and glycated haemoglobin. Diabetes Obes Metab 24:599–608
79. Zhang X, Belousoff MJ, Liang Y-L et al (2021) Structure and dynamics of semaglutide- and taspoglutide-bound GLP-1R-Gs complexes. Cell Rep 36:109374
80. Griffith DA, Edmonds DJ, Fortin J-P et al (2022) A small-molecule oral agonist of the human glucagon-like peptide-1 receptor. J Med Chem 65:8208–8226
81. Matyskiela ME, Lu G, Ito T et al (2016) A novel cereblon modulator recruits GSPT1 to the CRL4CRBN ubiquitin ligase. Nature 535:252–257
82. Monteleone S, Morao I, Fedorov DG et al (2025) Quantum mechanics-based ranking of predicted proteolysis targeting chimeras-mediated ternary complexes. ACS Med Chem Lett 16:420–427
83. Wurz RP, Rui H, Dellamaggiore K et al (2023) Affinity and cooperativity modulate ternary complex formation to drive targeted protein degradation. Nat Commun 14: 4177
84. Roy MJ, Winkler S, Hughes SJ et al (2019) SPR-measured dissociation kinetics of PROTAC ternary complexes influence target degradation rate. ACS Chem Biol 14:361–368
85. Tunjic TM, Weber N, Brunsteiner M (2023) Computer aided drug design in the development of proteolysis targeting chimeras. Comput Struct Biotechnol J 21:2058–2067
86. Gadd MS, Testa A, Lucas X et al (2017) Structural basis of PROTAC cooperative recognition for selective protein degradation. Nat Chem Biol 13:514–521

Chapter 7

Exploring Nucleic Acid Structure and Function with Molecular Dynamics and Fragment Molecular Orbital Methods

Tatsuya Ohyama, Hisae Tateishi-Karimata, Naoki Sugimoto and Teruki Honma

Abstract

In recent years, nucleic acid therapeutics have attracted increasing attention as a new modality for drug discovery and as a novel therapeutic approach for rare and intractable diseases. The G-quadruplex structure of nucleic acids is of interest as both a target in drug development and as a candidate for nucleic acid therapeutics, as it can inhibit replication, transcription, reverse transcription, translation, and elongation reactions. The G-quadruplex structures are known to undergo significant changes in both conformation and stability in response to environmental factors such as ionic concentration, pH, temperature, and molecular crowding. To develop nucleic acid therapeutics, including G-quadruplexes, effectively, it is necessary to consider not only interactions within the target nucleic acid but also those with binding nucleic acids, proteins, and various crowding molecules present in the cellular environment. In this study, in order to elucidate the interactions and behaviors of crowding molecules that affect nucleic acid stability, we performed molecular dynamics simulations and fragment molecular orbital (FMO) calculations of hairpin DNA and three types of G-quadruplexes, which are involved in transcription and translation in the absence and presence of tetraethylene glycol (EG4) as crowding molecules. The results of FMO calculations on multiple snapshots from MD simulations of DNA under dilute conditions showed that backbone interactions were similar across nucleic acid topologies, while base–base interactions showed distinct fluctuations in energy. Molecular dynamics simulations in the presence of EG4 molecules as crowding agents demonstrated that EG4 molecules, which are generally considered not to directly interact with target biomacromolecules, localized around DNAs for extended periods. From the FMO calculation, it was found that EG4 interacts with the backbone fragment through hydrogen bonding, whereas it interacts with the base fragment through CH–π interactions. These CH–π interactions were mainly observed with the G-quartets and with the bases stacked on the G-quartets in parallel- and hybrid-type G-quadruplexes that exhibited low fluctuation. On the other hand, in the anti-parallel-type G-quadruplex whose G-quartet was covered by highly fluctuating loop bases and in the hairpin DNA without a G-quartet, a few stable interactions were observed. These findings provide insight into how crowding molecules shape the intracellular environment surrounding nucleic acids.

Keywords FMO, Fragment molecular orbital method, DNA, G-quadrupex, Molecular dynamics simulations, ABINIT-MP, Hydrogen bonds, Aromatic π–π interactions, CH–π interactions

Alexander Heifetz (ed.), *Quantum Mechanics Applications in Drug Design*, Methods in Molecular Biology, vol. 3061,
https://doi.org/10.1007/978-1-0716-5420-0_7,

1 Introduction

Nucleic acids are not only responsible for storing genetic information in living organisms but also have attracted attention as a drug discovery modality in recent years, since various nucleic acid therapeutics have been developed that utilize the structure and properties of nucleic acids. Especially, G-quadruplexes formed by G-rich sequences are notable nucleic acid structures utilized both as drug discovery targets and as nucleic acid therapeutics. A G-quartet, which is a feature of G-quadruplexes, consists of four guanine bases bound by eight hydrogen bonds, as shown in Fig. 1, and the stacking of these G-quartets stabilizes the G-quadruplex. Even with the same G-rich sequence, it has been reported that G-quadruplexes can form

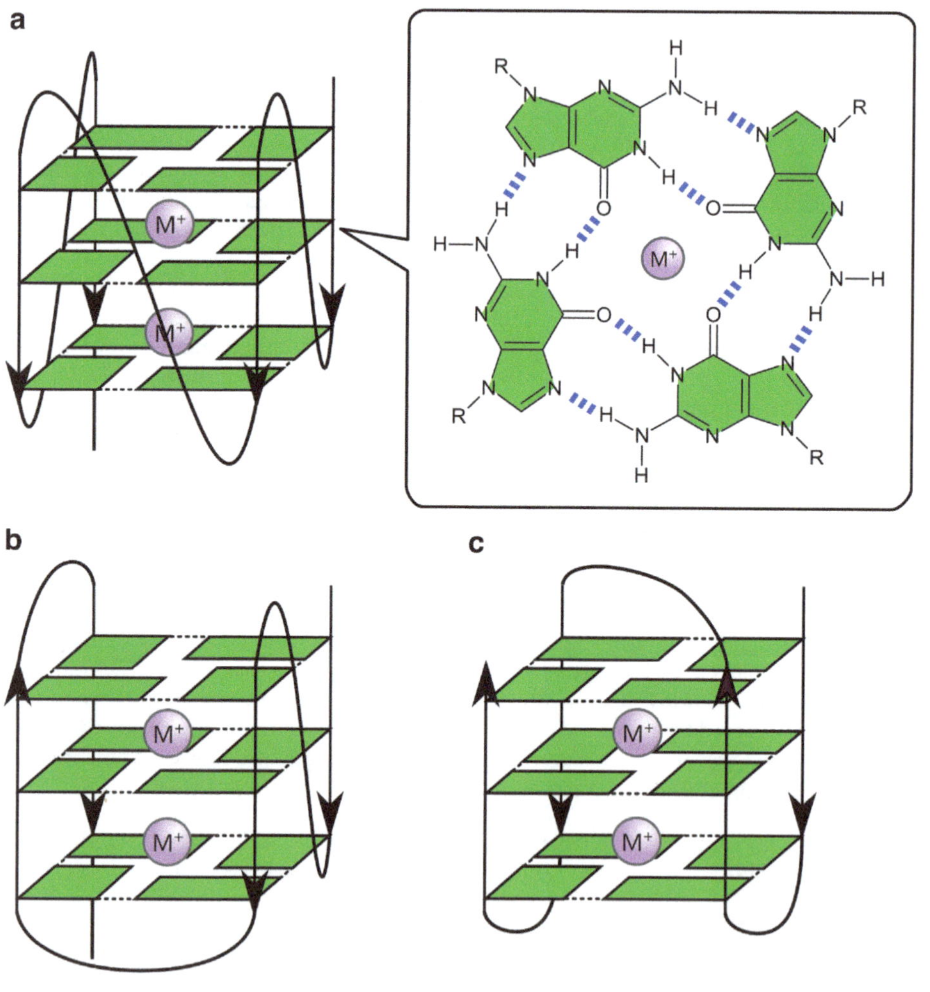

Fig. 1 Topologies for (**a**) parallel type G-quadruplex, (**b**) hybrid-type G-quadruplex, and (**c**) anti-parallel-type G-quadruplex. The right panel of (**a**) shows the G-quartet and its hydrogen bonds (blue lines) in a parallel-type G-quadruplex

three types of topology—parallel, hybrid, and anti-parallel—depending on the surrounding environment. Since stable G-quadruplexes inhibit replication [1], transcription [2], translation [3], reverse transcription [4], and telomerase elongation [5] reactions, several drug discovery studies have focused on utilizing this function. Thus, G-rich sequence regions that form G-quadruplexes are identified as drug targets aimed at inhibiting the above reactions. Although telomerase elongates telomeres at the ends of chromosomes, it is abnormally activated in cancer cells and promotes their proliferation [6]. Since telomeric regions with repeated TTAGGG sequences form G-quadruplexes, promoting stable G-quadruplex formation in these regions can inhibit elongation reactions, thereby suppressing the proliferation of cancer cells. Ligands with extensive π-planar surfaces, such as TMPyP4, BRACO-19, Phen-DC, and telomestatin, have been proposed as stabilizers of G-quadruplex structures [7]. In drug discovery strategies that employ transcriptional inhibition through G-quadruplex formation, G-rich sequence regions within oncogene promoters such as MYC, KRAS, BCL2, VEGF, PDGFR-β, RET, and EGFR are targeted. Furthermore, there are studies that attempt to induce G-quadruplex formation in G-rich sequence regions within the DNA or RNA of pathogenic bacteria [8] and viruses [9] to inhibit transcription. As a nucleic acid therapeutic, G-quadruplex is used in strategies that utilize its function by introducing nucleic acid sequences capable of forming G-quadruplex structures into a target without G-rich sequence regions. For example, previous research demonstrated that a designed DNA sequence comprising a region complementary to HIV-1 RNA and an additional G-rich region formed a G-quadruplex on the HIV-1 RNA, resulting in inhibition of the reverse transcription reaction [4, 10]. AS1411, which is a G-rich DNA aptamer composed of 26 nucleotides, selectively binds to nucleolin overexpressed in cancer cells and suppresses cancer cell proliferation by inhibiting DNA synthesis [11]. These nucleic acid therapeutics utilizing G-quadruplex structures require both the formation of a G-quadruplex and high stability. However, studies have reported that the stability of G-quadruplex structures varies depending on environmental factors such as temperature [12], ionic concentration [13], pressure [12, 14], pH [15], and molecular crowding conditions [16]. In this study, we focus on the atomic-level effects of the molecular crowding condition on G-quadruplexes.

The intracellular environment is packed with nucleic acids, proteins, and metabolites such as polysaccharides and osmolytes [17–19]. The environment is referred to as a molecular crowding condition. Under the environment, the stability and the behavior of proteins and nucleic acids differ from those under dilute conditions. For example, protein and enzyme reaction rates differ between dilute solutions and crowding conditions [20–22]. For nucleic acids, a crowding environment promotes the formation of G-quadruplexes and stabilizes these structures [23, 24]. These phenomena have been confirmed in experiments that reproduce the intracellular crowding environment by adding polyethylene glycol (PEG) [25–30], dextran [31, 32], ficoll [31, 32], or acetonitrile [25, 33]. These studies suggested that crowding molecules interact indirectly with biomolecules through both repulsive and attractive forces [22]. The repulsive forces were observed as the excluded volume effect [26, 34, 35]. On the other hand, the attractive interactions were observed by dielectric constant [5] and water activity in solution [27]. The conclusions of these studies are derived from the physical parameters of solutions with crowding molecules, and detailed atomic- or molecular-level structural information is limited. Several studies have analyzed the effects of crowding molecules on nucleic acids at the atomic level [10, 25, 36, 37], although few studies have depicted interactions in detail.

In this study, classical molecular dynamics (MD) simulations were performed on the noncanonical structure G-quadruplex structures and on single-stranded DNA with a hairpin-structure in which Watson–Crick base pairs are formed. For the G-quadruplex, we prepared parallel, hybrid, and anti-parallel types composed of the repeated TTAGGG sequence found in telomere regions. In our MD simulations, it was reaffirmed that the thermal fluctuations of the G-quadruplex are extremely small compared to single-stranded DNA, indicating very high stability. Interaction analysis using the fragment molecular orbital (FMO) method also showed that hydrogen bonds and aromatic π-π interactions among guanines within the G-quadruplex are stable. Furthermore, MD simulations under a crowding environment modeled with EG4, the main component of PEG200, revealed that some EG4 molecules are located around the DNAs. Interaction analysis showed that the major driving forces for the localization of EG4 molecules around DNA are strong hydrogen bonds with the backbone as well as hydrogen bonds and CH–π interactions with the bases. These results suggest that ligands targeting nucleic acids compete with crowding molecules, and that it is necessary for ligands to possess an affinity at least stronger than that of the crowding molecules. These findings will provide useful information for elucidating the mechanism of efficient biological responses and for developing an efficient drug delivery system.

2 Methods

2.1 Molecular Dynamics (MD) Simulations

We obtained a single-stranded hairpin DNA (PDB ID: 1AC7 [38]; Fig. 2a) from the PDB as a model of canonical structure nucleic acid, and parallel-type (PDB ID: 1KF1 [39]; Fig. 2b), hybrid-type (PDB ID: 2HY9 [40]; Fig. 2c), and anti-parallel-type (PDB ID: 2MBJ [41]; Fig. 2d) G-quadruplexes as models of noncanonical structure nucleic acids, to analyze the behaviors of nucleic acids and surrounding crowding molecules. Both 5′- and 3′-termini were capped with

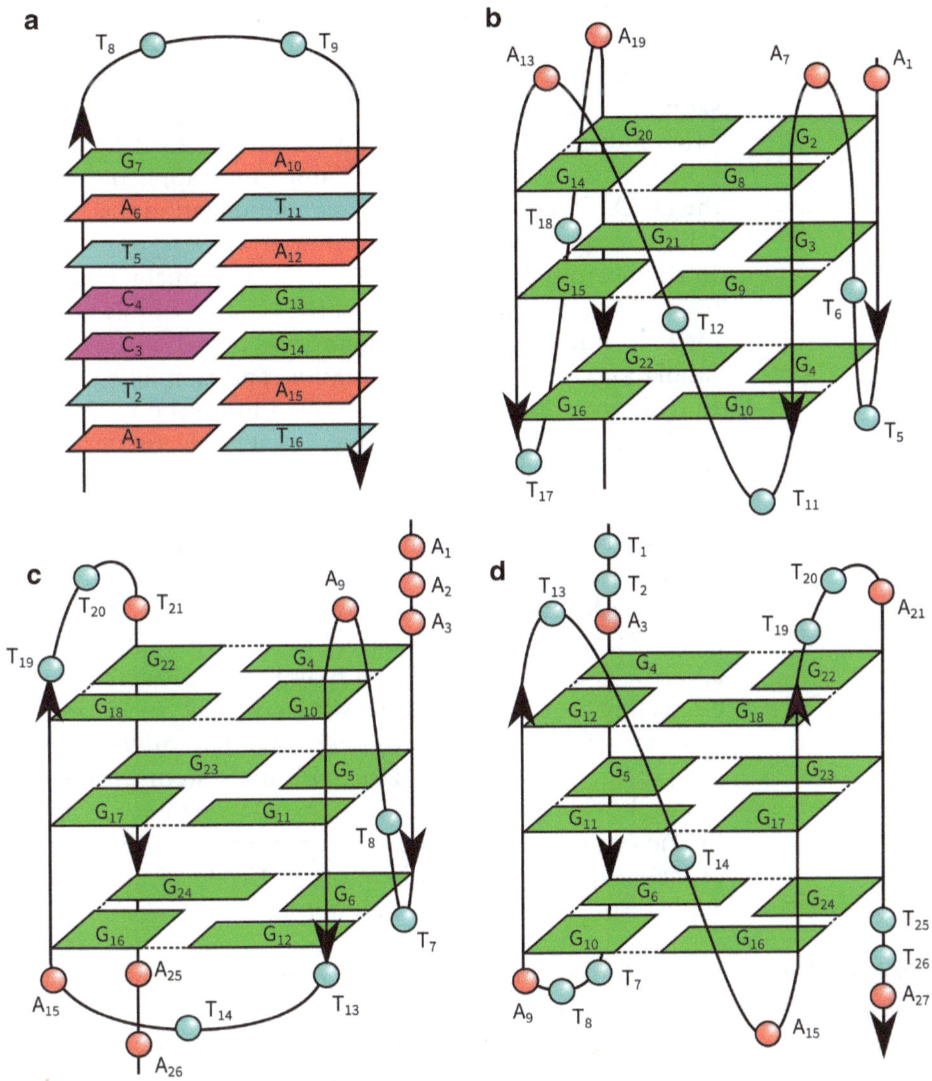

Fig. 2 Schematic structure of (**a**) hairpin DNA, (**b**) parallel-type G-quadruplex, (**c**) hybrid-type G-quadruplex, and (**d**) anti-parallel-type G-quadruplex. Red, magenta, green, and cyan spheres and parallelograms show adenine, cytosine, guanine, and thymine bases, respectively. The parallelograms indicate the bases that form base pairs

hydroxyl groups. Since the cavities of the G-quartets in the hybrid- and anti-parallel-type G-quadruplexes lacked cations, we introduced K^+ between the G-quartets of the hybrid-type and Na^+ at the center of the anti-parallel-type. System neutralization was performed by AmberTools [42] using cations equal to the number of phosphate groups. The system box was sized to maintain at least a 20 Å distance between nucleic acid and box wall, filled with water, and adjusted to 0.1 M ionic strength with K^+ or Na^+ and Cl^-. Parmbsc1 [43] and TIP3P [44] force fields were applied to nucleic acids and water, respectively. Three replicas of these systems were prepared. For each replica, structure optimization of the solvent and entire system, heating under the NVT ensemble, and NPT ensemble equilibration and sampling were sequentially performed using Gromacs Ver. 5.1.2 [45]. During solvent optimization and the heating process, positional restraints of 4184 kJ mol^{-1} nm^{-2} were applied to each atom of the DNA. Steepest descent energy minimization was performed for 10,000 steps during both solvent and whole-system optimization. In the heating process, the system was gradually heated to 298 K over 100 ps. Equilibration under the NPT ensemble was performed for 1 ns under 1 atm, and sampling was carried out for 50 ns. In the simulations for heating, equilibration, and sampling, covalent bonds to hydrogen atoms were constrained using the LINCS algorithm [46], and the time step was set to 2 fs. The electrostatic and van der Waals cutoffs were both set to 1.0 nm. In systems with crowding molecules, tetraethylene glycol (EG4), the main component of PEG200, was used as the crowding molecule. To provide sufficient degrees of freedom for water molecules and EG4, the concentration of EG4 was set to 0.5 M. EG4 was constructed using Discovery Studio Visualizer [47], and the General Amber Force Field (GAFF) [48, 49] was applied. The atomic charges of EG4 for GAFF were determined by calculating the restrained electrostatic potential (RESP) in Gaussian09 software [50]. The force fields were converted with ACPYPE software [51]. This system, as was done with the system without crowding molecules described above, was prepared with three replicas, followed by the optimization of the solvent and the entire system, heating process, equilibration, and sampling. No positional restraints were applied to EG4 throughout all steps.

2.2 Fragment Orbital Method Calculations

Interactions within DNA, as well as those between DNA and crowding molecules, were calculated at the MP2/6-31G(d) level using the fragment molecular orbital (FMO) method implemented in ABINIT-MP [52–54]. Target structures for statistical interaction analysis were generated by extracting DNA and associated water molecules within 5 Å, from snapshots taken every 5 ns along the MD trajectory. Since cations inside the G-quartet cavity are in close proximity to surrounding atoms and may cause convergence issues, they were excluded. Additionally, ions in the solvent were excluded because they contribute noise to the statistical analysis of interactions. Nucleotides were divided into backbone (deoxyribose and phosphate groups) and base fragments. Each water molecule was defined as an individual fragment. In addition, the interaction energies were divided into electrostatic (ES), exchange repulsion (EX), charge transfer, and others (CT+mix), and dispersion (DI) energies using pair interaction energy decomposition analysis (PIEDA) [55, 56]. For the system including crowding molecules, target structures were generated by extracting molecules within 5 Å of the DNA from snapshots at 50 ns of the MD trajectory. The system was divided into fragments for the backbone and bases of DNA, EG4 molecules, and water molecules. Interaction energies were divided into ES, EX, CT+mix, and DI components.

3 Results and Discussion

Previous studies have primarily employed molecular dynamics simulations of G-quadruplexes to evaluate and validate force fields [57, 58]. Accordingly, this study does not focus on the details of MD simulation protocols. Instead, we make use of structures obtained from MD simulations and focus our analysis on two types of interactions: those within DNA and those between DNA and surrounding crowding molecules.

3.1 Interactions in G-Quadruplex

To investigate differences in DNA behavior under dilute conditions, we measured the RMSD of C4′ atoms during the 30–50 ns of our MD simulations. For hairpin DNA with Watson–Crick base pairs, the RMSDs were 1.87 ± 0.33, 1.96 ± 0.35, and 1.86 ± 0.36 Å (Fig. 3a). For parallel G-quadruplex DNA, the values were 1.79 ± 0.21, 1.76 ± 0.23, and 2.00 ± 0.23 Å (Fig. 3b). For the hybrid G-quadruplex, the RMSDs were 2.58 ± 0.18, 2.00 ± 0.13, and 2.60 ± 0.15 Å (Fig. 3c). The anti-parallel G-quadruplex showed values of 1.55 ± 0.10, 1.63 ± 0.11, and 1.50 ± 0.17 Å (Fig. 3d). The standard deviations for hairpin DNA were 0.33–0.36 Å, while those for the G-quadruplexes were smaller, at 0.10–0.23 Å. This suggests that the G-quadruplex structures are highly stable. Such stability arises from the eight hydrogen bonds within each G-quartet and the aromatic π-π interactions between G-quartets.

To further analyze these interactions, we investigated internal DNA interactions in snapshots extracted every 5 ns from MD simulations of hairpin DNA and the parallel-type G-quadruplex, which is a representative G-quadruplex. We examined the distributions of distance and interaction energy for interactions classified as "Base pair" (base pair), "Stacking pair" (stacked bases), "Loop sequence" (nucleotides in loop regions), and "Other." The total interaction energy among backbone fragments ("TOTAL" in "Backbone fragment" in Fig. 4) was largely dominated by repulsive interactions in both hairpin DNA and the parallel-type G-quadruplex. These repulsions are electrostatic interactions between negatively charged phosphate groups, as indicated by the major contribution of the ES component. Since the DNA backbone is exposed to solvent, these interactions are usually neutralized or suppressed by surrounding cations and solvent molecules. Furthermore, the 5′- and 3′-termini were capped with hydroxyl groups in this study. The terminal nucleotides, which lack phosphate groups, exhibited interaction energies around 0 kcal/mol.

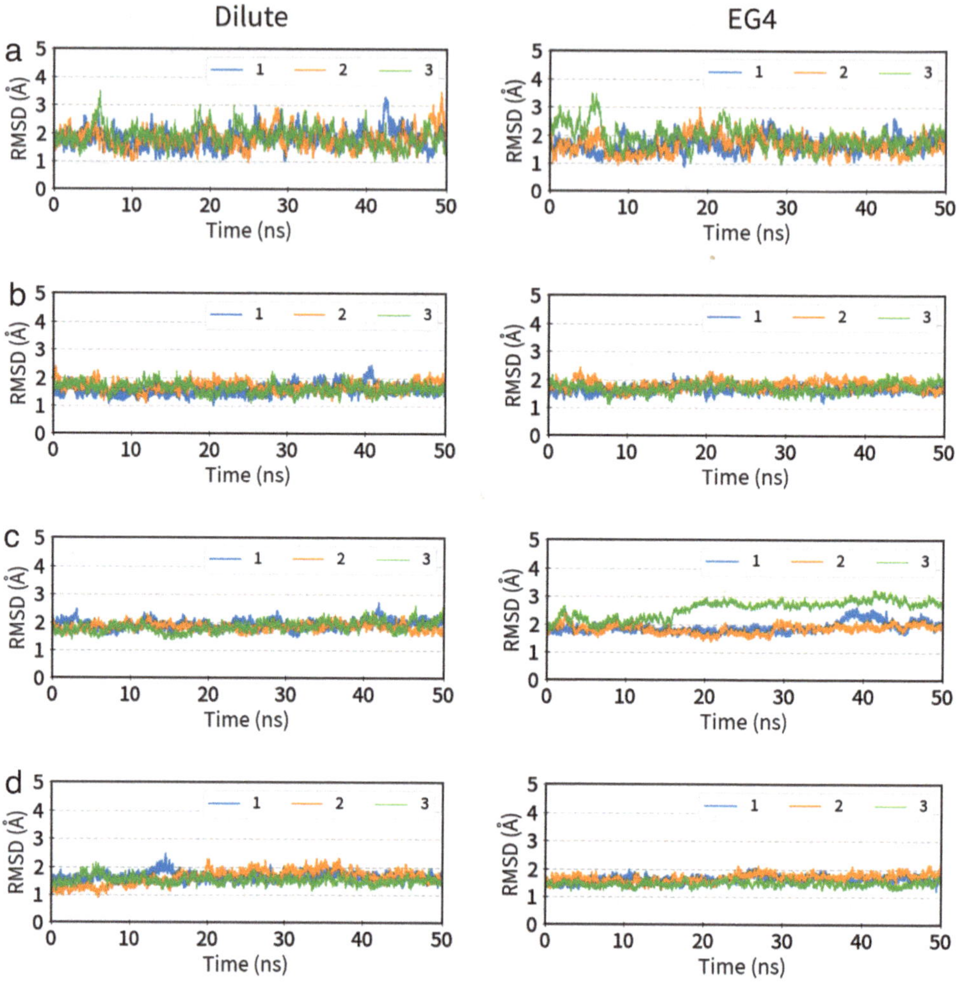

Fig. 3 Change in root mean square deviation (RMSD) of C4′ atoms in (**a**) hairpin DNA, (**b**) parallel-type G-quadruplex, (**c**) hybrid-type G-quadruplex, and (**d**) anti-parallel-type G-quadruplex. The DNA structures obtained after optimization of the entire system were used as reference structures. The C4′ atoms at the 5′- and 3′-termini were excluded. The blue, orange, and green lines represent the RMSD of DNA within each replica, respectively

The differences in interactions between hairpin DNA and the G-quadruplex were mainly evident among the base fragments ("Base fragments" in Fig. 4a, b). In hairpin DNA, the "Base pair" ES component (blue plot) was divided into two clusters. In the ES component of the hairpin DNA, the "Base pair" (blue plot) is divided into two clusters. The A-T base pairs were clustered from −36.2 to −12.4 kcal/mol, while the C-G base pairs ranged from −69.5 to −41.8 kcal/mol. In contrast, in the case of the parallel-type G-quadruplex, since the base pairs consist only of G-G, they form a single cluster. Significant differences were also observed in the "Stacking

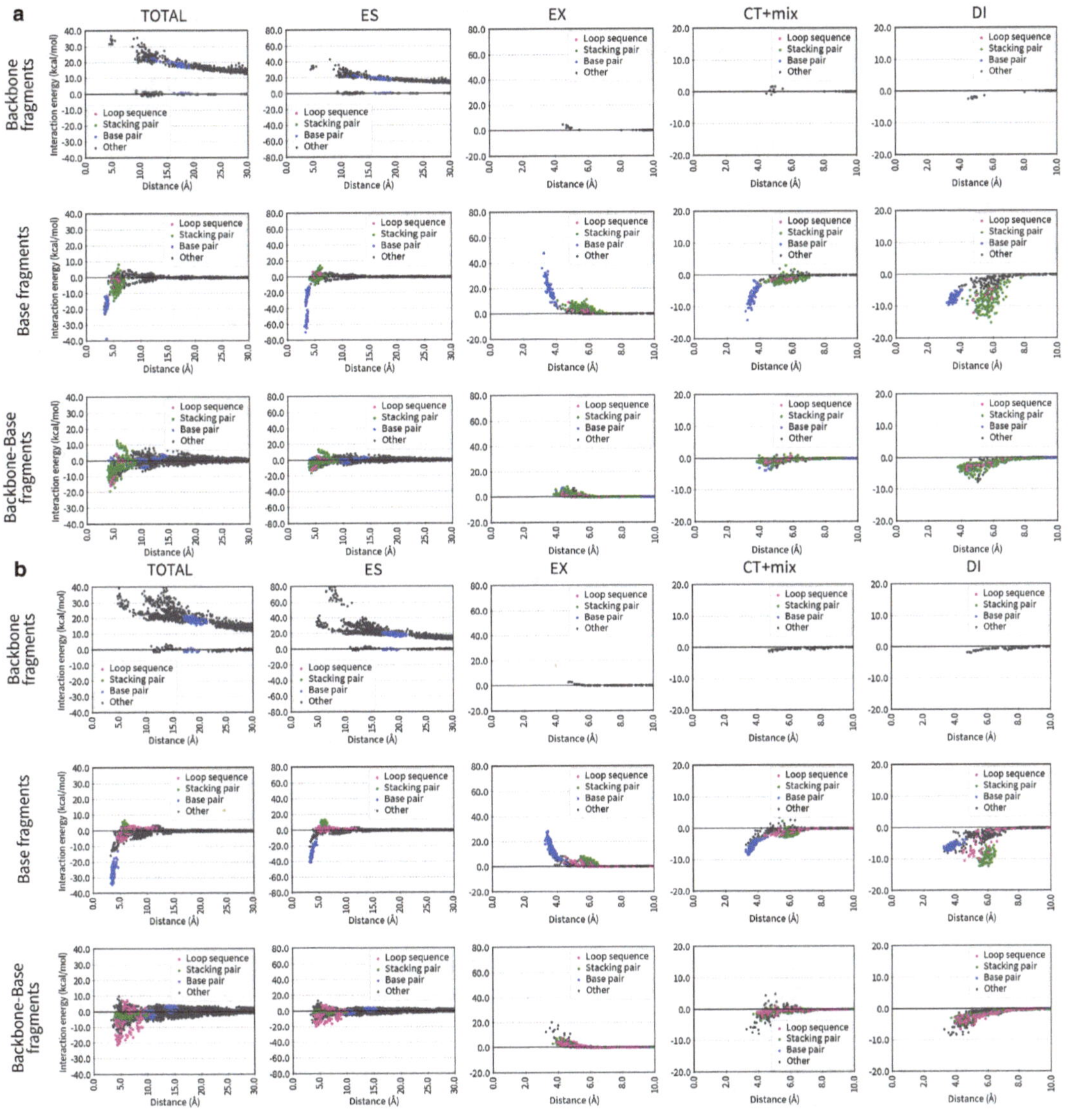

Fig. 4 Distribution of distances and interactions (TOTAL, ES, EX, CT+mix, and DI components) between DNA backbone fragments and between DNA base fragments in (**a**) hairpin DNA and (**b**) parallel-type G-quadruplex

pair" of the DI component. In hairpin DNA, the interactions of the "Stacking pair" (green plot) are distributed between −15.15 and −1.25 kcal/mol at distances ranging from 4.07 to 7.76 Å. In the parallel-type G-quadruplex, these interactions were distributed in a narrower range, from −12.32 to −5.65 kcal/mol, with distances from 5.34 to 6.49 Å. However, even in the parallel-type G-quadruplex, variations are observed in the aromatic π-π interactions between bases of the "Loop sequence".

These results indicate that the primary difference in interactions between hairpin DNA and the parallel-type G-quadruplex is the magnitude of fluctuations in these interactions. Minimal fluctuations in these interactions are a key factor in the sustained binding and stability of the G-quadruplex.

3.2 Interactions between G-Quadruplex and Cosolute

Polyethylene glycol (PEG) is widely used as a crowding agent to reproduce molecular crowding conditions. PEG200 is a mixture of short ethylene glycol oligomers and is mostly composed of tetraethylene glycol (EG4). We investigated the interactions between EG4 and nucleic acids using MD simulations combined with FMO calculations. Our simulation results (Fig. 5) reveal that EG4 molecules do not distribute uniformly in solution. Instead, they form clusters and frequently localize around DNA. In particular, it is observed that EG4 molecules are present around the G-quadruplex. To quantify the distribution of the EG4 molecules, the minimum distance from the DNA to each EG4 molecule was measured over 20,000 frames across three molecular dynamics simulations for 30–50 ns. The data presented in Fig. 6 are histograms of the minimum distances for 369 EG4 molecules, corresponding to 123 molecules per system

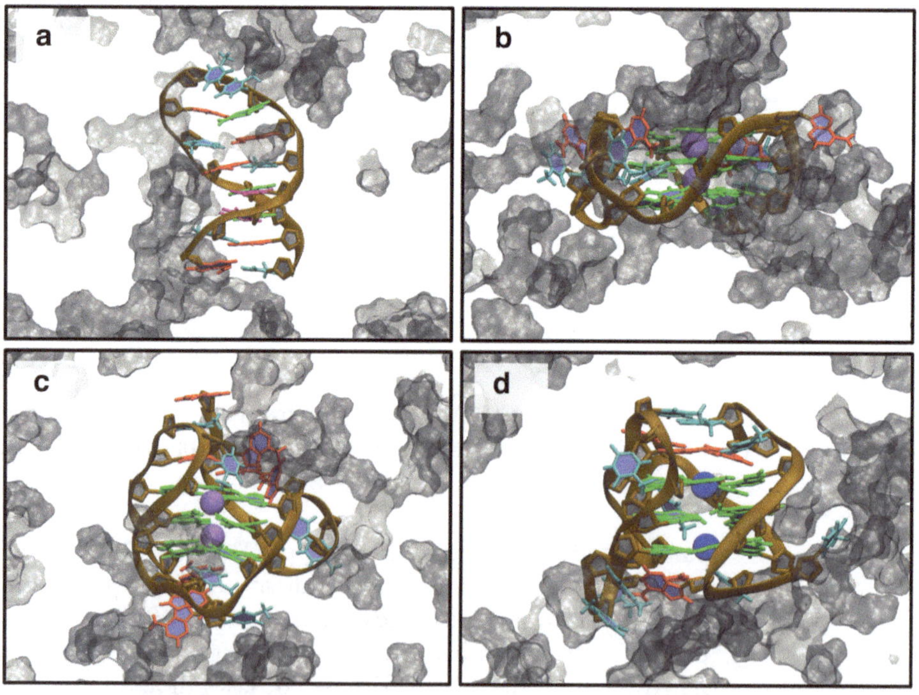

Fig. 5 Cross-sectional snapshots at 50 ns from the MD simulations for (**a**) hairpin DNA, (**b**) parallel-type G-quadruplex, (**c**) hybrid-type G-quadruplex, and (**d**) anti-parallel-type G-quadruplex. The brown ribbon, and red, magenta, green, and cyan stick representations indicate the DNA backbone, adenine base, cytosine base, guanine base, and thymine base, respectively. The gray surface representation shows the molecular surface of EG4

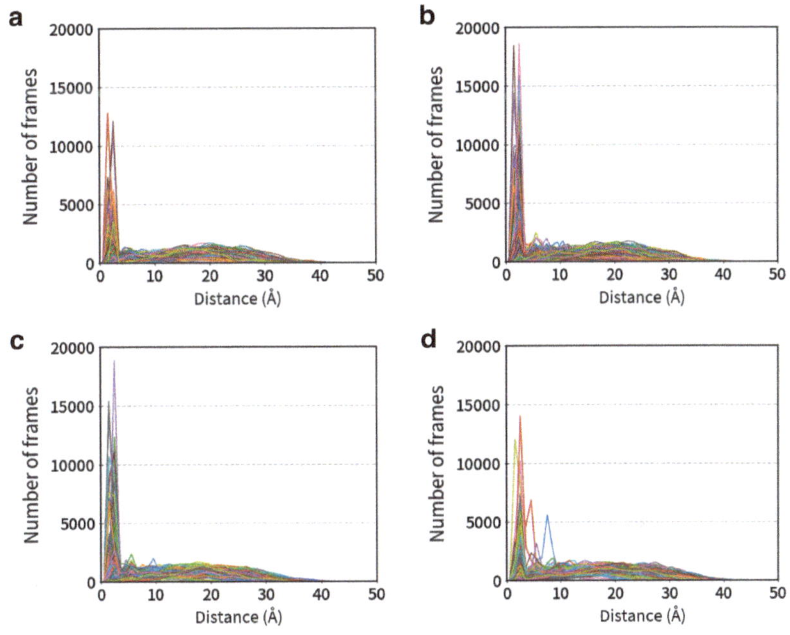

Fig. 6 Distribution of the minimum distances between DNA and EG4 molecules. DNA structures include (**a**) hairpin DNA, (**b**) parallel-type G-quadruplex, (**c**) hybrid-type G-quadruplex, and (**d**) anti-parallel-type G-quadruplex. Each line denotes one EG4 molecule during 30–50 ns (20,000 frames) of an MD trajectory. A total of 369 EG4 molecules, corresponding to 123 EG4 molecules in the system across three MD simulations, were analyzed

over three replicates. Across all systems, we detected sharp peaks at 1–3 Å. At greater distances, peaks are smaller, and the frequency in each histogram bin remains below 2000 frames. Maximum values for the sharpest peaks were observed as follows: 12,819 for hairpin DNA, 18,640 for parallel-type G-quadruplex, 18,848 for hybrid-type G-quadruplex, and 14,064 for anti-parallel-type G-quadruplex. These results indicate that several EG4 molecules remain in close proximity to DNA for long periods.

To examine the driving forces for EG4 molecules to remain on DNA for a long time, we extracted snapshots in which the greatest number of EG4 molecules were located within 5 Å of the DNA and performed FMO calculations. Figure 7 shows the interaction energies and distance distributions for backbone fragment—EG4 molecule (purple) and base fragment—EG4 molecule (cyan) pairs. For the interaction energy TOTAL, calculated as the sum of ES, EX, CT+mix, and DI components, the strongest interaction pairs show values of −31.6 kcal/mol for the EG4 molecule with the backbone fragment (hairpin DNA; purple arrow in Fig. 7a) and −18.6 kcal/mol with the base fragment (hybrid-type G-quadruplex; cyan arrow in Fig. 7c). For other DNA topologies, the strongest interaction pairs also exhibit

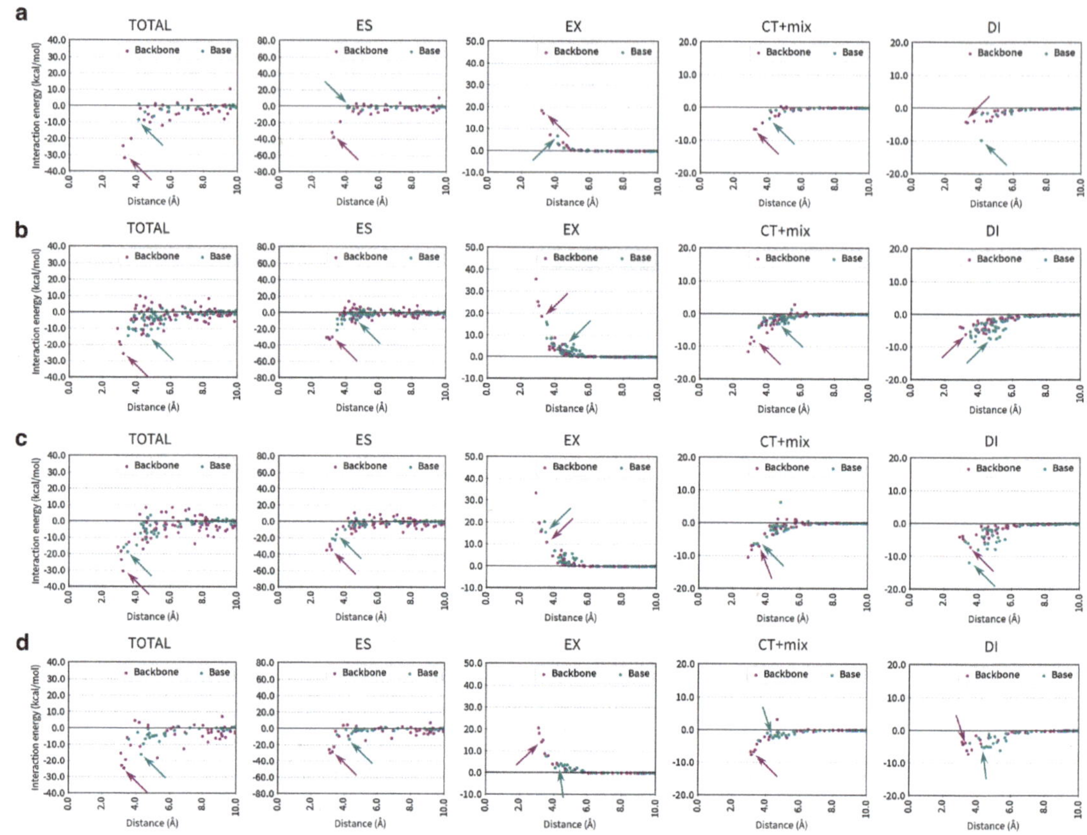

Fig. 7 Distribution of distances and interactions (TOTAL, ES, EX, CT+mix, and DI components) between DNA and EG4 molecules in the presence of (**a**) hairpin DNA, (**b**) parallel-type G-quadruplex, (**c**) hybrid-type G-quadruplex, and (**d**) anti-parallel-type G-quadruplex. Purple and cyan points show the interaction pair of EG4 molecules with the backbone and base fragments, respectively. Arrows indicate interaction pairs with the strongest interaction in TOTAL energy

interaction energies close to these values. In the components constituting these interactions, the ES and CT+mix components are predominant for backbone fragments and the DI component for base fragments. Since a previous study has systematically analyzed the database storing FMO calculation data and reported that the ES component and the CT+mix component are dominant in hydrogen bonds, most interactions with backbone fragments can be attributed to hydrogen bonds [59]. Figure 8 illustrates representative structures of the strongest interaction pairs. The left panel shows the structure of the strongest interaction pair with the backbone fragment (marked by purple arrows in Fig. 7), and the right panel shows the structure of the strongest interaction pair with the base fragment (marked by cyan arrows in Fig. 7). The DNA backbone mainly forms hydrogen bonds between phosphate groups and EG4 hydroxyls, which are only present at the ends of molecule (left panel in Fig. 8). Therefore, any

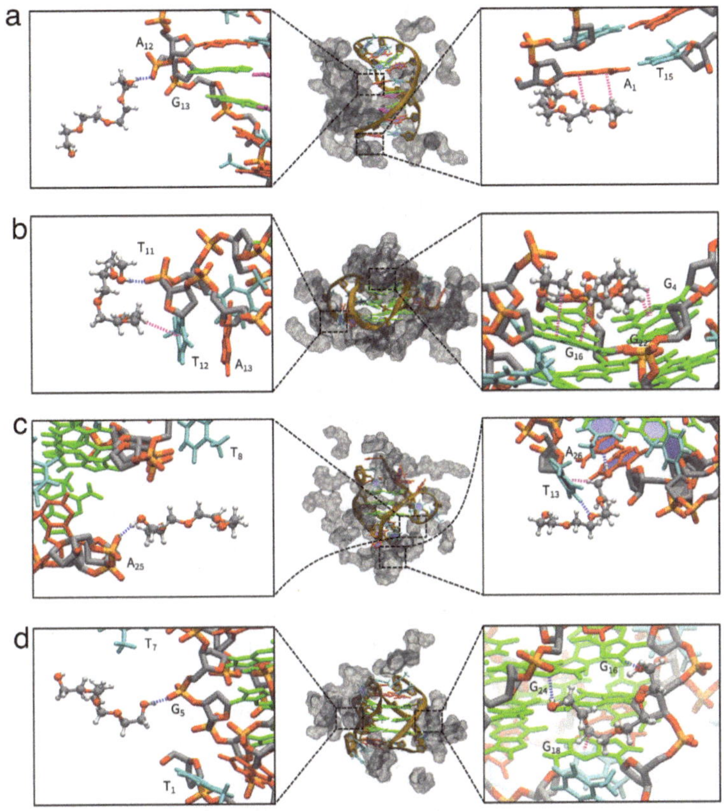

Fig. 8 Structures of the interaction pairs with the largest magnitude of the interaction energy TOTAL, as shown in Fig. 7, for (**a**) hairpin DNA, (**b**) parallel-type G-quadruplex, (**c**) hybrid-type G-quadruplex, and (**d**) anti-parallel-type G-quadruplex. The right and left panels show the strongest interaction pairs of EG4 molecules with backbone and base fragments, respectively. The brown ribbon, and red, magenta, green, and cyan stick representations indicate the DNA backbone, adenine base, cytosine base, guanine base, and thymine base, respectively. The gray surface representation shows the molecular surface of EG4. Blue and magenta lines show hydrogen bonds and CH–π interactions, respectively

single EG4 molecule can interact with at most two phosphate groups. In the native intracellular crowded environment, molecules such as other nucleic acids, proteins, and metabolized small molecules—which possess numerous sites capable of forming hydrogen bonds with the nucleic acid backbone—are densely crowded. Compared to these molecules, the FMO calculation results reaffirmed that EG4 is a molecule with the weak interactions with the DNA backbone.

On the other hand, for the interactions with the base fragment, not only hydrogen bonds with large contributions from the ES and CT+mix components, but also CH–π interactions with a large contribution from the DI component were observed. Typical CH–π interactions have energies of approximately 1.5–2.5 kcal/mol, and their strengths can accumulate [60]. However, the CH–π interaction is characterized by a cumulative increase in interaction energy and no mutual repulsive interactions other than steric hindrance [61]. Especially in topologies such as parallel-type G-quadruplex with exposed large π-planes of G-quartets, CH–π interactions with EG4 molecules constitute significant driving forces (Fig. 8b). Some interactions of a single EG4 molecule such as hydrogen bonds and CH–π interactions were also observed with bases partially stacked with a G-quartet in the hybrid-type G-quadruplex (Fig. 8c). On the other hand, in the anti-parallel-type G-quadruplex, a EG4 molecule forms hydrogen bonds and CH–π interactions with the G-quartet from the groove side of the G-quadruplex, rather than interacting with the G-quartet or bases stacked on the G-quartet (Fig. 8d). The reason is that the G-quartet in the anti-parallel-type G-quadruplex is completely covered by bases from the highly flexible loop region, resulting in the inability of the EG4 molecule to establish interactions. In the interactions with bases in the hairpin DNA, CH–π interactions, as shown in Fig. 8a, were observed. However, since similar interactions are not observed, as shown in Fig. 7a, this interaction is a rare case. Since hairpin DNA has no G-quartet and exhibits large fluctuations, as shown in Fig. 3a, it is expected that the EG4 molecule cannot establish stable interactions with the bases. These results provided a detailed depiction of the interactions between DNAs and crowding molecules. Because G-quadruplex ligands are intended to bind specifically to nucleobase sites, many compounds capable of aromatic π–π interactions—such as doxorubicin [62], proflavine [63], telomestatin [7], and BRACO-19 [7]—have been developed. Our results indicate that it is necessary to consider the topology of the G-quadruplex depending on the surrounding environment, even for the same sequence or G-quadruplex. Furthermore, our results suggest that CH–π interactions of EG4 may inhibit the binding of compounds targeting nucleic acid bases, indicating that it is necessary to develop ligands with affinities at least higher than those of interactions with crowding molecules.

4 Conclusion

In this chapter, the fundamental interactions of G-quadruplexes used as nucleic acid therapeutics and the interactions between crowding molecules and G-quadruplexes under molecular crowding conditions were investigated using MD simulations and FMO calculations. Since the phosphate groups on the backbone of nucleic acids are negatively charged, strong interactions between these phosphate groups can obscure important interactions in analyses at the nucleotide level. On the other hand, as shown in this chapter, the FMO method is a powerful tool that enables focused analysis of interactions involving important fragments by dividing nucleic acids into backbone and base fragments. In addition, analysis using the decomposition of interaction energies (PIEDA) and structural information enables discussion of the types of interactions involved. Cations in the cavity of the G-quartet were excluded from our calculations due to convergence problems. However, inclusion of cations reduces strong electrostatic interactions between backbone groups, resulting in more stabilized G-quadruplexes.

Detailed investigation of the behavior of crowding molecules is also necessary in the development of nucleic acid therapeutics. In this chapter, to mimic the molecular crowding environment, EG4 was used in the MD simulations. In the molecular crowding environment, various proteins, nucleic acids, and other metabolites are crowded and localized around the genes from which they are expressed and the enzymes responsible for their degradation [64]. It is expected that the properties of crowding molecules may differ among cellular compartments such as the nucleus, nucleolus, cytoplasm, and mitochondria. Therefore, to gain deeper insights into the behavior of crowding molecules surrounding the G-quadruplex, it is ideal to reproduce the system using the actual molecules present in biological systems. At least, it is necessary to investigate the effects of concentration using acetonitrile, ficoll, and dextran, which are commonly used in reproducible experiments. A deeper understanding of the effects of crowding molecule interactions is useful for the in silico development of both G-quadruplexes as drug targets and nucleic acid therapeutics that bind to G-quadruplexes.

Acknowledgments

The authors express their deep gratitude to Professor Shigenori Tanaka for his numerous insights. The authors deeply appreciate the support and assistance received from all involved.

References

1. Takahashi S, Brazier JA, Sugimoto N (2017) Topological impact of noncanonical DNA structures on Klenow fragment of DNA polymerase. Proc Natl Acad Sci U S A 114:9605–9610. https://doi.org/10.1073/pnas.1704258114
2. Tateishi-Karimata H, Isono N, Sugimoto N (2014) New insights into transcription fidelity: thermal stability of non-canonical structures in template DNA regulates transcriptional arrest, pause, and slippage. PLoS One 9:e90580. https://doi.org/10.1371/journal.pone.0090580
3. Endoh T, Sugimoto N (2013) Unusual -1 ribosomal frameshift caused by stable RNA G-quadruplex in open reading frame. Anal Chem 85:11435–11439. https://doi.org/10.1021/ac402497x
4. Tateishi-Karimata H, Muraoka T, Kinbara K, Sugimoto N (2016) G-Quadruplexes with tetra(ethylene glycol)-modified deoxythymidines are resistant to nucleases and inhibit HIV-1 reverse transcriptase. ChemBioChem 17:1399–1402. https://doi.org/10.1002/cbic.201600162
5. Yaku H, Murashima T, Tateishi-Karimata H et al (2013) Study on effects of molecular crowding on G-quadruplex-ligand binding and ligand-mediated telomerase inhibition. Methods

64:19–27. https://doi.org/10.1016/j.ymeth.2013.03.028

6. Hu M-H, Lin X-T, Liu B, Tan J-H (2020) Dimeric aryl-substituted imidazoles may inhibit ALT cancer by targeting the multimeric G-quadruplex in telomere. Eur J Med Chem 186:111891. https://doi.org/10.1016/j.ejmech.2019.111891
7. Monchaud D, Teulade-Fichou M-P (2008) A hitchhiker's guide to G-quadruplex ligands. Org Biomol Chem 6:627–636. https://doi.org/10.1039/B714772B
8. Cebrián R, Belmonte-Reche E, Pirota V et al (2022) G-Quadruplex DNA as a target in pathogenic bacteria: efficacy of an extended naphthalene diimide ligand and its mode of action. J Med Chem 65:4752–4766. https://doi.org/10.1021/acs.jmedchem.1c01905
9. Xu J, Huang H, Zhou X (2021) G-quadruplexes in neurobiology and virology: functional roles and potential therapeutic approaches. JACS Au 1:2146–2161. https://doi.org/10.1021/jacsau.1c00451
10. Tateishi-Karimata H, Ohyama T, Muraoka T et al (2017) Newly characterized interaction stabilizes DNA structure: oligoethylene glycols stabilize G-quadruplexes CH–π interactions. Nucleic Acids Res 45:7021–7030. https://doi.org/10.1093/nar/gkx299
11. Lourenço P, Moreira D, Miranda A et al (2025) Exploring the G-quadruplex formation of AS1411 derivatives. Molecules 30:1673. https://doi.org/10.3390/molecules30081673
12. Tariq N, Kume T, Feroze UN, Macgregor RB (2022) The pressure dependence of the stability of the G-quadruplex formed by d(TGGGGT). Life 12:765. https://doi.org/10.3390/life12050765
13. Bhattacharyya D, Mirihana Arachchilage G, Basu S (2016) Metal cations in G-quadruplex folding and stability. Front Chem 4:38. https://doi.org/10.3389/fchem.2016.00038
14. Takahashi S, Sugimoto N (2013) Effect of pressure on the stability of G-quadruplex DNA: thermodynamics under crowding conditions. Angew Chem Int Ed 52:13774–13778. https://doi.org/10.1002/anie.201307714
15. Navarro A, Benabou S, Eritja R, Gargallo R (2020) Influence of pH and a porphyrin ligand on the stability of a G-quadruplex structure within a duplex segment near the promoter region of the SMARCA4 gene. Int J Biiol Macromol 159:383–393. https://doi.org/10.1016/j.ijbiomac.2020.05.062
16. Gao C, Mohamed HI, Deng J et al (2023) Effects of molecular crowding on the structure, stability, and interaction with ligands of G-quadruplexes. ACS Omega 8:14342–14348. https://doi.org/10.1021/acsomega.3c01169
17. Ellis RJ (2001) Macromolecular crowding: an important but neglected aspect of the intracellular environment. Curr Opin Struct Biol 11:114–119. https://doi.org/10.1016/S0959-440X(00)00172-X
18. Ellis RJ (2001) Macromolecular crowding: obvious but underappreciated. Trends Biochem Sci 26:597–604. https://doi.org/10.1016/S0968-0004(01)01938-7
19. Nakano S, Miyoshi D, Sugimoto N (2014) Effects of molecular crowding on the structures, interactions, and functions of nucleic acids. Chem Rev 114:2733–2758. https://doi.org/10.1021/cr400113m
20. Balcells C, Pastor I, Vilaseca E et al (2014) Macromolecular crowding effect upon in vitro enzyme kinetics: mixed activation–diffusion control of the oxidation of NADH by pyruvate catalyzed by lactate dehydrogenase. J Phys Chem B 118:4062–4068. https://doi.org/10.1021/jp4118858
21. Wang Y, Sarkar M, Smith AE et al (2012) Macromolecular crowding and protein stability. J Am Chem Soc 134:16614–16618. https://doi.org/10.1021/ja305300m
22. Aumiller WM, Davis BW, Hatzakis E, Keating CD (2014) Interactions of macromolecular crowding agents and cosolutes with small-molecule substrates: effect on horseradish peroxidase activity with two different substrates. J Phys Chem B 118:10624–10632. https://doi.org/10.1021/jp506594f
23. Tateishi-Karimta H, Sugimoto N (2014) Control of stability and structure of nucleic acids using cosolutes. Methods 67:151–158. https://doi.org/10.1016/j.ymeth.2013.11.002
24. Trajkovski M, Endoh T, Tateishi-Karimata H et al (2018) Pursuing origins of (poly)ethylene glycol-induced G-quadruplex structural modulations. Nucleic Acids Res 46:4301–4315. https://doi.org/10.1093/nar/gky250
25. Buscaglia R, Miller MC, Dean WL et al (2013) Polyethylene glycol binding alters human telomere G-quadruplex structure by conformational selection. Nucleic Acids Res 41:7934–7946. https://doi.org/10.1093/nar/gkt440
26. Knowles DB, LaCroix AS, Deines NF et al (2011) Separation of preferential interaction and excluded volume effects on DNA duplex and hairpin stability. Proc Natl Acad Sci U S A 108:12699–12704. https://doi.org/10.1073/pnas.1103382108
27. Miyoshi D, Karimata H, Sugimoto N (2006) Hydration regulates thermodynamics of G-quadruplex formation under molecular crowding conditions. J Am Chem Soc

128:7957–7963. https://doi.org/10.1021/ja061267m
28. Ou Z, Wang Y, Gao Y et al (2017) Targeting human telomeric and c-myc G-quadruplexes with alkynylplatinum(II) terpyridine complexes under molecular crowding conditions. J Inorg Biochem 166:126–134. https://doi.org/10.1016/j.jinorgbio.2016.11.012
29. Takahashi S, Yamamoto J, Kitamura A et al (2019) Characterization of intracellular crowding environments with topology-based DNA quadruplex sensors. Anal Chem 91:2586–2590. https://doi.org/10.1021/acs.analchem.8b04177
30. Baltierra-Jasso LE, Morten MJ, Laflör L et al (2015) Crowding-induced hybridization of single DNA hairpins. J Am Chem Soc 137:16020–16023. https://doi.org/10.1021/jacs.5b11829
31. Ferreira LA, Uversky VN, Zaslavsky BY (2017) Role of solvent properties of water in crowding effects induced by macromolecular agents and osmolytes. Mol BioSyst 13:2551–2563. https://doi.org/10.1039/C7MB00436B
32. Kumar R, Sharma D, Kumar V, Kumar R (2018) Factors defining the effects of macromolecular crowding on dynamics and thermodynamic stability of heme proteins in-vitro. Arch Biochem Biophys 654:146–162. https://doi.org/10.1016/j.abb.2018.07.018
33. Morgan RK, Molnar MM, Batra H et al (2018) Effects of 5-hydroxymethylcytosine epigenetic modification on the stability and molecular recognition of VEGF i-motif and G-quadruplex structures. J Nucleic Acids 2018:1–14. https://doi.org/10.1155/2018/9281286
34. Minton AP (2001) The influence of macromolecular crowding and macromolecular confinement on biochemical reactions in physiological media. J Biol Chem 276:10577–10580. https://doi.org/10.1074/jbc.R100005200
35. Petraccone L, Pagano B, Giancola C (2012) Studying the effect of crowding and dehydration on DNA G-quadruplexes. Methods 57:76–83. https://doi.org/10.1016/j.ymeth.2012.02.011
36. Nakano M, Tateishi-Karimata H, Tanaka S et al (2015) Thermodynamic properties of water molecules in the presence of cosolute depend on DNA structure: a study using grid inhomogeneous solvation theory. Nucleic Acids Res 43:10114–10125. https://doi.org/10.1093/nar/gkv1133
37. Jaiswal AK, Srivastava R, Pandey P, Bandyopadhyay P (2018) Microscopic picture of water-ethylene glycol interaction near a model DNA by computer simulation: concentration dependence, structure, and localized thermodynamics. PLoS One 13:e0206359. https://doi.org/10.1371/journal.pone.0206359
38. van Dongen MJ, Mooren MM, Willems EF, et al (1997) Structural features of the DNA hairpin d(ATCCTA-GTTA-TAGGAT): Formation of a G-A base pair in the loop. Nucleic acids research 25:1537–1547. https://doi.org/gka257
39. Parkinson GN, Lee MPH, Neidle S (2002) Crystal structure of parallel quadruplexes from human telomeric DNA. Nature 417:876–880. https://doi.org/10.1038/nature755
40. Dai J, Punchihewa C, Ambrus A et al (2007) Structure of the intramolecular human telomeric G-quadruplex in potassium solution: A novel adenine triple formation. Nucleic Acid Research 35:2440–2450. https://doi.org/10.1093/nar/gkm009
41. Lim KW, Ng VCM, Martin-Pintado N et al (2013) Structure of the human telomere in Na+ solution: An antiparallel (2+2) G-quadruplex scaffold reveals additional diversity. Nucleic Acids Research 41:10556–10562. https://doi.org/10.1093/nar/gkt771
42. Case DA, Aktulga HM, Belfon K et al (2023) AmberTools. Journal of Chemical Information and Modeling 63:6183–6191. https://doi.org/10.1021/acs.jcim.3c01153
43. Ivani I, Dans PD, Noy A, et al (2016) Parmbsc1: A refined force field for DNA simulations. Nature Methods 13:55–58. https://doi.org/10.1038/nmeth.3658
44. Jorgensen WL, Chandrasekhar J, Madura JD et al (1983) Comparison of simple potential functions for simulating liquid water. The Journal of Chemical Physics 79:926–935. https://doi.org/10.1063/1.445869
45. Abraham MJ, Murtola T, Schulz R et al (2015) GROMACS: High performance molecular simulations through multi-level parallelism from laptops to supercomputers. SoftwareX 1–2:19–25. https://doi.org/10.1016/j.softx.2015.06.001
46. Hess B, Bekker H, Berendsen HJC, Fraaije JGEM (1997) LINCS: A linear constraint solver for molecular simulations. Journal of Computational Chemistry 18:1463–1472. https://doi.org/10.1002/(SICI)1096-987X(199709)18:121463::AID-JCC43.0.CO;2-H
47. BIOVIA (Dassault Systèmes) (2017). Discovery studio visualizer. Release 2017. Dassault Systèmes, San Diego
48. Wang J, Wolf RM, Caldwell JW et al (2004) Development and testing of a general Amber force field. Journal of Computational Chemistry 25:1157–1174. https://doi.org/10.1002/jcc.20035

49. Wang J, Wang W, Kollman PA et al (2006) Automatic atom type and bond type perception in molecular mechanical calculations. Journal of Molecular Graphics and Modelling 25:247–260. https://doi.org/10.1016/j.jmgm.2005.12.005
50. Frisch MJ, Trucks GW, Schlegel HB, et al (2013). Gaussian09 Revision D.01.
51. Sousa da Silva AW, Vranken WF (2012) ACPYPE - AnteChamber PYthon Parser interfacE. BMC Research Notes 5:367. https://doi.org/10.1186/1756-0500-5-367
52. Tanaka S, Mochizuki Y, Komeiji Y et al (2014) Electron-correlated fragment-molecular-orbital calculations for biomolecular and nano systems. Physical Chemistry Chemical Physics 16:10310–10344. https://doi.org/10.1039/c4cp00316k
53. Fedorov DG, Jensen JH, Deka RC et al (2008) Covalent bond fragmentation suitable to describe solids in the fragment molecular orbital method. Journal of Physical Chemistry A 112:11808–11816. https://doi.org/10.1021/jp805435n
54. Fukuzawa K, Kurisaki I, Watanabe C et al (2015) Explicit solvation modulates intra- and inter-molecular interactions within DNA: Electronic aspects revealed by the ab initio fragment molecular orbital (FMO) method. Computational and Theoretical Chemistry 1054:29–37. https://doi.org/10.1016/j.comptc.2014.11.020
55. Chiba M, Fedorov DG, Kitaura K (2007) Time-dependent density functional theory based upon the fragment molecular orbital method. The Journal of Chemical Physics 127:104108. https://doi.org/10.1063/1.2772850
56. Green MC, Fedorov DG, Kitaura K, et al (2013) Open-shell pair interaction energy decomposition analysis (PIEDA): Formulation and application to the hydrogen abstraction in tripeptides. The Journal of Chemical Physics 138:074111. https://doi.org/10.1063/1.479061610.1039/c4cp00099d
57. Le HT, Dean WL, Buscaglia R, et al (2014) An Investigation of G - Quadruplex Structural Polymorphism in the Human Telomere Using a Combined Approach of Hydrodynamic Bead Modeling and Molecular Dynamics Simulation. The Journal of Physical Chemistry B 118:5390–5405. https://doi.org/10.1039/c0cp02559c
58. Castelli M, Doria F, Freccero M et al (2022) Studying the Dynamics of a Complex G-Quadruplex System: Insights into the Comparison of MD and NMR Data. Journal of Chemical Theory and Computation 18:4515–4528. https://doi.org/10.1021/acs.jctc.2c00291
59. Takaya D, Watanabe C, Nagase S et al (2021) FMODB: The World's First Database of Quantum Mechanical Calculations for Biomacromolecules Based on the Fragment Molecular Orbital Method. Journal of Chemical Information and Modeling 61:777–794. https://doi.org/10.1021/acs.jcim.0c01062
60. Nishio M, Umezawa Y, Fantini J et al (2014) CH–π hydrogen bonds in biological macromolecules. Physical Chemistry Chemical Physics 16:12648. https://doi.org/10.1039/c4cp00099d
61. Kumari M, Balaji PV, Sunoj RB (2011) Quantification of binding affinities of essential sugars with a tryptophan analogue and the ubiquitous role of C–H···π interactions. Physical Chemistry Chemical Physics 13:6517. https://doi.org/10.1039/c0cp02559c
62. Pérez-Arnaiz C, Busto N, Leal JM, García B (2014) New Insights into the Mechanism of the DNA/Doxorubicin Interaction. The Journal of Physical Chemistry B 118:1288–1295. https://doi.org/10.1021/jp411429g
63. Yoshida N, Kiyota Y, Hirata F (2011) The electronic-structure theory of a large-molecular system in solution: Application to the intercalation of proflavine with solvated DNA. Journal of Molecular Liquids 159:83–92. https://doi.org/10.1016/j.molliq.2010.04.019
64. Alfano C, Fichou Y, Huber K, et al (2024) Molecular Crowding: The History and Development of a Scientific Paradigm. Chemical Reviews 124:3186–3219. https://doi.org/10.1021/acs.chemrev.3c00615

Check for updates

Chapter 8

Protein–Protein Interactions Analyzed by the Fragment Molecular Orbital Method

Shigenori Tanaka

Abstract

This chapter describes the applications of the fragment molecular orbital (FMO) method to analyzing the protein–protein interactions (PPIs). Employing the protein complexes related to influenza virus, measles virus, SARS-CoV-2 and HER2, the intermolecular interactions between amino acid residues are quantitatively evaluated. Recent advances in computational methodology, such as the systematic identification of interaction hotspots, the incorporation of conformational fluctuations and the clustering-based statistical averaging are addressed.

Keywords Fragment molecular orbital (FMO) method, Inter-fragment interaction energy (IFIE), Protein–protein interaction (PPI), Molecular dynamics (MD), Structural clustering, Singular value decomposition (SVD)

1 Introduction

Understanding protein–protein interactions (PPIs) [1–3] is fundamental to elucidating the molecular mechanisms that govern intracellular recognition and signal transduction, including phenomena such as viral infection and antigen–antibody binding. Recent advances in structural biology, particularly in cryo–electron microscopy (cryo-EM), together with the rapid progress in AI-based biomolecular modeling, have markedly deepened our understanding of PPIs through high-resolution molecular simulations [4, 5]. Traditionally, docking simulations employing scoring functions and molecular dynamics (MD) simulations based on empirical force fields have been the primary tools for quantitative characterization at the molecular level. However, to achieve higher accuracy in describing intermolecular interactions, quantum chemical approaches that explicitly incorporate electronic degrees of freedom are increasingly recognized as indispensable.

Alexander Heifetz (ed.), *Quantum Mechanics Applications in Drug Design*, Methods in Molecular Biology, vol. 3061, https://doi.org/10.1007/978-1-0716-5420-0_8,

This article introduces an example of PPI analysis using the fragment molecular orbital (FMO) method [6–8], a useful method for performing quantum chemical calculations on biomolecular systems. Given the structure of a protein complex, FMO calculations can be performed to determine the inter-fragment interaction energies (IFIEs) between the amino acid residues that make up the complex. Analyzing these IFIEs at the residue level or at the residue group level can comprehensively and systematically determine which interactions between residues are important for the PPIs. However, the huge number of their possible combinations necessitates some form of information compression. Furthermore, because the molecular structure of protein complexes can fluctuate significantly, it is important to perform calculations not only for statically stable structures but also for various fluctuating structures and consider their statistical average. Below, we illustrate examples of FMO calculations for proteins related to the influenza virus, measles virus, SARS-CoV-2 and HER2 (human epidermal growth factor receptor 2) with their interaction partners or receptors, and demonstrate that systematic PPI analysis is possible by combining these FMO calculations with methods such as MD calculations, structural clustering, and singular value decomposition (SVD).

2 Methods

2.1 Structure Preparation

We use the crystal structures of protein complexes from the Protein Data Bank (PDB) if available, and perform the necessary structure preprocessing for MD and FMO calculations. Structure preprocessing is performed mainly employing the Molecular Operating Environment (MOE) [9], a software suite integrating molecular modeling and computational chemistry applications, as follows (corresponding functional modules are indicated in parentheses):

1. Obtaining the protein complex structure from the PDB.
2. Deleting unnecessary molecules.
3. Structure modification, and completion of missing residues and loop regions (Structure Preparation).
4. Residue replacement if necessary (Protein Builder).
5. Hydrogen addition (Protonate 3D).
6. Structure optimization (Energy Minimization).

In the absence of structural information for a protein complex, we generate the complex structure. For example, in the case of the measles virus (MV) demonstrated below, the structures of the individual proteins, MV-H (hemagglutinin), and bat SLAM (signaling lymphocyte-activation molecule), were created using AlphaFold2

[10] based on their respective amino acid sequences. The individual protein structures were then superimposed using MOE [9] onto the complex structure of MV-H and cotton-top tamarin SLAM (PDB code: 3ALZ) [11], generating three-dimensional models of the MV-H and bat SLAM complex. The residue at position 187 in MV-H was subsequently mutated from Asn (N) to Tyr (Y) (N187Y) using MOE. Finally, homology modeling was performed on the N-terminal regions of SLAM to refine any unnatural structural features. In the procedure described above, AlphaFold2 was executed on Google Colab [12] without a template, and the structure with the highest pLDDT (predicted local distance difference test) score was selected from the five generated models. For structural optimization using MOE, the AMBER10:EHT force field was employed.

2.2 MD Calculation and Structural Clustering

When necessary, we perform structure optimization to remove interatomic collisions and minimize the system energy, where simulations are performed using the equilibration procedure to bring the optimized structure closer to the state at room temperature and ambient pressure, followed by the actual calculation procedure. During equilibration, atomic constraints other than hydrogen atoms are imposed and pressure control is gradually conducted to prevent sudden changes in the protein structure, bringing the simulation conditions closer to those of the actual ones.

In MD simulation using GROMACS [13], the Amber ff14SB or ff19SB force field [14] is usually employed as the protein force field, and TIP3P model [15] water molecules are placed in a rectangular simulation box as the solvent, with periodic boundary conditions. The MD time step is 2 fs. Temperature control is performed using Bussi's method [16], and pressure control is performed using Berendsen's [17] and Parrinello-Rahman's [18] methods. The LINCS method is used to constrain chemical bonds. The reference temperature of the system is 300 K, the reference pressure is 1 bar, and the cutoff distances for van der Waals and electrostatic interactions (with the Ewald sum) are 1.0 nm.

For structural clustering, we use the Python program TTClust [19] to examine the structural distributions of protein complexes in MD simulations by dividing them into several groups. TTClust calculates the alignment and the root mean square deviation (RMSD) for each frame of the input trajectory for the selected atomic group and generates a distance matrix. Then, using Ward's method, which minimizes intra-cluster variance and clearly separates different conformations, we classify the structures into an arbitrary number of clusters. Additionally, the frame structure with the smallest average RMSD relative to other frames within each cluster is output for each cluster.

2.3 FMO Calculation and Singular Value Decomposition (SVD)

FMO calculations are performed for the structures extracted from PDB or after MD simulations, using the electron-correlated Møller–Plesset (MP) perturbation theory. The calculation method usually employed is FMO2-MP2/6-31G* with the FMO calculation program ABINIT-MP [7, 8], and the calculation results are analyzed using BioStation Viewer [7, 8], where the inter-fragment interaction energies (IFIEs) E_{ij} play a pivotal role.

We perform the CP (CANDECOMP/PARAFAC) decomposition or the SVD [20–23] for the IFIEs as

$$E_{ij} = \sum_{\mu} \lambda_{\mu} u_{\mu i} v_{\mu j} \quad (1)$$

where λ_{μ} is the (decremental) eigenvalue for mode μ, and $u_{\mu i}$ and $v_{\mu j}$ refer to the associated eigenvectors describing the contributions from each fragment (amino-acid residue in protein complex) of two monomers (chains) so that $U=(u_{\mu i})$ and $V=(v_{\mu j})$ construct unitary matrices.

Furthermore, for the three-body interactions, we carry out the CP decomposition as [23, 24]

$$E_{ijk} = \sum_{\mu} \Lambda_{\mu} u_{\mu i} v_{\mu j} w_{\mu k} \quad (2)$$

as well, where Λ_{μ} is the (decremental) eigenvalue for mode μ, and $u_{\mu i}$, $v_{\mu j}$, and $w_{\mu k}$ refer to the associated eigenvectors to form unitary matrices, $U=(u_{\mu i})$, $V=(v_{\mu j})$ and $W=(w_{\mu k})$. The crucial residues for the three-body interactions are thus extracted as important contributors to the higher-eigenvalue modes in a systematic way. The SVD (CP decomposition) is performed with the TensorLy library (http://tensorly.org/stable/index.html) [25].

The FMO-IFIE-SVD analysis above has been carried out without hydration. It is well known, however, that the solvation effects play vital roles for PPIs. The incorporation of solvation and Coulombic screening effects into the FMO methodology has been discussed in the literature [21, 26, 27] and will be an important issue in future research.

3 Applications

3.1 Influenza Virus

The first application of FMO calculations to protein complexes was to the influenza hemagglutinin (HA) protein in complex with its antibody. In 2008, using the 4096 processors of the Earth Simulator, a supercomputer at the time, an FMO-MP2/6-31G calculation was completed in 53.4 min [28] for the HA antigen-antibody system

(PDB code: 1EO8) of the influenza H3N2 A/Aichi/2/68 virus, which contained 921 amino acids, 14,086 atoms, and 54,262 electrons. The results for IFIEs were then used to predict viral mutations [29]. Subsequently, electron-correlated FMO calculations were performed on the HA trimer-Fab antibody complex system (PDB code: 1KEN; 2351 residues) with the MP3 perturbation method on the Earth Simulator, and a detailed PPI analysis was performed [30]. In recent years, FMO calculations on a scale of 10,000 fragments, including surrounding water molecules, have become possible.

3.2 Measles Virus Hemagglutinin Complexes

Measles virus (MV), belonging to the genus *Morbillivirus* of the family Paramyxoviridae, is the causative agent of measles in humans. Measles is an acute and highly contagious disease that often induces immunosuppression. Measles is still a serious disease in children, especially in developing countries [31]. As MV is serologically monotypic, effective vaccination programs are expected to achieve global eradication. The extensive vaccination campaign promoted by the World Health Organization (WHO) has dramatically decreased the number of measles cases. However, as the occurrence of diseases and deaths is still recorded, especially in Asia and Africa, the risk of MV transmission needs to be considered in the current globalized world.

Cellular receptor molecules on host cells are needed for viral entry. They determine the virus tropism of cells and tissues, and the host specificities of the virus; they are also deeply related to pathogenesis. For MV, three distinct receptor molecules, signaling lymphocyte-activation molecule (SLAM), Nectin-4, and CD46, have been identified. SLAM is a principal receptor for virus entry and is expressed selectively in immune cells such as monocytes, dendritic cells, and activated T and B cells [32, 33]. Nectin-4 is expressed at the adherens junction of epithelial cells, and is used for virus release from the human body [34–36]. In addition to SLAM and Nectin-4, only some vaccine MV strains can use CD46 as a receptor, which is a complement-regulatory protein expressed in all nucleated human cells [37, 38]. On the other hand, MV possesses two viral envelope glycoproteins, hemagglutinin (H) and fusion (F) proteins, on its surface. H binds to the receptors and then initiates fusion between the F protein and host cell membrane in the initial step of MV invasion [39]. The H proteins form homophilic dimers, and each H protein contains the stalk and head domains. The head domain exhibits six-bladed β-propeller folds [40]. The crystal structures of respective complexes of the measles virus H protein (MVH) and the three receptors mentioned above were already determined [11, 41, 42]. These studies demonstrated that MVH uses the same β4-β5 groove to bind to all three receptors. In contrast, there are considerable differences among the structures of the three receptors binding to the MVH [11, 41, 42].

Employing the crystal structures of three receptors, SLAM, CD46, and Nectin-4, in complex with the measles virus hemagglutinin (MVH), we elucidated the details of binding energies between the amino acid residues of MVH and those of the receptors with the FMO-MP2/6-31G* method, computationally [43]. The calculated inter-fragment interaction energies (IFIEs) revealed a number of significantly interacting amino acid residues of MVH that played essential roles in binding to the receptors. As predicted from previously reported experiments, some important amino acid residues of MVH were shown to be common, but others were specific to interactions with the three receptors. Particularly, some of the (nonpolar) hydrophobic residues of MVH were found to be attractively interacting with multiple receptors, thus indicating the importance of the hydrophobic pocket for intermolecular interactions (especially in the case of Nectin-4). In contrast, the electrostatic interactions tended to be used for specific molecular recognition. Furthermore, we carried out FMO calculations for in silico experiments of amino acid mutations, finding reasonable agreements with virological experiments concerning the substitution effect of residues. Thus, our study demonstrated that the electron-correlated FMO method is a powerful tool to search exhaustively for amino acid residues that contribute to interactions with receptor molecules.

A theoretical scheme to systematically describe collective (network-like) interactions between molecular fragments was proposed within the framework of the FMO method combined with the singular value decomposition (SVD) of the IFIE matrix [20, 21, 24]. For a comprehensive description of PPIs in the context of molecular recognition, we applied the proposed method to a complex of MVH and human-like (cotton-top tamarin) SLAM receptor (Fig. 1), thus finding a usefulness for efficiently eliciting the correlated interactions among the amino acid residues involved in the two proteins.

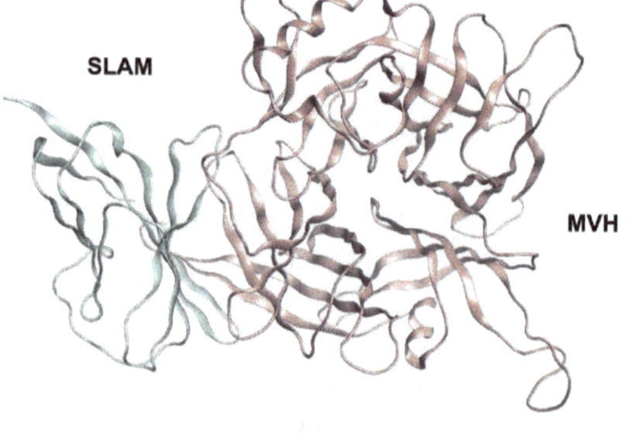

Fig. 1 Complex structure of MVH (measles virus hemagglutinin) and cotton-top tamarin SLAM (PDB code: 3ALZ)(Reprinted with permission from [21]. Copyright 2020 Elsevier)

Additionally, collective interaction networks by amino-acid residues important for mutation experiments were clearly visualized [21].

Through the FMO calculation for MVH-SLAM complex [43], we obtained an IFIE matrix composed of the residue components of MVH and those of SLAM, where 416 and 110 fragments were considered for MVH and SLAM, respectively. Then, performing the SVD for the IFIE matrix, we obtained the singular vectors for MVH and SLAM, each with the corresponding singular values, which represent the modes of the collective interaction pattern. The first singular vectors of MVH and SLAM are depicted in Fig. 2 for the complex form, thus showing particularly important interactions for complex formation. It is noted that the signs of the components of singular vectors do not necessarily correspond to those of the electric charges or the interaction energies associated with the amino-acid residues, due to the arbitrariness of the total phase (positive/negative) of the singular vectors.

From the calculated results, we detected D505, D507, D530, and R533 on the MVH side as important residues, which corresponded well to experimental findings [44–46]. On the other hand, we observed important residues of E75, K77, and E123 on the SLAM side, as was remarked in the previous study [43]. For singular vectors of the top five rank 1–5 with higher singular values, the main components were predominantly composed of charged residues, showing the importance of electrostatic interactions, especially in the case of MVH-SLAM complex [43]. It was also noted that the interaction patterns given by each component of singular vectors 1–5 varied slightly but distinctively, thus showing their proper meanings. For the first singular vectors, we observed the tendency that the positively and negatively charged residues had the opposite signs and interacted with the opposite-sign partners at the MVH-SLAM interface (Fig. 2). Interestingly, the relative phase (positive versus negative signs) of the dominant components was different between the singular vectors of 2 and 3, while the similar residues (e.g., K77, E123, D505, D507, and R533) were involved, thus representing the complex mixture of attractive and repulsive interactions among charged residues. As for the rank 6–10 singular vectors, on the other hand, we found important contributions from neutral (polar and hydrophobic) residues, some of which were related to the hydrogen bond or dispersion interactions. In addition to the interactions associated with polar residues such as Ser and Thr, we observed those by Tyr, Phe, Leu and Ile in the eigenvectors 6–10. These charged and neutral residues significantly affected the binding affinity of the MVH-SLAM complex when they were mutated [43].

A novel feature of the SVD methodology proposed in our work [21] was to explicitly elicit the correlated or cooperative IFIEs shared among multiple residue fragments. For example, as seen in Fig. 2, the cluster formed by E503, D505, D507, and R533 in MVH and

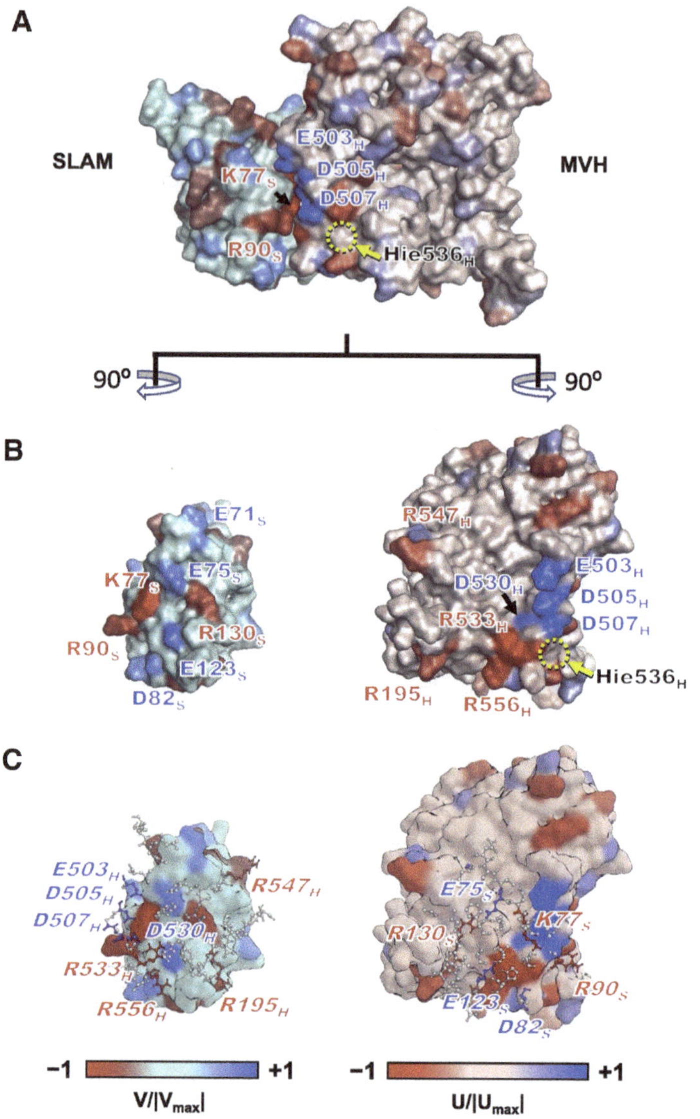

Fig. 2 (**A**) Residues with large magnitude in the first singular vector for the complex of MVH (right) and SLAM (left). Blue and red colors indicate positive and negative values, respectively, whose hue depth refers to the magnitude. The suffixes "H" and "S" indicate the residues of MVH and SLAM, respectively. The position of H536 (Hie536) is highlighted by the yellow dotted circle. (**B**) Interaction surfaces of MVH (right) and SLAM (left) depicted by opening out the interfaces of the complex with 90 degrees rotation. (**C**) Interacting residues of the target molecules are depicted by the ball-and-stick representation on the surface of partner proteins(Reprinted with permission from [21]. Copyright 2020 Elsevier)

that by E75, K77, R90, and E123 in SLAM cooperatively interacted with each other, thus yielding a collective interaction network for the formation of a protein complex. Such a novel viewpoint would be essential when we describe the indirect interactions between fragments [47] as well as their direct interactions, which often play an important role in biomolecular complex formation but were difficult to detect by earlier analyses [43] based on the individual IFIE values for each residue pair. Thus, the proposed approach can provide a novel FMO-based tool for the PPI analysis as an alternative to other attempts based on different viewpoints, such as the 3D Scattered Pair Interaction Energies (3D-SPIEs) [48], the PIEPRN (Pair Interaction Energy and Protein Residue Network) [49, 50], and VIINEC (Visualization of the Interfacial Electrostatic Complementarity) [51].

Moreover, as an example of morbillivirus adaptation to SLAM receptors from different host species, we conducted a detailed computational analysis of MV adaptation to utilize bat SLAM [52]. The first step in this calculation procedure was to create a structure of the complex between the MVH protein and bat SLAM (see Sect. 2) and to perform classical MD calculations on it to take into account the dynamical fluctuation effects around the stable structure. Structures were then extracted from the trajectory of the MD calculations, and the FMO calculations were performed on the extracted structures to evaluate the inter-residue interactions. We could understand why and how particular mutations modify SLAM–MVH interactions through these computational analyses.

These analyses demonstrated that the N187Y mutation decreased the average IFIE of the SLAM–MVH interaction (from -659.5 ± 13.7 kcal/mol to -698.8 ± 14.7 kcal/mol), enhancing attractive interactions. Furthermore, MVH-Tyr187 formed a new hydrogen bond with a threonine near the N-terminus of bat SLAM (SLAM-Thr25). Additionally, MVH-Tyr187 formed a CH/π interaction with tryptophan at position 112 of SLAM (SLAM-Trp112). Furthermore, the aspartic acid at position 507 of the MVH protein (MVH-Asp507) significantly interacted with the lysine at position 72 of SLAM (SLAM-Lys72). In contrast, in MVH without the N187Y mutation, SLAM-Lys72 interacted mainly with aspartic acid at position 505 (MVH-Asp505), instead of MVH-Asp507. These two residues (MVH-Asp505 and MVH-Asp507) play key roles in MVH binding to cotton-top tamarin SLAM (PDB code: 3ALZ) [11] and are highly conserved in morbillivirus H proteins other than the MBaMV (myotis bat morbillivirus) H protein. This shift in interacting residues, induced by the N187Y mutation in the H protein, further supports its role in enhancing the SLAM–MVH interaction.

Many previous studies have shown that morbilliviruses can change their ability to utilize SLAM through a single amino-acid substitution in the H protein. However, detailed molecular analyses of how these point mutations alter the interactions between the H protein and SLAM at the molecular level remain limited. This analysis presents one such example, demonstrating that a single amino-acid change can influence multiple intermolecular interactions. Importantly, this study underscores the potential of computational approaches in elucidating the mechanisms of viral adaptation to new host receptors.

3.3 SARS-CoV-2 Spike Protein Complexes

Recently, the advances in cryogenic electron microscope (cryo-EM) technology [53, 54] have significantly spurred the structure determinations of huge biomolecular systems such as membrane proteins, protein–protein, and protein–DNA (or RNA) complexes, which were highly intractable previously. This trend necessarily demands, in the context of computational structural biology, the analysis of complicated interactions between many amino acid residues or nucleotides involved in the molecular complexes.

In a recent study on SARS-CoV-2 spike protein composed of a trimer of chains A–C (three monomers) with more than a thousand residues each, the electron-correlated FMO calculations at the perturbative Møller–Plesset second-order (MP2) to fourth-order (MP4) levels were carried out [55] on the Fugaku supercomputer. It is striking that, since FMO-MP3 calculations [56] on these systems can be performed only at about twice the computational cost compared with FMO-MP2 on the Fugaku supercomputer, we can obtain the MP2.5 level evaluations [57] for the IFIEs [7, 8] in terms of a scaled MP3 approach, which are comparable in accuracy to very accurate CCSD(T) level values. These accurate results on the IFIEs have then provided useful information about the molecular recognition in biomolecular complexes.

Then, our problem consists of the fact that there are too many inter-residue interactions to be analyzed. For example, if we consider the inter-residue interactions between two protein monomers with a thousand (10^3) residues each, we should take account of $(10^3)^2 = 10^6$ interaction pairs. Furthermore, if we simultaneously consider the inter-residue interactions in a trimer with a thousand residues each, the number of combinations of three-body interactions amounts to $(10^3)^3 = 10^9$, whose comprehensive analysis gives a big challenge in data science. It is noted in this context that we cannot confine our consideration to the short-range interactions between nearby residues due to the involvement of long-range electrostatic interactions. The aim of the study is thus to analyze the FMO-based inter-residue interactions (or IFIEs) in huge trimer or higher complexes, in which the information of many combinations of three-body interactions is efficiently compressed to extract the essential interactions crucial for molecular recognition and structural stability of complexes.

Employing the trimers of SARS-CoV-2 spike protein as a prototypical system, we performed the FMO2-MP2.5/6-31G* calculations [55] to obtain the IFIEs between all the amino-acid residues contained in the trimer, where "FMO2" means the two-body expansion approximation [7, 8] conventionally adopted in the FMO method. We then used a data compression technique based on the singular value decomposition (SVD) [20–24] in which the important residues in each monomer (chain) were selected at the preliminary level of dimer interaction analysis and employed as bases to find the crucial interaction networks at the trimer level. While higher accuracy analyses based on a higher-cost basis set and FMO3 or higher expansion method [7, 8] would be optionally feasible, the primary aim of this work was to provide a basic computational framework for coping with a huge number of IFIE combinations. With the aid of the proposed methodology, we thus elucidated the differences in trimer interactions between the closed and open structures of SARS-CoV-2 spike proteins, which would be useful for understanding the infection mechanism of COVID-19 viruses.

We carried out the three-body (trimer) SVD analysis on the basis of the FMO-IFIE values for closed and open structures of SARS-CoV-2 spike proteins [24]. First, for each pair combination of the three chains of A, B, and C, we performed the dimer SVD based on Eq. (1) in order to select the important residues for protein–protein interactions. We retained the top-10 eigenmodes with higher eigenvalues, and selected the top-10 residues with higher amplitudes in each eigenvector. Removing the duplication of residues, we thus chose 63, 56, and 59 individual residues for chains A, B, and C, respectively, in the case of the closed form, and 69, 60, and 63 individual residues for chains A, B, and C, respectively, in the case of the open form. Next, the trimer SVD was carried out in terms of Eq. (2), where only the key residues selected above were employed to construct the basis space.

We first focused on the second and third eigenmodes for the closed structure. For the second eigenmode, the contributions of Glu1031 and Arg1039 were dominant in all three chains of A, B, and C. For the third eigenmode, the contributions of Arg1091 and Asp1118 were dominant in all three chains of A, B, and C. These facts thus implied that these key residues belonging to the central helix, β-hairpin, and β-sheet domains were essential for keeping the stability of the trimer structure, primarily due to the salt bridges among charged residues. Similarly, we observed the dominance of the same residues in the cases of the first and second eigenmodes for the open structure as well.

As for the first eigenmode for the closed structure, we found the important residues Arg983, Asp745, Asp737, and Arg44 in the chain A, Arg765, Arg457, Lys458, Asp737, and Lys417 in the chain B, and Arg319 and Asp614 in the chain C [24]. In contrast to the second and third eigenmodes addressed above, this eigenmode illustrated the collective residue interactions over widely spread domains of the three chains, including the receptor binding domain (RBD), $S1^D$ domain, N-terminal domain (NTD), central β-strand, downward helix, and heptad repeat region, reflecting the significance of long-range electrostatic interactions. It is thus supposed that this interaction mode represents the cooperation between the motion of RBD and the stability of the whole complex, which should be essential for the viral infection through the binding of RBD to the ACE2 receptor. In particular, Asp614 in chain C is known [58] to play a key role, whose mutation into Gly614 (D614G) significantly enhances the infection efficiency through the structural changes of the spike protein. In addition, the mutations of Lys417 belonging to RBD were considered to be crucial [59, 60] in variants of concern [61]. Then, this interaction mode corresponds to the third eigenmode for the open structure, where Lys786 and Lys854 in chain A, Arg765 and Arg1019 in chain B, and Asp614 and Arg646 in chain C show dominant interactions. It is remarkable that the interactions associated with Arg44, Arg319, Lys417, Arg457, Lys458, Asp737, and Asp745 are substantially reduced in this case through the transformation from the closed to open structures [55]. Overall, such a finding of essential interaction networks would be helpful for the elucidation of those allosteric effects induced by the occurrence of amino-acid mutations.

By using the decomposition technique proposed in this work, one can easily identify a small number of important interaction networks in biomolecular complexes (e.g., trimers) from a huge number of combinations of inter-residue interactions, which may amount to the order of billions if naively treated. In this FMO-based study, we applied the proposed prescription to the trimer complexes of SARS-CoV-2 spike protein, and succeeded in the efficient data compression concerning the extraction of the inter-residue interaction network crucial for the structural stability and dynamics associated with the viral transmission into host cells. It is then remarked in the framework of FMO methodology that the intended applications incorporating the pair interaction energy decomposition analysis (PIEDA) [62] and FMO3 expansion [7] are fairly straightforward. For example, a similar SVD analysis focusing on the electron-correlated, dispersion interaction part would be interesting in order to discuss the significance of relatively short-range intermolecular interactions such as π–π, CH–π, and cation–π interactions. Furthermore, the proposed methodology on the basis of tensor decomposition [22, 23] has more general applicability, also including the analysis of experimental or observational results with high complexities.

4 Future Directions

As seen in the applications discussed in the previous section, most conventional FMO calculations for protein complexes have been single-point calculations on snapshots of energetically stable structures. However, recent studies have begun to investigate the effects of structural fluctuations, such as in protein-ligand systems, by generating dynamic trajectories using molecular dynamics (MD) and then performing FMO calculations on a number of snapshot structures selected from the trajectories [63]. Because protein complexes are generally flexible systems with large structural fluctuations, this approach is thought to be effective for analyzing protein–protein interactions.

Below, as a future direction for research on FMO-based analysis of PPIs, we illustrate an example of calculations performed on a large number of structures obtained by MD simulations. Structures obtained by MD are statistically distributed based on their energy and temperature (Boltzmann distribution), which may demand huge computational costs for FMO calculations. However, by clustering the large number of structures into several characteristic structural groups, selecting representative structures from each group, performing FMO calculations for them, and weighting the results according to the size of the cluster, we can perform analyses that take structural fluctuations into account while keeping computational costs relatively low. Here, we present an example of MD-FMO analysis using the HER2-Pertuzumab complex.

Human epidermal growth factor receptor 2 (HER2) [64, 65] is a protein involved in normal cell proliferation and cell death. It is present on the cell surface and controls signal transduction from the extracellular to the intracellular environment. It is overexpressed in cancer cells, such as breast cancer and gastric cancer, and is involved in their proliferation and malignant transformation. Pertuzumab, on the other hand, is an antibody molecule that specifically binds to HER2 and exhibits pharmacological activity [66]. By inhibiting HER2 dimerization, it can be used as a molecularly targeted drug to suppress the abnormal activation of cancer cell signaling pathways. A detailed understanding of the binding between antibody proteins and target proteins that cause disease will provide useful information for designing future inhibitors and considering drug resistance due to amino acid mutations.

Experiments [67] have shown that the introduction of the N52A or N53A point mutation into the original Pertuzumab (wild-type; WT) reduces the binding strength to HER2 by more than 1000-fold. To simulate this phenomenon, we introduced the above point mutations into the crystal structure of wild-type Pertuzumab-HER2 using molecular modeling software MOE [9]. Then, we performed MD calculations of approximately 10 ns for each structure to generate dynamic structures. From these, we selected structures at equal time intervals (1 ns increments) and performed FMO calculations. However, the protein-protein IFIEs obtained thus failed to reproduce the experimental results above. Further analysis of these preliminary results revealed that one of the main reasons for this discrepancy was the accidental (fluctuating) dissociation of the Lys64-Asp255 ionic interaction, one of the main factors that mediates Pertuzumab-HER2 binding, during the early stages of MD for WT. Therefore, longer MD runs and structural clustering are suggested to be effective in obtaining statistically selected structures suitable for FMO calculations.

In the following study, we used the wild-type (WT) HER2-Pertuzumab complex (Fig. 3) registered in the Protein Data Bank (PDB code: 1S78) and introduced the N52A or N53A mutation into it using the molecular modeling software MOE. Ten 100-ns MD simulations were performed using GROMACS [13] for the WT and two mutant structures with various initial atomic velocities, and 10,000 structures were extracted from each of them at equal time intervals (0.1 ns intervals). These structures were then clustered using the Python program TTClust [19] using the Ward method. The backbones of His245-Gly265 (in HER2) and Ser54-Ser74 (in Pertuzumab) were specified as the target atomic groups for clustering. FMO2-MP2/6-31G* calculations were then performed on the representative structure (with the smallest RMSD from other structures) of each cluster to obtain the IFIE values between HER2 and Pertuzumab.

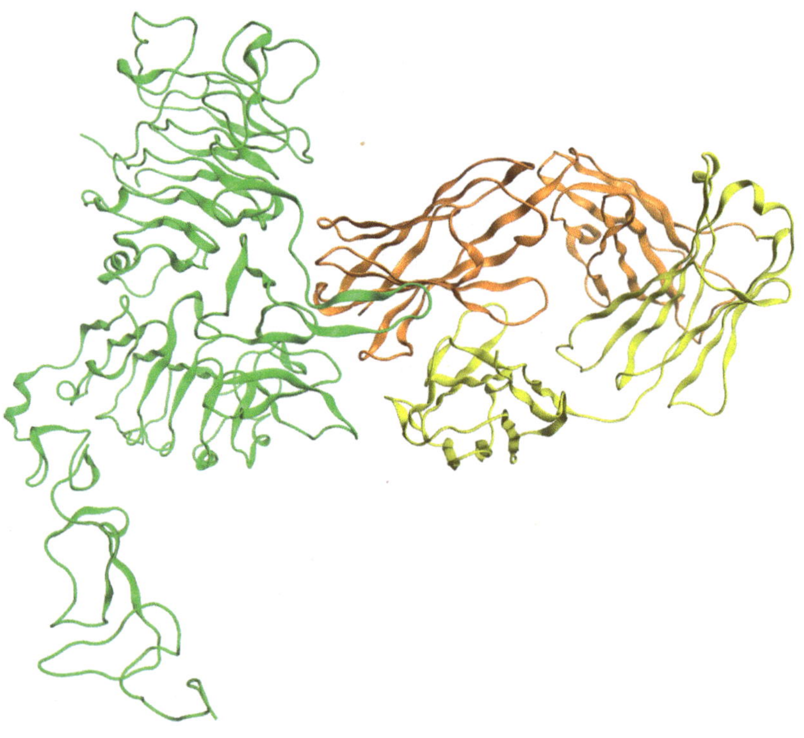

Fig. 3 Crystal structure of the extracellular domain of HER2 with the antigen-binding fragment of Pertuzumab (PDB code: 1S78). HER2 is colored in green. Pertuzumab is colored by chain, with the light chain in yellow and the heavy chain in orange

For multiple MD trajectories of WT and mutants, we focused on the displacement of the HER2 loop region (Asn253-Phe257) and divided the structures into several groups to analyze their interactions. For 100 ns × 10 trajectories (3 × 1000 frames × 10 = 30,000 frames) of WT, N52A, and N53A, we aligned both ten residues before and after HER2 Asp255 and Pertuzumab heavy chain Lys64 (His245-Gly265 and Ser54-Ser74), calculated the RMSD, and created a distance matrix. Based on this result, we classified the structures into three clusters using Ward's method, a hierarchical clustering method.

The clustering by TTClust successfully separated all the WT and mutant structures into three clusters. Figure 4 shows a superposition of the representative structure for each cluster (the structure with the smallest average RMSD for all frame pairs within the cluster). The crystal structure is shown in white, the structure group closest to the (closed form) loop region of the crystal structure is shown as Cluster 1 (red), the structure group whose loop region is furthest from Pertuzumab is shown as Cluster 3 (green), and the structure group in between is shown as Cluster 2 (blue).

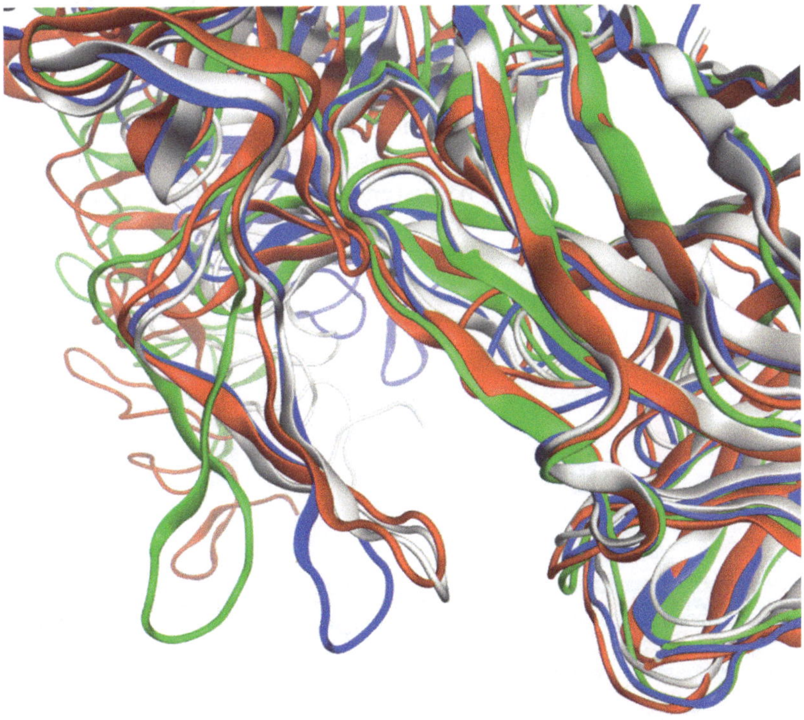

Fig. 4 Superposition of representative structures of HER2-Pertuzumab complex from each cluster (1: red, 2: blue, 3: green) with the white structure representing the crystal structure

Figure 5a shows the proportion of structures included in the three clusters for the wild type (WT) and mutants (N52A and N53A). Compared to the WT, the mutants have a lower proportion of structures in Cluster 1, indicating that the HER2 loop region tends to be more divergent. Figure 5b shows the temporal evolution of cluster species during ten 100-ns MD simulations for WT and mutants. The clusters transition between Cluster 1 and Cluster 2, and between Cluster 2 and Cluster 3. It can be seen that the clusters change from Cluster 1 to Cluster 2 and then to Cluster 3 as the HER2 loop region moves away from Pertuzumab.

To investigate the changes in intermolecular interactions between the clusters in WT and mutants, FMO calculations were performed on ten structures from each of the three clusters for WT, N52A, and N53A, for a total of 90 structures ($3 \times 3 \times 10 = 90$). To extract structures from the MD simulations for FMO calculations, the structures with the smallest RMSD relative to the representative structures of the three clusters (Fig. 4) were selected from each 100-ns trajectory. (If a 100-ns trajectory did not contain all three cluster structures, the trajectory with the largest proportion of the cluster structures not included was selected from the other nine trajectories, and the two structures with the smallest RMSD were extracted.) The results obtained were analyzed to determine the average IFIEs (total interaction energies between HER2 and Pertuzumab) and their standard deviations for each structure and cluster.

As shown in Table 1, when comparing the average IFIEs of WT and mutants, WT exhibits greater attractive interaction energy in each cluster, particularly in Cluster 1. This result is thought to be

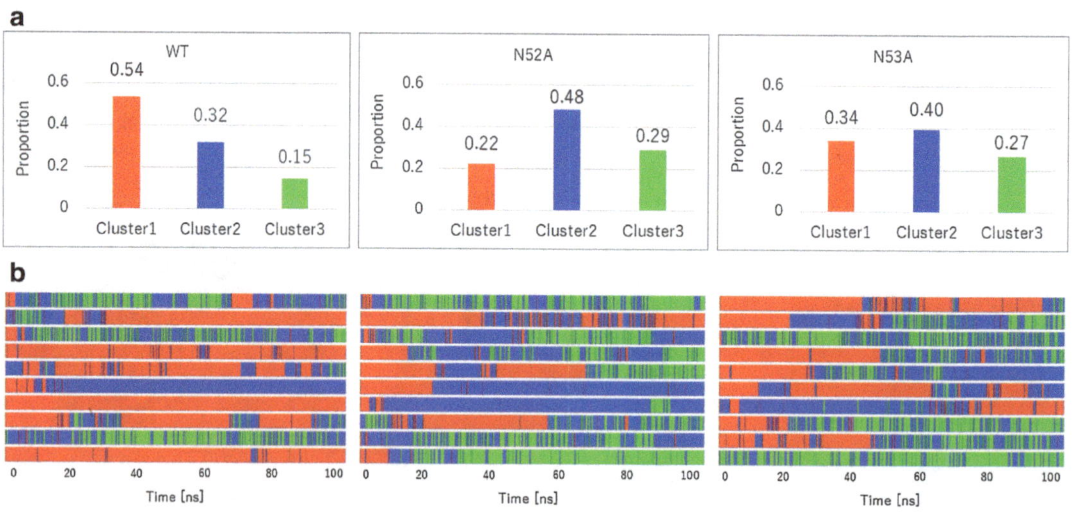

Fig. 5 (**a**) Proportions (0–1) of clusters (1: red, 2: blue, 3: green) during the MD simulations for WT and mutants of HER2-Pertuzumab complex. (**b**) Transitions of clusters during the ten 100 ns MD simulations for WT and mutants (N52A and N53A in Pertuzumab) corresponding to (**a**)

Table 1
Average IFIEs (total interaction energies) and their standard deviations (in units of kcal/mol) between HER2 and Pertuzumab for each cluster in both wild type (WT) and mutants (N52A and N53A)

	Cluster 1	Cluster 2	Cluster 3
WT	−885.34 ± 67.21	−745.18 ± 65.36	−765.56 ± 85.70
N52A	−805.77 ± 72.77	−744.84 ± 81.58	−764.65 ± 68.52
N53A	−845.18 ± 62.33	−708.45 ± 56.21	−687.35 ± 59.53

consistent with previous research [67] showing that the binding activities of point mutants N52A and N53A are reduced compared to WT. Furthermore, when comparing the average IFIEs of each cluster for WT and mutants, the attractive interaction energy between HER2 and Pertuzumab in Clusters 2 and 3 was reduced compared to Cluster 1. This can be interpreted as a decrease in attractive interactions at the binding interface between the HER2 loop region and Pertuzumab.

The average values of IFIE sums between HER2 and Pertuzumab for 30 structures of WT and mutants were −798.69 kcal/mol for WT, −771.75 kcal/mol for N52A, and −747.00 kcal/mol for N53A. Furthermore, weighted averages based on the proportion of structures belonging to each cluster (Fig. 5a) obtained by structural clustering over 100 ns × 10 MD simulations yielded −823.15 kcal/mol for WT, −764.25 kcal/mol for N52A, and −749.24 kcal/mol for N53A. Thus, the mutants exhibit a reduced attractive interaction energy between HER2 and Pertuzumab compared to the WT through the MD+FMO simulations. This decrease is due to the fact that the proportion of dynamic structures belonging to Cluster 1 is lower in the mutants compared to the WT, and that attractive interactions tend to be stronger in the WT, even when limited to structures belonging to Cluster 1. Detailed residue-level analysis of this mechanism will be reported elsewhere [68].

5 Summary

This review article has presented some examples of PPI analysis using the FMO method with its recent progress. Recent advances in HPC (high-performance computing) technology have made it possible to perform first-principles FMO calculations on protein complexes with more than thousands of residues. This has spurred the performance of IFIE analyses on various biomolecular complexes of medical and pharmaceutical interest, and has been utilized in the interpretation of experimental results, prediction of viral mutations, and drug design. Using supercomputers such as Fugaku, it is now possible to perform not only snapshot calculations of single-point structures but also statistical calculations for fluctuating structures generated by MD. Combined with information compression techniques such as structural clustering and SVD, this has ushered in an era of more efficient and comprehensive PPI analysis.

Acknowledgments

This work was supported by the MEXT Quantum Leap Flagship Program, Japan (Grant no. JPMXS0120330644).

References

1. Fields S, Song O (1989) A novel genetic system to detect protein–protein interactions. Nature 340:245–246
2. Gavin AC, Bösche M, Krause R et al. (2002) Functional organization of the yeast proteome by systematic analysis of protein complexes. Nature 415:141–147
3. Roux KJ, Kim DI, Raida M, Burke B (2012) A promiscuous biotin ligase fusion protein identifies proximal and interacting proteins in mammalian cells. J Cell Biol 196:801–810
4. Wodak SJ, Bonvin AMJJ, Honorato RV et al. (2023) Protein–protein docking: from interaction to interactome. Annu Rev Biophys 52:89–114
5. Honorato RV, Koukos PI, Jiménez-García B et al. (2024) Structural biology in the clouds: HADDOCK2.4 web server protocols. Nat Protoc 19:394–423
6. Fedorov DG, Kitaura K (eds) (2009) The fragment molecular orbital method. CRC Press, Boca Raton
7. Tanaka S, Mochizuki Y, Komeiji Y, Okiyama Y, Fukuzawa K (2014) Electron-correlated fragment-molecular-orbital calculations for biomolecular and nano systems. Phys Chem Chem Phys 16:10310–10344
8. Mochizuki Y, Tanaka S, Fukuzawa K (eds) (2021) Recent advances of the fragment molecular orbital method. Springer, Singapore
9. Molecular Operating Environment (MOE) (2024) 2024.0610 ed. Chemical Computing Group ULC, Montreal, QC, Canada, 910-1010
10. Jumper J, Evans R, Pritzel A, Green T, Figurnov M, Ronneberger O et al. (2021) Highly accurate protein structure prediction with AlphaFold. Nature 596:583–589
11. Hashiguchi T, Ose T, Kubota M, Maita N, Kamishikiryo J, Maenaka K et al. (2011) Structure of the measles virus hemagglutinin bound to its cellular receptor SLAM. Nat Struct Mol Biol 18:135–141
12. Mirdita M, Schütze K, Moriwaki Y, Heo L, Ovchinnikov S, Steinegger M (2022) ColabFold: making protein folding accessible to all. Nat Methods 19:679–682
13. Berendsen HJC, van der Spoel D, van Drunen R (1995) GROMACS: a message-passing parallel molecular dynamics implementation. Comput Phys Commun 91:43–56
14. Tian C, Kasavajhala K, Belfon KAA, Raguette L, Huang H, Migues AN et al. (2020) ff19SB: amino-acid-specific protein backbone parameters trained against quantum mechanics energy surfaces in solution. J Chem Theory Comput 16:528–552
15. Jorgensen WL, Chandrasekhar J, Madura JD, Impey RW, Klein ML (1983) Comparison of simple potential functions for simulating liquid water. J Chem Phys 79:926–935
16. Bussi G et al. (2007) Canonical sampling through velocity rescaling. J Chem Phys 126:014101
17. Berendsen HJC et al. (1984) Molecular dynamics with coupling to an external bath. J Chem Phys 81:3684–3690
18. Parrinello M, Rahman A (1981) Polymorphic transitions in single crystals: a new molecular dynamics method. J Appl Phys 52:7182–7190
19. Tubiana T et al. (2018) TTClust: a versatile molecular simulation trajectory clustering program with graphical summaries. J Chem Inform Model 58:2178–2182
20. Maruyama K, Sheng Y, Watanabe H, Fukuzawa K, Tanaka S (2018) Application of singular value decomposition to the inter-fragment interaction energy analysis for ligand screening. Comput Theor Chem 1132:23–34
21. Tanaka S, Watanabe C, Honma T, Fukuzawa K, Ohishi K, Maruyama T (2020) Identification of correlated inter-residue interactions in protein complex based on the fragment molecular orbital method. J Mol Graph Model 100:107650

22. Appellof CJ, Davidson ER (1981) Strategies for analyzing data from video fluorometric monitoring of liquid chromatographic effluents. Anal Chem 53:2053–2056
23. Kilmer ME, Horesh L, Avron H, Newman E (2021) Tensor-tensor algebra for optimal representation and compression of multiway data. Proc Natl Acad Sci 118:e2015851118
24. Okuwaki K, Akisawa K, Hatada R, Mochizuki Y, Fukuzawa K, Komeiji Y, Tanaka S (2022) Collective residue interactions in trimer complexes of SARS-CoV-2 spike proteins analyzed by fragment molecular orbital method. Appl Phys Express 15:017001
25. Kossaifi J, Panagakis Y, Anandkumar A, Pantic M (2019) TensorLy: tensor learning in Python. J Mach Learn Res 20:1
26. Okiyama Y, Fukuzawa K, Komeiji Y, Tanaka S (2020) Taking water into account with the fragment molecular orbital method. In: Heifetz A (ed) Quantum mechanics in drug discovery, Methods in molecular biology, vol. 2114. Humana Press, New York, pp 105–122, Chap 7
27. Tanaka S, Watanabe C, Okiyama Y (2013) Statistical correction to effective interactions in the fragment molecular orbital method. Chem Phys Lett 556:272–277
28. Mochizuki Y, Yamashita K, Murase T, Nakano T, Fukuzawa K, Takematsu K, Watanabe H, Tanaka S (2008) Large scale FMO-MP2 calculations on a massively parallel-vector computer. Chem Phys Lett 457:396–403
29. Takematsu K, Fukuzawa K, Omagari K, Nakajima S, Nakajima K, Mochizuki Y, Nakano T, Watanabe H, Tanaka S (2009) Possibility of mutation prediction of influenza hemagglutinin by combination of hemadsorption experiment and quantum chemical calculation for antibody binding. J Phys Chem B 113:4991–4994
30. Yoshioka A, Fukuzawa K, Mochizuki Y, Yamashita K, Nakano T, Okiyama Y, Nobusawa E, Nakajima K, Tanaka S (2011) Prediction of probable mutations in influenza virus hemagglutinin protein based on large-scale ab initio fragment molecular orbital calculations. J Mol Graph Model 30:110–119
31. Bulter D (2015) Measles by the numbers: a race to eradication. Nature 518:148–149
32. Tatsuo H, Ono N, Tanaka K, Yanagi Y (2000) SLAM (CDw150) is a cellular receptor for measles virus. Nature 406:893–897
33. Schwartzberg PL, Mueller KL, Qi H, Cannons JL (2009) SLAM receptors and SAP influence lymphocyte interactions, development and function. Nat Rev Immunol 9:39–46
34. Tahara M, Takeda M, Shirogane Y, Hashiguchi T, Ohno S, Yanagi Y (2008) Measles virus infects both polarized epithelial and immune cells by using distinctive receptor-binding sites on its hemagglutinin. J Virol 82:4630–4637
35. Noyce RS, Bondre DG, Ha MN, Lin LT, Sisson G, Tsao MS, Richardson CD (2011) Tumor cell marker PVRL4 (nectin 4) is an epithelial cell receptor for measles virus. PLoS Pathog 7:e1002240
36. Muhlebach MD, Mateo M, Sinn PL, Prufer S, Uhlig KM, Leonard VH, Navaratnarajah CK, Frenzke M, Wong XX, Sawatsky B et al. (2011) Adherens junction protein nectin-4 is the epithelial receptor for measles virus. Nature 480:530–533
37. Naniche D, Wild TF, Rabourdin-Combe C, Gerlier D (1992) A monoclonal antibody recognizes a human cell surface glycoprotein involved in measles virus binding. J Gen Virol 73:2617–2624
38. Dorig RE, Marcil A, Chopra A, Richardson CD (1993) The human CD46 molecule is a receptor for measles virus (Edmonston strain). Cell 75:295–305
39. Lin LT, Richardson CD (2016) The host cell receptors for measles virus and their interaction with the viral hemagglutinin (H) protein. Viruses 8:250
40. Hashiguchi T, Kajikawa M, Maita N, Takeda M, Kuroki K, Sasaki K, Kohda D, Yanagi Y, Maenaka K (2007) Crystal structure of measles virus hemagglutinin provides insight into effective vaccines. Proc Natl Acad Sci U S A 104:19535–19540
41. Santiago C, Celma ML, Stehle T, Casasnovas JM (2010) Structure of the measles virus hemagglutinin bound to the CD46 receptor. Nat Struct Mol Biol 17:124–129
42. Zhang X, Lu G, Qi J, Li Y, He Y, Xu X, Shi J, Zhang CW, Yan J, Gao GF (2013) Structure of measles virus hemagglutinin bound to its epithelial receptor nectin-4. Nat Struct Mol Biol 20:67–72
43. Xu F, Tanaka S, Watanabe H, Shimane Y, Iwasawa M, Ohishi K, Maruyama T (2018) Computational analysis of the interaction energies between amino acid residues of the measles virus hemagglutinin and its receptors. Viruses 10:236
44. Vongpunsawad S, Oezgun N, Braun W, Cattaneo R (2004) Selectively receptor-blind measles viruses: identification of residues necessary for SLAM- or CD46-induced fusion and their localization on a new hemagglutinin structural model. J Virol 78:302–313
45. Masse N, Ainouze M, Neel B, Wild TF, Buckland R, Langedijk JPM (2004) Measles virus (MV) hemagglutinin: evidence that attachment sites for MV receptors SLAM and CD46 overlap on the globular head. J Virol 78:9051

46. Navaratnarajah CK, Vongpunsawad S, Oezguen N, Stehle T, Braun W, Hashiguchi T, Maenaka K, Yanagi Y, Cattaneo R (2008) Dynamic interaction of the measles virus hemagglutinin with its receptor SLAM. J Biol Chem 283:11763–11771
47. Anzaki S, Watanabe C, Fukuzawa K, Mochizuki Y, Tanaka S (2014) Interaction energy analysis on specific binding of influenza virus hemagglutinin to avian and human sialosaccharide receptors: importance of mutation-induced structural change. J Mol Graph Model 53:48–58
48. Lim H, Chun J, Jin X, Kim J, Yoon JH, No KT (2019) Investigation of protein-protein interactions and hot spot region between PD-1 and PD-L1 by fragment molecular orbital method. Sci Rep 9: 16727
49. Heifetz A, Morao I, Babu MM, James T, Southey MWY, Fedorov DG, Aldeghi M, Bodkin MJ, Townsend-Nicholson A (2020) Characterizing interhelical interactions of g-protein coupled receptors with the fragment molecular orbital method. J Chem Theor Comput 16:2814–2824
50. Heifetz A, Sladek V, Townsend-Nicholson A, Fedorov DG (2020) Characterizing protein-protein interactions with the fragment molecular orbital method. Methods Mol Biol 2114:187–205
51. Ozono H, Ishikawa T (2021) Visualization of the interfacial electrostatic complementarity: a method for analysis of protein–protein interaction based on ab initio quantum chemical calculations. J Chem Theory Comput 17:5600–5610
52. Hyodo A, Seki F, Fukuda K, Tashiro K, Kitai Y, Akahori Y, Watabe H, Katoh H, Osaki R, Takaya D, Kawashita N, Fukuhara H, Ikegame S, Yoshikawa T, Eunsil P, Morikawa S, Yamaguchi R, Lee B, Maenaka K, Shirai T, Fukuzawa K, Tanaka S, Takeda M (2025) Evolutionary and structural basis of slam utilization in morbilliviruses–its implications for host range and cross-species transmission. PLoS Pathog 21:e1012990
53. Callaway E (2015) The revolution will not be crystallized: a new method sweeps through structural biology. Nature 525:172–174
54. Zivanov J, Nakane T, Forsberg BO, Kimanius D, Hagen WJH, Lindahl E, Scheres SHW (2018) New tools for automated high-resolution cryo-EM structure determination in RELION-3. eLife 7:e42166
55. Akisawa K, Hatada R, Okuwaki K, Mochizuki Y, Fukuzawa K, Komeiji Y, Tanaka S (2021) Interaction analyses on SARS-CoV-2 spike protein based on fragment molecular orbital calculations. RSC Adv 11:3272–3279
56. Mochizuki Y, Yamashita K, Fukuzawa K, Takematsu K, Watanabe H, Taguchi N, Okiyama Y, Tsuboi M, Nakano T, Tanaka S (2010) Large-scale FMO-MP3 calculations on the surface proteins of influenza virus, hemagglutinin (HA) and neuraminidase (NA). Chem Phys Lett 493:346–352
57. Pitonak M, Neogrady P, Cerny J, Grimme S, Hobza P (2009) Scaled MP3 non-covalent interaction energies agree closely with accurate CCSD(T) benchmark data. Chem Phys Chem 10:282–289
58. Gobeil SMC et al. (2021) D614G mutation alters SARS-CoV-2 spike conformation and enhances protease cleavage at the S1/S2 junction. Cell Rep 34:108630
59. Akisawa K, Hatada R, Okuwaki K, Kitahara S, Tachino Y, Mochizuki Y, Komeiji Y, Tanaka S (2021) Fragment molecular orbital based interaction analyses on complexes between SARS-CoV-2 RBD variants and ACE2. Jpn J Appl Phys 60:090901
60. Khan A, Zia T, Suleman M, Khan T, Ali SS, Abbasi AA, Mohammad A, Wei DQ (2021) Higher infectivity of the SARS-CoV-2 new variants is associated with K417N/T, E484K, and N501Y mutants: an insight from structural data. J Cell Physiol 236:7045
61. Harvey WT et al. (2021) SARS-CoV-2 variants, spike mutations and immune escape. Nat Rev Microbiol 19:409–424
62. Fedorov DG, Kitaura K (2007) Pair interaction energy decomposition analysis. J Comput Chem 28:222–237
63. Takaba K, Watanabe C, Tokuhisa A, Akinaga Y, Ma B, Kanada R, Araki M, Okuno Y, Kawashima Y, Moriwaki H, Kawashita N, Honma T, Fukuzawa K, Tanaka S (2022) Protein–ligand binding affinity prediction of cdk2 inhibitors by dynamically averaged FMO-based interaction energy. J Comput Chem 43:1362–1371
64. Gravaos C, Jimeno A (2008) HER2 in gastric cancer: a new prognostic factor and a novel therapeutic target. Ann Oncol 19:1523–1529
65. Hynes NE, Lane HA (2005) ERBB receptors and cancer: the complexity of targeted inhibitors. Nat Rev Cancer 5:341–354
66. Franklin MC et al. (2004) Insights into ErbB signaling from the structure of the ErbB2-pertuzumab complex. Cancer Cell 5:317–328
67. Vajdos FF et al. (2002) Comprehensive functional maps of the antigen-binding site of an Anti-ErbB2 antibody obtained with shotgun scanning mutagenesis. J Mol Biol 320:415–428
68. Enomoto Y, Mori Y, Fukuzawa K, Tanaka S (2026) Statistical protein-protein interaction analysis of HER2-Pertuzumab complex by the fragment molecular orbital method. J Comput Chem 47:e70382

Check for updates

Chapter 9

Untangling the Complexity of Binding Across Biomolecular Interfaces with Interaction Network Models

Vladimir Sladek and Dmitri G. Fedorov

Abstract

The biological functions of large biomolecules such as proteins are often realized through very specific interactions with partner molecules by forming complex protein–protein interaction interfaces. There is a vast amount of pairwise interactions between residues of two proteins as well as water. Residue interaction network (RIN) models are very useful in analyzing the complexity of binding, in particular, when hotspots responsible for molecular recognition are identified with the singular value centrality (C^{SV}) technique, that delivers a single score value per residue quantifying its importance in the binding. The C^{SV}/RIN methodology is described and its practical usage is illustrated on representative complexes.

Keywords Protein–protein interactions, Residue interaction networks, Water bridges, Hotspot prediction, Dimensionality reduction

1 Introduction

Proteins are linear hetero polymers in which monomer units are standard amino acid residues. The overall shape of the protein is to a large extent [1], but not exclusively [3, 2], governed by the number and specific order of the amino acids, also known as the sequence. The structure reflects a concerted interplay of interactions at different levels, known as the primary, secondary, tertiary, and quaternary structures. While the secondary and tertiary structures are adopted by single peptide strands and chains, respectively, the quaternary structure arises due to aggregation of two or more polypeptides to form a multimer.

The biological importance of proteins can hardly be overstated. They govern catalysis (enzymes), cell signaling, and immune reactions, to name a few roles of proteins. In many, if not most cases, their actions are manifested through their ability to bind specifically to other molecules, be it other proteins, nucleic acid chains (DNA and RNA), (poly)saccharides, or small molecules [4, 5]. The importance

Alexander Heifetz (ed.), *Quantum Mechanics Applications in Drug Design*, Methods in Molecular Biology, vol. 3061,
https://doi.org/10.1007/978-1-0716-5420-0_9,

of this field is reflected in several reviews addressed from different perspectives [6–8]. When proteins interact with other big molecules, the contact areas can be quite large. The buried areas in protein–protein interactions (PPI) are from 300 to 6000 Å^2 [5, 9]. Lo Conte et al. [10] reported that buried areas in the range of 1200 to 2000 Å^2 are the standard size of PPI interfaces, with small and large ones below and above these limits. Hwang et al. [11, 12] arrived at similar values. The buried area refers to the molecular surface area that ceases to be accessible to the solute (water) when the PPI interface is formed [13]. Specifically, there is no easy direct exchange of water molecules between the buried area and the bulk water, but trapped water can be a part of the binding interface.

Protein contact surfaces contain a large number of interactions between residues. The average is around 60 with the maximum of 803 residues [5]. For larger proteins, the protein–protein interaction interface (PPIi) is seldom continuous, but rather consists of several contact patches [13–15]. This is often reflected in the core–rim model of the patches, where the core is formed by hotspots (residues that directly bind to the partner protein), while the rim is made up of residues with supportive role [16]. Typically, hydrophobic residues (Ala, Cys, Ile, Leu, Met, Phe, and Val) tend to be more inside the protein and hydrophilic residues (Asn, Asp, Glu, Lys, and Arg) are on the surface. The residue occurrence frequencies in the composition of PPIi are in general more uniform [5, 17].

A PPIi is seldom flat and fully dehydrated. In other words, some parts of the PPIi that are not contact patches can form gaps [18], and residual water can be trapped inside. This water is, however, not necessarily residual in the sense of "forgotten" or otherwise idle. On contrary, it can have significant structural implications toward the shape and stability of the complex [19–26]. This formation of water-mediated contacts between residues can lead to energetic stabilization of protein aggregates.

One can rather straightforwardly deduce from the above that a detailed study of protein interfaces, a subject tightly tied to their quaternary structure, is not a trivial task. In this context, it is noteworthy that the quaternary structure was an early discovery among the other levels of macromolecular assembly [13, 27, 28]. The works of Svedberg in the mid 1920s led to the realization that some proteins consist of several subunits, predating the Pauling et al. description of the secondary structures and the determination of proteins amino acid sequences, both made in the 1950s.

Over the subsequent decades considerable effort was put towards developing a more structure based perspective on protein–protein interactions [29]. While initially leaning heavily towards experimental investigations, computational tools started to gain traction in the recent decades [30, 31]. Computational models allowed to investigate even experimentally undetermined complexes [32]. Gathering a certain volume of data is critical for discerning some general patterns and principles of PPI formation and stability. Presently, big-data-driven approaches from the realm of machine learning, e.g., AlphaFold 3 [33], are capable to predict quaternary structures of protein aggregates or complexes of proteins with other large molecules such as DNA/RNA, etc.

As it appears, big-data-driven approaches for generating sensible structures of proteins and/or their complexes (or complexes with/of other large biomolecules) seem to be a viable option to gather structural data [30]. However, in order to make sense of this data, one needs methods to analyze it, preferably in a consistent, reproducible, more-or-less fast, and ideally quantitative way. A considerable effort was dedicated towards development of databases [34] and tools for hotspot identification and/or prediction [35–46]. It may seem that there is already a sufficiently wide selection of tools. Yet, it should be stated that hotspot prediction is not the same as analysis of existing PPI interfaces. The rather complex nature of the PPIi was outlined before, with the core–rim composition and the role of water. Janin et al. [13] specifically stress that core–rim model and the data from mutational assays should not be interpreted to say that the rim plays no role in the interaction. As the data suggest, the rim contributes less to the free energy of binding (or dissociation) than the hotspots in the core. Schreiber [47] adds that many interface residues have a dual role—in addition to being a binding partner for the other protein, they also stabilize the molecular surface before complexation, as also discussed elsewhere [16].

The methods presented in this chapter are capable of addressing such issues in a deeper way than "just" hotspot prediction. The interpretation of structural and experimental data (e.g., mutational) is of nonnegligible value if one wishes to understand the PPIi to the extent that would make attempts to modulate it feasible. Also, the role of water-bridged interactions is seldom elucidated in hotspot predictions. In fact, whether residues with predominantly water-bridged contacts to the partner protein should be regarded as hotspots or not is an open question.

It was argued that the development of PPI modulators may be a promising drug development venue [35, 48]. It is a rather new, but rapidly expanding field of research attracting growing attention [49–54]. The need for (computational) analytical tools capable of providing deeper insight into residue roles in PPIi does exist. Below, more details are provided on the principles and the practical deployment of the methods.

It should be noted that, to some extent, methods developed for PPIs are applicable to domain–domain interactions within one multi-domain protein. Although the focus is on protein–protein complexes, interactions of proteins with other large biomolecules (DNA, RNA,...) are also governed by much the same principles.

2 Residue Interaction Network Models

2.1 An Overview of RIN Models

It was mentioned before that proteins, and, for that matter also other large biomolecules, are heteropolymers. Their sequence is given by the covalently bound amino acid residues. The higher order secondary, tertiary, and quaternary structures are determined by an intricate interplay of weaker noncovalent interactions. These can be backbone–backbone, side chain–side chain, and backbone–side chain interactions.

The pioneering work utilizing modeling with network/graph formalism can be traced to Kannan et al. [55]. They considered the proximity of residues as an indicator of interaction, building a graph/network representing the overall topology of all residue–residue interactions and then infer some important residue clusters. Others adopted similar graph-theoretic formalism on network models of proteins and/or their complexes in attempts to elucidate several aspects of protein structure [56–63] and the rise of communication/allosteric effects in these molecules [60, 64–74]. Several software tools were presented to aid with specific problems in this field [75–82].

Estrada et al. [83–86] and others [87–90] focused on identification of general protein properties. The protein contact networks (PCN) were declared to be a new paradigm in chemistry [91] and a useful tool to designing new proteins with altered functions [92]. Besides the more general term residue interaction networks favored in this chapter, sometimes other terms, such as protein residue networks (PRN), protein structure networks, or protein contact networks are used as well. A visual RIN representation of a protein tetramer is shown in Fig. 1.

So what is a RIN actually? It is a graph, sometimes called network, representing the interactions between all residues within a protein or a protein complex. Other molecules can be treated by the same formalism (DNA, polysaccharides, glycopeptides, and their complexes). The graph $G(V, E)$ is a mathematical object carrying information about the connections between its constituents. These are vertices V, sometimes called nodes, and their connections are represented by edges E, also called links. For RIN, vertices represent individual residues, and edges represent the contacts between them. A convenient and often used method to represent the graph is through its adjacency matrix $\mathbf{A}$. This is a square matrix with nonzero elements A_{ij} if there is an edge between vertices i and j, see Fig. 2. The diagonal elements are always zero. Adjacency matrices of undirected graphs are symmetric. RIN models are symmetric because there is symmetry rooted in the basic physics of the interaction, be it distance between residues or some other property discussed later. They are also simple, as there is no "self interaction" of one residue

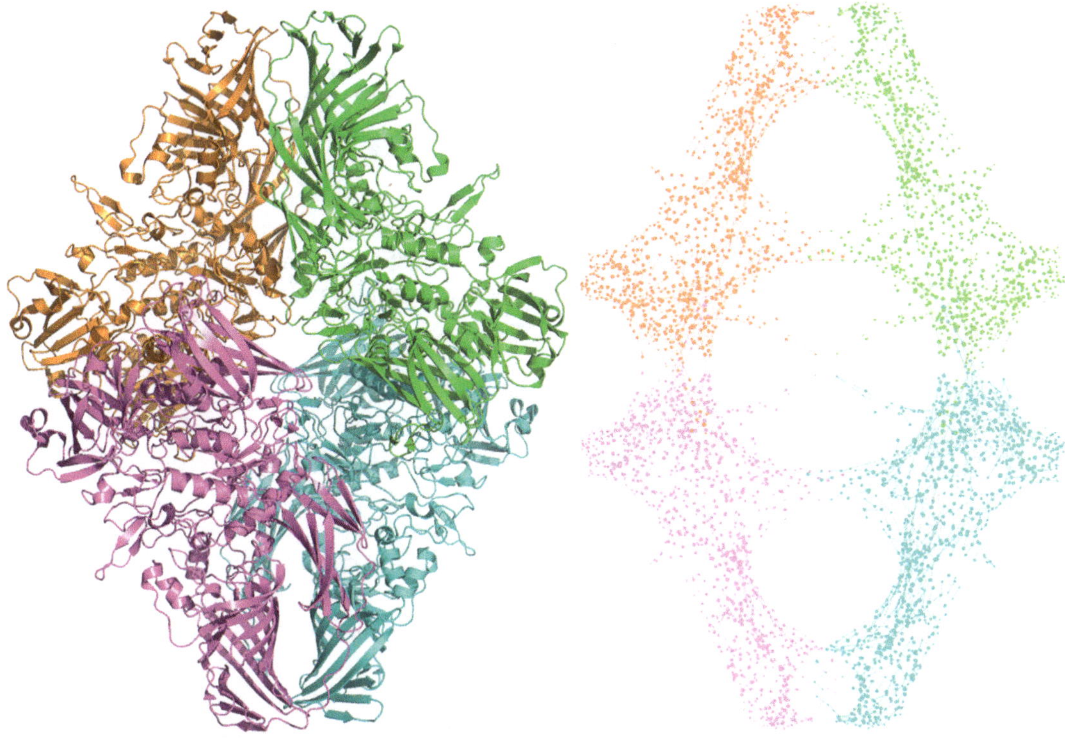

Fig. 1 A tetramer of the *E. coli* beta-galactosidase in complex with allolactose (PDB ID: 1jz8). On the left is the cartoon representation and the RIN model on the right. Note that the protein–protein interaction interface includes rather few residues in direct contact with the other protein despite quite large contact surfaces in the 3D structure

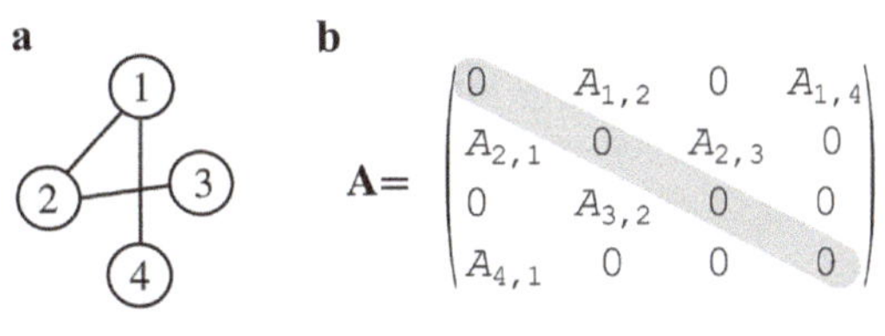

Fig. 2 A graph and its adjacency matrix. (**a**) A simple *undirected* graph with four nodes. (**b**) The adjacency matrix of the graph. The nonzero matrix elements are either 1 for unweighted graphs or are the edge weights in weighted graphs. Note the zeros on the main diagonal and that $A_{ij} = A_{ji}$

with itself and no multi-edges (multiple interactions) between any residue pair.

The earlier RIN models, cited before, define the existence of an edge from residue proximity. Typically, if two residues are closer than some threshold [59], they are assumed to be in contact, or in other words, to have a mutual interaction. Implicitly, such interaction is often interpreted as attractive or stabilizing. This, however, is not necessarily always true. The concept of frustration in biomolecules

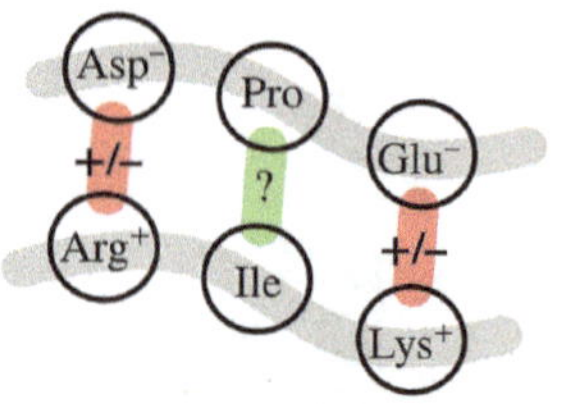

Fig. 3 Two segments of protein sequence in geometric proximity. Some residues are close together because they attract each other (red dash) and some (green dash) are brought to proximity as a result of the interactions of their neighbors

[93] was introduced to describe such phenomena. Following this concept, one can think of biomolecular structures to be in a (at least locally) stable state in which stabilizing and destabilizing "forces" are balanced. The result can be some amount of stress in the structure. Consider the trivial example depicted in Fig. 3. All three pairs Asp–Arg, Pro–Ile, and Glu–Lys are close to each other. In distance based RINs one could assign an edge to each of these pairs. However, clearly, the Pro–Ile interaction is of different "quality" than the other two. At minimum, it is not clear whether there is attraction, repulsion, or no interaction at all in that pair. In some cases, it may therefore be insufficient to draw conclusions about residue interactions based solely on distance criteria.

2.2 Energy-Based RIN Models

Residue interaction network models based on pair interaction energy (PIE–RIN) are advantageous in that they overcome the problem of misassigned interactions present in distance-based models. The work of Vijayabaskar et al. [94] was perhaps the first one to introduce PIE–RIN. Comparisons to distance-based RINs (D-RIN) were made elsewhere [61, 95]. The pair interaction energies (PIE) can be calculated by various means. The seemingly simplest is to use force field parameters to evaluate the Coulomb electrostatic and the van der Waals terms. However, the solvent screening effect is not accounted for in such simple approach. One would have to evaluate the generalized Born equations to assess the effect of the solvent/solute environment. Another way is to use quantum chemical calculations in ready-to-use packages that can account for electrostatic screening via various models (perhaps most notably the polarizable continuum model PCM, but other models beyond that are available). The fragment molecular orbital (FMO) method [96–105] was chosen to study some properties of PIE-RINs [95, 106].

The FMO method uses an expansion of the energy of a molecular system into fragment energies, and higher order terms, which are the two-, three-, and higher-order interaction energies between fragments. The fragments are chosen to minimize the error of fragmentation, so that fragments typically differ from conventional amino acid residues by a shift of one CO group in the backbone, while keeping the correct side chain [107]. Interactions for FMO fragments can be recomputed for conventional residues when the partition analysis (PA) is applied after FMO calculations [108].

The FMO energy of an N-fragment system can be written as

$$E=\sum_{I=1}^{N}\left(E'_I+\Delta E_I^{\text{solv}}\right)+\sum_{I=1}^{N}\sum_{J=1}^{I-1}\Delta E_{IJ}+\sum_{I=1}^{N}\sum_{J=1}^{I-1}\sum_{K=1}^{J-1}\Delta E_{IJK}+\ldots \quad (1)$$

combining monomer energies E'_I, solvation energies ΔE_I^{solv}, and many-body contributions ΔE_{IJ} and ΔE_{IJK}. The order at which one truncates the expansion determines the FMO energy, so the second-order FMO2 energy would contain only the first two sums, FMO3 also the third one, and so on. In FMO2, ΔE_{ij} are the fragment–fragment (i.e., residue–residue) interaction energies used in PIE–RIN models. ΔE_{IJ} is invariant with respect to an index permutation so that RINs as defined here are undirected graphs. The advantage of using FMO over force fields for PIE evaluation lies also in the fact that the energies are evaluated in a self-consistently constructed embedding potential accounting for the charges/polarization of the whole system. Hence, also each residue is correctly polarized.

The PA energy expression is formally the same as Eq. 1 with the three-body terms ΔE_{IJK} equal to zero, but the atomic definitions of fragments may be different, and all energy values in PA differ from FMO in general. When a FMO/PA calculation is performed, there are two sets of PIEs, one for shifted residues (fragments) and another one for conventional residues (called segments; there are other methodological differences between fragments and segments, for example, fragments have integer charges and segments have fractional charges).

2.3 Edge Weights and Network Analysis

Distance and PIE based RIN concepts are discussed above. A word on how to use the distance and/or energy values as edge weights is due. First, one clarification about edge weights in graphs should be done. Two categories of edge weights exist, based on cost or importance. In the former case, a lower value implies a stronger or "better" connection, as it is associated with a lower cost/penalty for traversing along the edge in a path in the graph. A road network with toll highways may be an example. An electrical circuit, where edges represent resistance of some segment is another. Contrary to this, if edges represent road lanes, then a higher number may imply a better connection between points in the road network. Stronger friendships in social networks may also indicate a stronger connection.

The shortest path is a path between two vertices with the minimal sum of weights of traversed edges. Negative edge weight values are not permitted to avoid the negative cycle trap. Many centrality algorithms use shortest paths. A centrality is a ranking metric used to answer the question "Which are the important vertices in the graph?" Different centralities are used to reflect different properties a node is expected to possess to be considered important. Naturally, graph centrality algorithms based on evaluations of shortest paths require a weight-as-cost interpretation. Other methods can use the weight-as-importance interpretation [109, 110]. The concept of centrality has its limitations. Firstly, a centrality optimal for some type of problems may yield poor results for other problems. Secondly, the ranking is often not designed to provide relative importance throughout all vertices, albeit, just to point out the most important ones [111, 112].

The construction of PIE–RIN networks must account for these limitations. Firstly, a lower negative interaction energy between residues implies a stronger connection between the vertices. Conventionally, albeit not always, RINs account for attractive interactions. However, negative weights are not permitted in shortest path-based analysis. Therefore, often the weight is defined as $w_{ij} = |\Delta E_{ij}|^{-1}$in such cases. For the singular value centrality the weights can be directly set as $w_{ij} = \Delta E_{ij}$, as this one does not rely on shortest paths. Generally, it is not advantageous to analyze an overly dense network. A threshold is, therefore, applied to ΔE_{ij} to be accepted as an edge in the graph. Only such edges i, j are defined, for which $\Delta E_{ij} \leq E_{\text{lim}}$ with $E_{\text{lim}} = -3$ kcal mol^{-1} (this value is commonly used to pick important interactions [35]). Other E_{lim} and additional rules can be chosen if one wishes to analyze repulsive interactions (which is rather uncommon).

RIN models are particularly well suited to reveal topologically important residues with respect to higher order organizational forms of protein structures (tertiary, quaternary, and in part secondary). The choice of the centrality may play a role in it. The efficiency centrality [113, 114] seems to work well. This centrality was proposed to identify nodes most responsible for the "small world" character of graphs. It was noted before that RINs do behave like small worlds [56–58, 60, 62, 63, 66, 87], meaning that (allosteric) communication between any residue pair involves only relatively short paths, i.e., few intermediate residues. The efficiency centrality for a given node k is defined as the relative loss of the efficient small world behavior of the whole network as a result of removal of that particular node.

$$C^{\text{eff}}(k) = 1 - \frac{E[G'_k]}{E[G]} \tag{2}$$

Here, G is the original graph and G'_k is the graph after the node k is removed (and all its adjacent edges). The efficiency E of a network G with N nodes is evaluated according to the Latora and Marchiori [115] formula using the lengths (distances) of the shortest paths between all node pairs d_{ij}.

$$E[G]=\frac{1}{N(N-1)}\sum_{i\neq j\in G} d_{ij}^{-1} \tag{3}$$

It was mentioned before that a convenient way to store information about the network topology is through its adjacency matrix. The topology in real-world graphs can be rather complex. Matrix reduction and/or decomposition techniques are more and more commonly applied to address such problems in many research fields where complex data is stored in matrices. This led to the definition of another type of centrality—the singular value centrality [116], inspired by the work of Tanaka et al. [117, 118]. The singular value decomposition (SVD) method allows us to rewrite a matrix in a particular way. Let us start with a general case of a nonsquare matrix with n_A rows and n_B columns.

$$\mathbf{A}=\mathbf{U}\boldsymbol{\Sigma}\mathbf{V}^{\mathrm{T}} \tag{4}$$

The size of $\mathbf{U}$ is $n_A\times m$, the size of $\mathbf{S}$ is $m\times m$ and the size of $\mathbf{V}^{\mathrm{T}}$ is $m\times n_B$, where $m=\min\{n_A, n_B\}$. The columns $\mathbf{u}^{(i)}$ of $\mathbf{U}$ are called left singular vectors of $\mathbf{A}$ and the rows $\mathbf{v}^{\mathrm{T}(i)}$ of $\mathbf{V}^{\mathrm{T}}$ are the right singular vectors. The columns in $\mathbf{U}$ are orthogonal and so are the rows in $\mathbf{V}^{\mathrm{T}}$. Depending on the field of application, these may be called principal components, features, or motifs. The matrix $\boldsymbol{\Sigma}$ is a diagonal matrix, and its diagonal elements $\sigma_{i=1,\ldots,m}$ are called the singular values. Typically, they are ordered in such a way that $\sigma_i\geq\sigma_{i+1}$.

To emphasize the structure, we write

$$\mathbf{U}=[\mathbf{u}^{(1)},\mathbf{u}^{(2)},\ldots,\mathbf{u}^{(m)}],\quad \mathbf{V}^{\mathrm{T}}=\begin{bmatrix}\mathbf{v}^{\mathrm{T}(1)}\\ \mathbf{v}^{\mathrm{T}(2)}\\ \vdots\\ \mathbf{v}^{\mathrm{T}(m)}\end{bmatrix} \tag{5}$$

Alternatively, one may rewrite Eq. 4 using dyadic products

$$\mathbf{A}=\sum_{l=1}^{m}\sigma_l\mathbf{u}^{(l)}\otimes\mathbf{v}^{\mathrm{T}(l)} \tag{6}$$

or for matrix elements

$$A_{ij}=\sum_{l=1}^{m}\sigma_l u_i^{(l)} v_j^{\mathrm{T}(l)} \tag{7}$$

Singular values act like *weights* in the sum, and since $\sigma_l\geq\sigma_{l+1}$, the terms with low/small l in Eqs. 6 and 7 contribute the most to the sum. Note that each singular vector $\mathbf{u}^{(l)}$ has n_A components $u_i^{(l)}$ and $\mathbf{v}^{\mathrm{T}(l)}$ has n_B components $v_j^{\mathrm{T}(l)}$.

We can define the new column vectors

$$\tilde{\mathbf{u}}^{(i)} \equiv \begin{bmatrix} \sigma_1^{1/2} u_i^{(1)} \\ \sigma_2^{1/2} u_i^{(2)} \\ \vdots \\ \sigma_m^{1/2} u_i^{(m)} \end{bmatrix}, \quad \tilde{\mathbf{v}}^{(i)} \equiv \begin{bmatrix} \sigma_1^{1/2} v_i^{\mathrm{T}(1)} \\ \sigma_2^{1/2} v_i^{\mathrm{T}(2)} \\ \vdots \\ \sigma_m^{1/2} v_i^{\mathrm{T}(m)} \end{bmatrix} \tag{8}$$

There are n_A of the $\tilde{\mathbf{u}}^{(i)}$ vectors and n_B of the $\tilde{\mathbf{v}}^{(i)}$ vectors. The modified matrices are

$$\tilde{\mathbf{U}} = \begin{bmatrix} \tilde{\mathbf{u}}^{\mathrm{T}(1)} \\ \tilde{\mathbf{u}}^{\mathrm{T}(2)} \\ \vdots \\ \tilde{\mathbf{u}}^{\mathrm{T}(n_A)} \end{bmatrix}, \quad \tilde{\mathbf{V}} = [\tilde{\mathbf{v}}^{(1)}, \tilde{\mathbf{v}}^{(2)}, \ldots, \tilde{\mathbf{v}}^{(n_B)}] \tag{9}$$

Then the matrix elements and the original matrix can be written as

$$A_{ij} = \tilde{\mathbf{u}}^{\mathrm{T}(i)} \cdot \tilde{\mathbf{v}}^{(j)} \tag{10}$$

$$\mathbf{A} = \tilde{\mathbf{U}}\tilde{\mathbf{V}} \tag{11}$$

To use these techniques for the decomposition of square and symmetric matrices where $\mathbf{A}^{\mathrm{T}} = \mathbf{A}$, it is useful to realize that $\mathbf{A}^{\mathrm{T}} = \mathbf{V}\Sigma\mathbf{U}^{\mathrm{T}}$. It follows that in such case $\mathbf{U}$ and $\mathbf{V}$ are identical up to a sign.

For an adjacency matrix, which is symmetric in RIN models, each row (or column) can be associated with a node in the network. For example, the k-th row (column) contains information about all edges of the vertex k, see Fig. 2. Let us construct a centrality that would assess the nodes topological relevance to the graphs overall structure. First, a metric is needed to identify nodes in the graph that have large edge weights and good connectedness. Since $\sigma_l \geq \sigma_{l+1}$, and assuming Eq. 6, then the vectors $\mathbf{u}^{(l)}$ (or $\mathbf{v}^{\mathrm{T}(l)}$) with lower l are more dominant in $\mathbf{A}$. From this follows that nodes for which large components occur in these lower l motifs are more dominant. The information about each node can be aggregated into one object by constructing the vectors $\tilde{\mathbf{u}}^{(l)}$ and $\tilde{\mathbf{v}}^{(l)}$, see Eq. 8. Let us define the singular value (SV) centrality as

$$C^{SV}(i) \equiv |\tilde{\mathbf{u}}^{(i)}| \tag{12}$$

The SVD decomposition is unique up to the sign. This, in turn, has no effect on the norm of the vectors $\tilde{\mathbf{u}}$ and $\tilde{\mathbf{v}}$ and, therefore, their norms are robust measures of node contribution to the network topology. The performance of this methodology was validated elsewhere [116]. The main conclusion from that work is that this node-ranking metric does correlate with both the edge weight(s) of edges adjacent to that node and with the number of edges. Good correlation with experimental data from mutational assays (e.g., Ala scans) was reported [116].

3 RIN Models and Techniques for PPI Analysis

Analyses of interactions between multiple proteins and/or protein domains in protein complexes are typically challenging tasks in molecular biology and computational chemistry. Nonetheless, such task to analyze residue interaction networks representing systems with multiple components can be approached with the same means as described in the previous sections, provided some simple assumptions are made. First, each subsystem (monomer) in the complex (e.g., protein, DNA molecule, and the water ensemble) can be represented by its own RIN graph. Such *subsystem graph* contains all residues constituting the monomer and all internal interaction between its residues. Second, the interactions between the subsystems can be represented by a RIN graph as well. Such *subsystem interaction graphs* will contain all the residues from both subsystems, but only those edges that connect residues from two different subsystems. The graph of the whole system can be constructed as a union of the subsystem graphs and the subsystem interaction graphs. An illustration of a graph union is shown in Fig. 4.

A few remarks about the systems structure are due. For a system with N subsystems there are N subsystem graphs G_k and $N(N-1)/2$ subsystem interaction graphs G_{kl}. The G_{kl} are bipartite graphs—one partition consists of nodes from monomer k, the second partition from monomer l, and there are only edges between the two monomers, not within. The overall RIN graph G can be built according to the subsystem interaction network construction (SINC) method

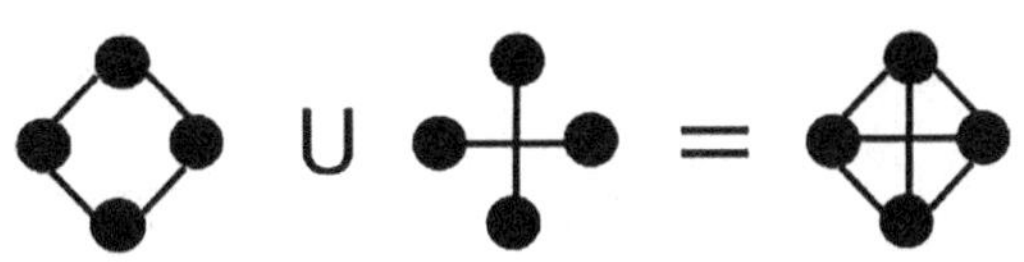

Fig. 4 Union of graphs. The two graphs are subgraphs of the resulting graph

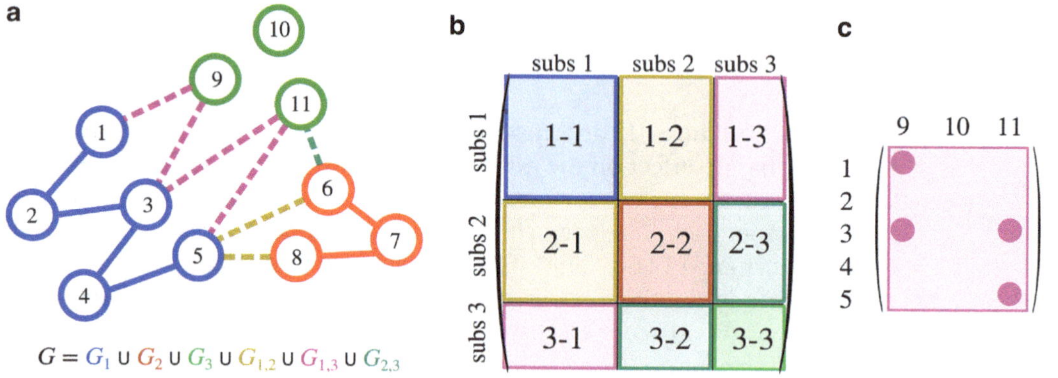

Fig. 5 SINC applied to interaction networks. (**a**) A three-subsystem RIN—the blue, red, and green subgraphs and same color edges represent the subsystem (monomer) graphs and internal interactions. The dashed edges are in the subsystem interaction graphs. These connect only nodes from different monomers. (**b**) The adjacency matrix of the overall graphs G. Color coding corresponds to the components. The diagonal matrix blocks contain intra-monomer interactions, the off-diagonal blocks contain edges between monomers. (**c**) The block $\mathbf{A}^{(1,3)}$ with nonzero elements shown as dots. Note, that for any k, l pair $\mathbf{A}^{(k,l)} = \mathbf{A}^{(l,k)T}$

[116] (a visualization of this process applied to RINs is shown in Fig. 5).

$$G = \left(\bigcup_{k}^{N} G_k \right) \cup \left(\bigcup_{l<k}^{N(N-1)/2} G_{kl} \right) \tag{13}$$

In PPI analysis, one is interested mainly in deciphering residue roles in the interaction interface. This can be quite large and complex, as mentioned in the Introduction. Let us discuss two distinct approaches to this problem reported in literature. The first is the network differential analysis (NDA) technique [82]. The idea behind NDA lies in the premise that if there is a system represented by G, such that $G = G_k \cup G_l \cup G_{kl}$ then the difference in centrality ranking of residues in isolated monomers (i.e., G_k and G_l) and in the supersystem G is affected mostly by the edges forming the interaction interface G_{kl}. Hence, subtracting the centrality (any kind) of a node i in the isolated state from the value in the complex should do the trick, like $\Delta C(i) = C_G(i) - C_{G_k}(i)$. However, due to the different number of nodes in G and the monomer graph $G_{k(l)}$ there is no guarantee of transferability of the centrality. The SINC method can again be employed to arrive at a more suitable definition

$$\Delta C(i) = C_G(i) - C_{G_k \cup G_l}(i) \tag{14}$$

A visual aid is shown in Fig. 6. The definition of this NDA centrality leaves some room for how the monomer graphs are

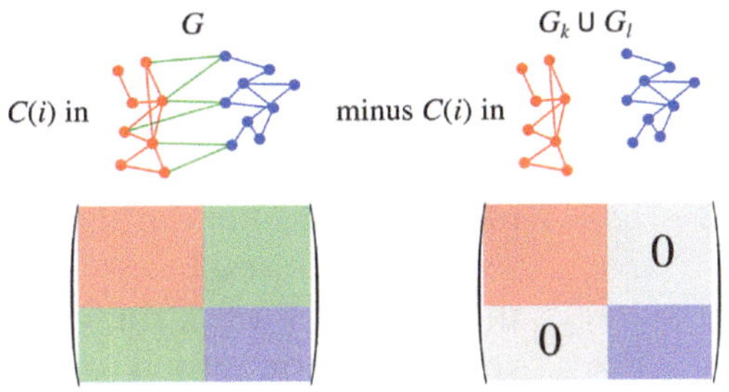

Fig. 6 NDA centrality is the centrality difference between two graphs. The $G_k \cup G_l$ graph on the right does not contain the edges connecting the two monomers, which are highlighted by the zeros in the adjacency matrix. It does have the same size (number of nodes) as G on the left

constructed. It is well known that proteins (and other large biomolecules) often undergo structural changes upon binding in a complex. The term induced fit is often used in this context. If one constructs the monomer RINs with these relaxed monomer structures, then the NDA analysis can reflect such geometry relaxation effects. If one does not like to account for geometry relaxation, one FMO calculation of the complex can be performed, and the elements in the adjacency matrix of $G_k \cup G_l$ can be zeroed out that correspond to the interactions between subsystems, see Fig. 6. Alternatively, one can perform three separate FMO calculations: one for the dimer, and one for each isolated monomer.

Some aspects to consider may be the following. First, in multicomponent complexes with more than two monomers the number of FMO calculations increases quite considerably if all pairs of monomer interaction interfaces are to be analyzed. The zeroing-out approach seems more viable in such case. On the other hand, if one just deletes the edges, the polarization of all the residues remains in the monomers identical to the polarization as in the dimer, because all data stems from one FMO calculation. Obviously, polarization of contact residues can considerably change in complex compared to isolated states even if no geometry relaxation is accounted for. Two factors contribute to the change of polarization: the presence of the partner residue from the other monomer, and the change of solvation on the contact surface. Obviously, isolated monomers are fully solvated from all sides. This fact would speak in favor of running three separate jobs. In either case, two separate centrality calculations are required.

NDA analysis for three separate FMO calculations (or just three separate.pdb structures) can be readily done in pyProGA [82], the PyMOL plugin to deal with RINs based on distance or PIEs. The PIEs can be from FMO or force field calculations (the latter values are evaluated directly in the plugin). It is accessible here: https://gitlab.com/Vlado_S/pyproga.

Equally, the singular value centrality can be used to analyze PPI interfaces. Recollect that rows and columns in the adjacency matrix of a graph can be associated with nodes—one row/column per node. Columns are just transposed rows for a symmetric matrix as adjacency matrices are. One can then perform SVD on $\mathbf{A}$ and construct the vectors $\widetilde{\mathbf{u}}^{(i)}$ for each vertex and define the C^{SV} as in Eq. 12. Due to the symmetry of $\mathbf{A}$ the $\widetilde{\mathbf{u}}^{(i)}$ and $\widetilde{\mathbf{v}}^{(i)}$ are equivalent, and the latter vectors are not needed to define the centrality. Such C^{SV} evaluation account for all the edges in a network. PPI interface analysis may require some adjustments.

Figure 5b shows that for multicomponent systems only parts of $\mathbf{A}$ hold information about the PPI interface interactions. Moreover, it is there twice, since it holds true that $\mathbf{A}^{(k,l)} = \mathbf{A}^{(l,k)\mathrm{T}}$ for the off diagonal matrix blocks relevant for PPI contacts. Within such block $\mathbf{A}^{(k,l)}$ one can still associate rows with residues in subsystem k and columns with residues in subsystem l. If one performs the SVD decomposition on such matrix block, because it is no longer symmetric, one has to use the $\widetilde{\mathbf{u}}^{(i)}$ vectors to define centrality for residues in monomer k and the $\widetilde{\mathbf{v}}^{(i)}$ vectors for residues in the l subsystem.

$$C^{\mathrm{SV}}(i) \equiv \begin{cases} |\widetilde{\mathbf{u}}^{(i)}| & \text{if } i \in G_k \\ |\widetilde{\mathbf{v}}^{(i)}| & \text{if } i \in G_l \end{cases} \tag{15}$$

It does not matter whether one uses the $\mathbf{A}^{(k,l)}$ block or the transposed $\mathbf{A}^{(l,k)}$ block as shown in the original paper [116]. Note again that the blocks are not true (full) adjacency matrices of any graph. The reason to do this instead of using the adjacency matrix of the bipartite G_{kl} graph, which would have two off-diagonal blocks $\mathbf{A}^{(k,l)}$ and $\mathbf{A}^{(l,k)}$ and zero diagonal blocks, is computational efficiency. Why crunch a roughly four times larger matrix when it can be done in a smaller computation? The other reason is that the singular value centrality defined more generally for asymmetric matrices can be used in other spheres of data science beyond applications to PPIs.

But what if two subsystems are covalently bound? Indeed, one can encounter such challenge either in cross-linked proteins or if one wishes to analyze domain–domain interactions in single-chain proteins. In fact, there is no formal hindrance that would prohibit the application of the presented methods to such problems. Several options are possible. The simplest approach would be to just ignore the interaction through the covalent bond, as the domain–domain interaction is not dependent on that one covalent linkage, but rather on numerous non bonded weak interactions as in the case of interactions of two protein chains. Alternatively, partition analysis based on FMO results yields residue–residue interaction energies for covalently linked residues on par with con-covalent interactions. Hence, the presence of such link between two interacting domains would not significantly alter the analysis. In fact, such links are as a rule kept in the overall RIN of the whole system.

3.1 Water-Mediated Contacts in PPI Interfaces

How to assess the role of water in protein–protein interaction interfaces deserves a special discussion. It was mentioned in the introduction that the importance of water in PPIs is accepted in the community [13, 19–26]. It was argued [116] that a sensible approach to this problem is to include the water in the RIN and to measure its relevance for the interface indirectly, i.e., evaluate the importance/centrality of the (protein) residues in the presence of water. Then, comparing residue centralities calculated with and without water can shed some light on the dominant "mode of action" for each residue. This strategy was proposed mostly to accommodate to the fact that there are no routine approaches to experimentally assess the water role directly. This method allows the quantification of contribution of water-bridged protein–protein contacts on per residue basis.

The RIN is constructed by adhering to the same principles as before—specifically, the condition $\Delta E_{ij} \leq E_{\text{lim}}$ is applied to water contacts as well. Hence, a water bridge is defined, if any given water has at least one interaction with a residue from monomer (protein) k and at least one interaction with a residue from monomer l. Let us label the water nodes with the index w, and the goal is to subject the tripartite network G_{klw} to the C^{SV} calculation. Formally, utilizing graph union, one can write $G_{klw} = G_{kl} \cup G'_{kw} \cup G'_{lw}$. The apostrophe in the subsystem interaction graphs with water indicates that they are modified in such way that if a water molecule has interactions with only one monomer, such edges are removed. Water–water edges are not present in G_{klw}, as these are not of interest. Figure 7 may speak a thousand words.

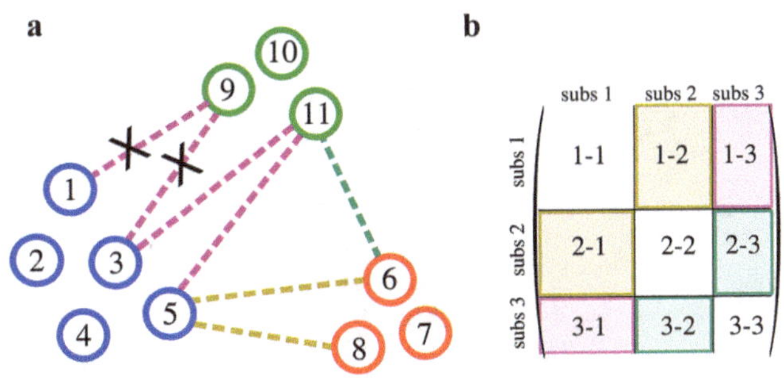

Fig. 7 A tripartite network allowing the assessment of water-bridged contacts in PPIs. (**a**) Let subsystems 1 (blue) and 2 (red) represent two proteins and subsystem 3 (green) represent water. Since water 9 has interactions only with one of the monomers it cannot fulfil the role of a water-bridge. Hence, its edges are removed (indicated by cross). All interactions between the protein residues are maintained. (**b**) Nonzero block in the adjacency matrix of G_{klw} are colored

Given that now a true adjacency matrix is used in the calculation of the singular value centrality, one can use the definition of C^{SV} form Eq. 12.

4 Practical Examples

Let us focus on the technical aspects of how to use available software tools to analyze complex PPIs in biological systems. A real-world example is used for which experimental data is available. A more detailed discussion of the results is in the original paper [116]. Tetracycline receptor protein (TetR) forms a complex with a synthetic RNA aptamer in water (PDB ID: 6sy4) [119]. TetR is a homodimer protein. In the published structure the protein chains A and B have 200 and 176 residues, respectively. The RNA aptamer has 38 residues (nucleotides), and there are 19 waters included. The waters are in interaction interfaces formed by the biomolecules, see Fig. 8. The discussion focuses on the interaction between residues in the protein chain A and the RNA, as experimental data are available for this.

In Table 1, the experimental findings of Grau et al. [119] and corresponding singular value centralities are compiled. Note that the centrality calculation is carried out only once in the wild type, while the effect of mutations has to be measured individually for each residue mutation.

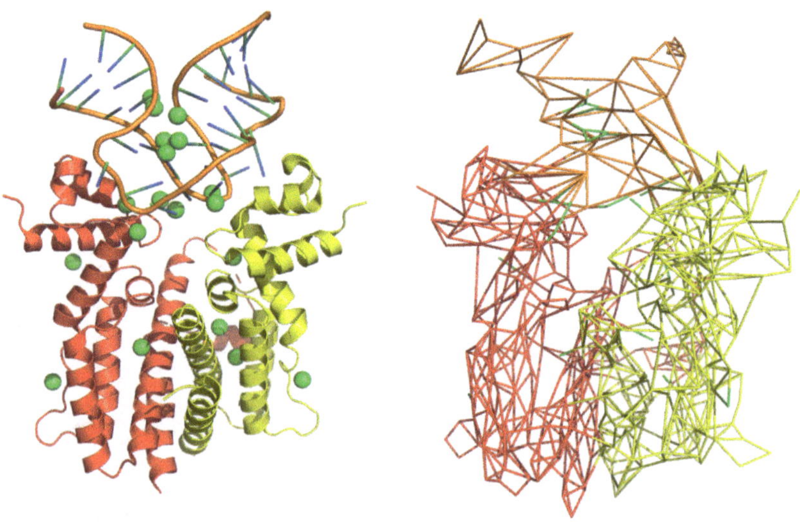

Fig. 8 The TetR protein homodimer in complex with an aptamer with water in the interaction interfaces (PDB ID: 6sy4). 3D rendering on the left and the RIN model on the right. The 916 edges in the model represent res–res interactions of 433 nodes: 200 protein chain A (red) + 176 protein chain B (yellow) + 38 nucleotides (orange) + 19 waters (green)

Table 1

Effects of Ala mutations on the TetR–RNA binding free energy[a] ΔG compared to wild type (WT) in TetR [119]. C^{SV} values are also given.[b] All energies are in kcal mol^{-1}

Mutation	ΔG	$C^{SV}_{1,3}$
Arg28A	nb	5.140
Gln38A	> −8.6[c] (−10.4)	1.074
Tyr42A	nb (−6.7)[d]	2.828

[a]Mutations in the TetR protein. ΔG were calculated from dissociation constants from EMSA (ITC values are in parentheses); nb means no binding observed. $\Delta G = -9.7$ and -11.3 kcal mol^{-1} for the WT in EMSA and ITC, respectively.

[b]Centrality values are for WT residues of chain A.

[c]Inaccurate EMSA value.

[d]Size exclusion chromatography (SEC) registered complex formation.

The purpose of this section is to introduce the reader to the practical usage of available software tools for this kind of PPI interface analysis. The focus is on the package `SVD_PPI_centrality.py` allowing the calculation of C^{SV}. It is made publicly available at https://gitlab.com/Vlado_S/singular-value-centrality.

Before usage, pay attention to the commentary in the Installation section in the `README` file. The execution itself is rather simple, just run the command

```
python  SVD_PPI_centrality.py param_inp.csvinp
```

in your command line environment. Running the command

```
python SVD_PPI_centrality.py -h
```

will display a simple help text in the terminal.

A set of keywords is available to the user to control the execution of the code. They are parsed from the input file `param_inp.csvinp`. Some are mandatory, others are specific to certain scenarios. Hence, a collection of examples representing common cases was compiled and added to the repository into the (compressed) directory `6sy4_examples` at the https://gitlab.com/Vlado_S/singular-value-centrality `GitLAB` page, discussed here from the simplest to more difficult ones, using the 6sy4 as the model system.

4.1 Case 1: FMO Data Including Defined MOLFRG or NOPSEG

All examples use files 6sy4.pdb, 6sy4.inp, and 6sy4.out. An .inp file is a GAMESS [120–122] input for the fragment molecular orbital calculation [123] and the .out is the corresponding output. The input was created using the .pdb file.

Since the goal is to analyze the interaction interface between monomers (subsystems), one has to define them first. This is achieved by assigning individual residues to the monomers. Each monomer should have its own ID value. Typically, an integer is used; monomer 1, 2,... Each monomer can have multiple residues in it. Luckily, the

FMO as implemented in GAMESS, does already provide the option to do this. The setup of the actual FMO calculation is described elsewhere [124]. In the subsystem analysis (SA) performed for the regular FMO, MOLFRG is used, whereas for PA, NOPSEG is employed (limited to DFTB). Here, the model calculation makes use of the FMO/PA method hence the NOPSEG keyword was set. The code can read this setting from the .out file and the user does not have to define the residue-to-subsystem membership again.

The contents of the .csvinp file used for this example are reproduced below.

```
param_inp_exmpl_1.csvinp

fmo_inp = 6sy4.inp; fmo_out = 6sy4.out
pdb_file = 6sy4.pdb
E_lim= -3.00
SEGMENTS_USE = True
draw_PPI = True
out_f = svd_ppi_out_exmpl_1.dat
print_nonzero_CSV = True
idw = 4
CSV_incl_wb = True
```

Let us discuss the keywords one by one. First, the .inp and .out files have to be defined by `fmo_inp` and `fmo_out`. Next, the energy threshold `E_lim` for edge creation $\Delta E_{ij} \leq E_{\mathrm{lim}}$ must be set. Remember, the ΔE_{ij} values are in the .out file for all ij combinations, but not all are used as edges in the RIN model. Lastly, the keyword `SEGMENTS_USE = True` tells the code to look for ΔE_{ij} values from the FMO/PA calculation (for FMO/PA jobs there are always also ΔE_{ij} values of the preceding FMO calculation in the same .out file, yet we are not interested in them for now). With just these keywords we should be able to successfully execute the code.

Additional keywords set some other options. The name of the output file from this C^{SV} calculation is set by `out_f` (if not set, a default name is used). The presence of `pdb_file` tells the code that the calculated centralities (summarized in `out_f`) will be added to a newly generated .pdb file with the beta-factor column replaced by the centralities. This can be used, e.g., in PyMOL [125] to color the residues by the magnitude of their respective C^{SV} values. These new .pdb files are always named as 'structure_CSV_*k*_*l*.pdb', where *kl* are the monomer IDs.

Even though the PPI interface can be large and quite complex, still, usually most of the residues in a protein do not participate in it. Hence, most residues have a centrality score of zero. To avoid excessively large listings of zeros and print only the nonzero values, let us set the keyword `print_nonzero_CSV = True`.

To analyze the effect of water-mediated contacts in the PPIs, `CSV_incl_wb = True` is used. To tell the program which subsystem is the one with waters, we set the water ID by `idw` = 4 (obviously, 4 applies to our example and the user has to set it to fit his definition).

Lastly, the keyword `draw_PPI = True` tells the program to draw schematic images of all the identified interfaces in the model system. Four subsystems (protein A, protein B, RNA, and water) have 6 possible interaction interfaces in total (all *kl* combinations).

Let us finally run the program and monitor its progress in the terminal. First, a short summary of the system is printed. In this case it reads

```
    We recognize 4 subsystems with IDs: [1, 2, 3, 4]
and the numbers of nodes in each SS are 1: 200, 2:
176, 3: 38, 4: 19 (433 nodes in total)
    Read 7286 atom coordinates
    Total of 433 residues, of which 376 amino acids.
```

The user should pay attention to this and carefully check if everything is as it should be. The program then proceeds to calculate C^{SV} values for each interaction interface—i.e., it runs through all *kl* combinations. If there is contact between monomers *k* and *l* a basic summary will be printed, if no contact, it will be skipped. It is reported by the line `A_k,l: has nonzero elements, carry on!` where `A_k,l` stands for the martix block in the adjacency matrix as shown in Fig. 5. Similar progress is reported for the calculation including water-mediated contact that follow the "dry" PPIs, if requested. A successful execution should end by printing the total run time, which is typically a few seconds.

All computed results are in the output file (`out_f`). The content is rather self explanatory and the data is tabulated in a form that should facilitate swift transfer to publication ready tables such as Table 2.

The schematic drawing of the interaction interfaces (`draw_PPI = True`) is shown in Fig. 9. At this point it is time to pay some attention to the interpretation of the acquired centralities. Firstly, Arg28 attains the highest centrality score and, according to the experimental data, is the only residue, which when mutated for Ala leads consistently to the loss of binding between the protein and the RNA aptamer. The calculation predicts the importance of Lys48, yet there is no experimental data available for Lys48 mutants. From the remaining residues the next highest centrality is assigned to Tyr42. Its mutation can be prohibitive to binding, albeit experimental data from another source suggest otherwise. Lastly, the mutation of Gln38 does not prevent complex formation; however, the binding free energy is lowered, indicating its complex stabilizing role. The centrality analysis indicates a similar role. Moreover, its C^{SV} is lower than that of Tyr42 and certainly of Arg28, following the trend of decreasing relevance with respect to complex stabilization. This, combined with more examples in the original work [116], leads to the

Table 2

Effect of water bridges on centrality C^{SV} shown by comparing protein–RNA (C^{SV}_{kl}) with protein–RNA-water C^{SV}_{klw} values for the tripartite graph $G_{k\cup l\cup w}$ in the TetR–RNA complex in water[a]

Subsustem	Residue[b]	C^{SV}_{kl}	C^{SV}_{klw}
1	Thr27*	0.980	2.716
1	Arg28	5.140	5.136
1	Gln38	1.074	1.052
1	Pro39	1.736	1.736
1	Tyr42	2.828	2.828
1	Asn47	2.193	2.150
1	Lys48	3.450	3.261
1	Arg49*	0	2.366
3	U14*	1.963	2.942
3	A15**	4.199	4.410
3	U16	3.318	3.318
3	A30	1.707	1.709
3	G31	4.566	4.570
4	w301		3.678
4	w106		2.726

[a]Here, water bridges are reported between the protein chain A (k=1) and the aptamer (l=3). Water forms subsystem w=4

[b]The number of asterisks is equal to the number of water-mediated contacts. Underlined residues were investigated experimentally, see Table 1

conclusion that C^{SV} data can be used as a reliable qualitative indicator of residue importance in interaction interfaces. To some extent semi-quantitative agreement with experiment can be reported.

Hence, singular value centrality can be used to perform reproducible and automated analysis of PPIs where normally human judgement/expertise was the only leading principle, e.g., selecting mutation candidates. It is not clear why Lys48 evaded the experimentalists attention.

Table 2 hints on the effect of water-mediated contacts in interaction interfaces. For some residues such contacts seem to be the dominant mode by which they contribute to the binding. Thr27 and Arg49 are good examples. Both residues attain considerably higher score when the tripartite graph with water (Fig. 7) is used for C^{SV} calculation (Eq. 12) instead of the bipartite graph (Fig. 5, Eq. 15) without water. Arg49 does not have any direct contact with the aptamer residues, but one water-bridge connects it to A15. Thr27 does have one contact to A15 with $\Delta E_{ij} = -3.5$ kcal mol^{-1}, but the

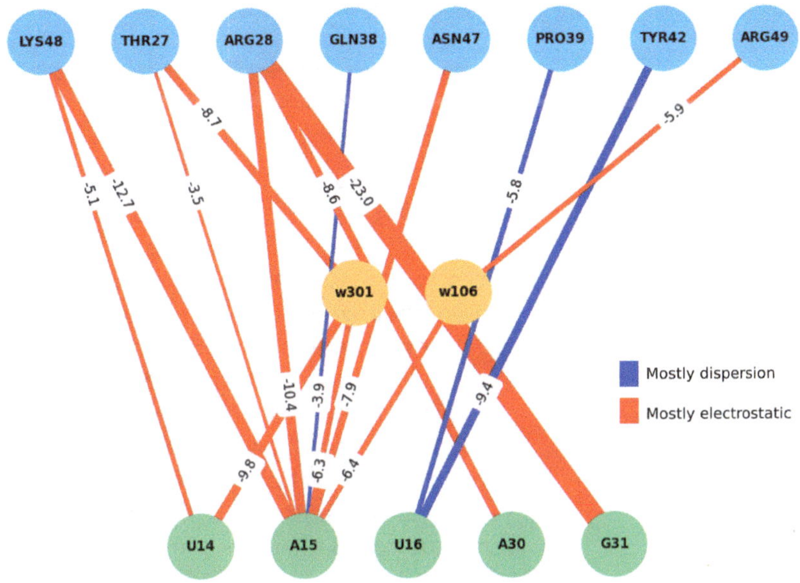

Fig. 9 The interaction interface between protein residues in chain A (blue) and the RNA residues (green) accounting for water-mediated (yellow) contacts in the TetR–RNA complex 6sy4. ΔE_{ij} values are shown (kcal mol– 1). Red lines indicate predominantly electrostatic interaction, blue dispersion interactions as categorized by the PIC factor [106]

water bridge of water301 with ΔE_{ij} values −8.7 and −6.3 kcal mol^{-1} seems to be a stronger connection. Hence, the almost tripling of the score by including water.

On the other hand, the presence of water does not automatically increase the C^{SV} score dramatically. For the aptamer nucleotide A15 such accounting for the water bridge led to a marginal increase in the score from about 4.2 to 4.4. By looking at Fig. 9 it becomes clear that this is because both water-mediated contacts to the protein are much weaker than the direct contacts of A15. On the other hand, for U14 the presence of the strongly bound water301 is a much more dominant factor compared to its single direct interaction with Lys48.

4.2 Case 2: Manual Definition of Subsystems

Let us use the same .inp and .out files, and suppose that for some reason the RIN model should be built with a different subsystem definition than in the previous example. The addition of the keyword `SS_dict` takes care of the residue-to-subsystem assignment within the `SVD_PPI_centrality.py` program, see below.

```
param_inp_exmpl_2.csvinp

fmo_inp = 6sy4.inp; fmo_out = 6sy4.out
pdb_file = 6sy4.pdb
E_lim= -3.00
```

```
SEGMENTS_USE = False
draw_PPI = True
out_f = svd_ppi_out_exmpl_2.dat
print_nonzero_CSV = True
idw = 4
CSV_incl_wb = True
SS_dict = {1: [(0,198), 199], 2: [(200, 375)],
3: [(376, 413)], 4: [(414,432)]}
```

The structure of the argument for the `SS_dict` keyword may look confusing. First, the keyword defines a python dictionary. There are four monomers (1, 2, 3, and 4)—the dictionary keys. To each a range of residues (nodes in RIN) is assigned—the dictionary values. The range itself must always be given as a python list, i.e., enclosed in square brackets [], where one could either list all residues explicitly or give an interval or combine the two. The pythonic numbering starting at zero is used. To illustrate this take a look at the assignment for subsystem 1 that says [(0,198), 199]. The list inside the square brackets has two elements; first (0,198), and second 199. Fig. 8 shows that monomer A (chain A of the protein) has 200 residues, so in pythonic indexing 0 to 199. The pair in round parentheses (0,198) indicates that residues (nodes) 0 to 198 (including 198) belong to monomer 1. The following number 199 tells the code that also residue 199 belongs to monomer 1. It would be of course easier to write 1: [(0,199)] instead of the split assignment, but this serves the purpose of demonstrating the input logic for this `SS_dict` keyword. The order in which the residue indices are given does not matter. In fact, setting 1: [(5,198), 199, (0,4)] produces the same result.

The remaining keywords are identical to the previous example, except setting `SEGMENTS_USE = False` and omitting `print_nonzero_CSV`. That sets it to its default False value. Note how the output is much longer, albeit no extra useful information is printed. The actual C^{SV} score can differ when compared to the previous case. One reason is that interactions ΔE_{ij} in FMO and PA are in general different, because fragments in FMO are not chosen as conventional residues, but in PA it is possible to use conventional residues. The missassigned CO group in proteins and the PO_3^- group in nucleotides can play a role [107] if the given groups interact, but "belong" to the neighboring residue in FMO logic (this is not a mistake and was set up this way to achieve better convergence in the SCF steps of the calculation). The partition analysis corrects this and the residues are identical to conventionally defined residues in protein/DNA biology.

4.3 Case 3: Distance Data to Define RIN

Let us build a distance-based RIN model of our system. For that the 6sy4.pdb file will suffice, but some small modifications have to be done. The .pdb files have only three chains defined A and B for protein monomers and C for the RNA molecule. Water molecules were assigned to these chains by the authors of the structure. However, for our purposes, the chain ID is used as the subsystem ID. Thus, in order to define a separate, fourth subsystem that contains only waters, the chain IDs of all waters are changed to W. This creates one additional problem, since the waters formerly in chains A and B have identical residue names and residue IDs. One cannot have residues with identical names and IDs in one chain, now chain W, hence one should change the residue IDs of waters formerly in chain B by adding +100. Compare the 6sy4.pdb in example three with those in previous examples.

A few more steps have to be completed before calculating the singular value centralities. The inter-residue distances have to be converted to edge weights. In distance based RINs (D-RIN) the convention is such, that shorter inter-residue distance indicates "stronger connection." Hence, a larger edge weight is expected in the C^{SV} algorithm for strongly interacting (nearby) residues and a lower weight for weakly interacting (distant) residue pairs. A small program can be used to extract the inter-residue distances and convert them to edge weights. For example_3 in step_1 in terminal run

```
python PDB_to_WG.py 6sy4.pdb iCM 6sy4
```

To get a full list of options and a simple help run

```
python PDB_to_WG.py -h
```

In this case the program is told to use the structure file `6sy4.pdb`. The `iCM` keyword means that the inverse of the center of mass distances between residues will be used as weights. Distant residue pairs will have low edge weights, nearby pairs a higher weight. Lastly, the keyword `6sy4` is the name base for three output files that will be generated: `6sy4_WG.csv`, `6sy4_LBL.csv`, `6sy4_SS.dict`. The first one contains the weights, the second file the names of the nodes (residues) and the last one contains the subsystem dictionary in a structure as discussed in the second example. The three files should be copied to the directory step_2.

Now everything is ready to calculate the singular value centralities from the D-RIN model of our complex. First, the .csvinp file has to be set up as indicated below

```
param_inp_exmpl_3.csvinp

PIE_csv_file = 6sy4_WG.csv
node_label_file = 6sy4_LBL.csv
SS_dict = 6sy4_SS.dict
include_rep = only
E_lim = 0.15
draw_PPI = True
out_f = svd_ppi_out_exmpl_3.dat
print_nonzero_CSV = True
idw = 4
CSV_incl_wb = True
```

Several new keywords had to be added. `PIE_csv_file` points to a file holding the edge weights. `node_label_file` points to a file with node (residue) labels/names. `SS_dict` is not a dictionary now, instead it points to a file containing the subsystem-defining dictionary. Since, in contrast to attractive ΔE_{ij}, the edge weights are now positive numbers, the option `include_rep = only` is used. It is a bit confusing, but the name of this keyword is motivated by the possibility to analyze also repulsive (positive) ΔE_{ij} interactions. Hence, for any network with positive weights one has to use this keyword. The choice of a suitable `E_lim` deserves some thought. Let us use the value 0.15, which is the inverse of roughly 6.7 Å. This, combined with `include_rep = only` will build a `RIN` model with edges connecting residue pairs, for which the centers of mass are closer than 6.7 Å. The meaning of the remaining keywords was explained in examples 1 and 2.

Let us execute the program to calculate the centralities

```
python SVD_PPI_centrality.py param_inp_exmpl_3.
csvinp
```

and several files are saved in the results directory. In contrast to the PIE-RIN model, the settings as used to identify water-mediated contacts besides the subsystems 1 and 3 (prot chain A and RNA), also between subsystems 1 and 2 (prot chains A and B), and subsystems 2 and 3 (prot chain B and RNA), see data in the output file `svd_ppi_out_exmpl_3.dat` after `Aklw = A_1_3_4`, `Aklw = A_1_2_4`, and `Aklw = A_2_3_4`, respectively. To compare results with the PIE–RIN model, one can analyze just the 1–3–4 interface, see Table 3.

Immediately noticeable is the larger size of the interaction interface in the D-RIN model compared to the PIE–RIN model, in terms of number of participating residues, compare to Table 2. Residues with experimentally validated influence on the interaction interface are Arg28, Tyr42, and Gln38 with importance in that order, see Table 1. All three were identified as important by the singular value centrality calculated in the PIE–RIN model with the same order of importance. With the D-RIN model, set up in this example, the method fails to identify Arg28 and Tyr42. However, Gln38 is identified as participating in direct contact to the RNA aptamer. Perhaps the reason for failing to identify Arg28 and Tyr42 is an insufficient distance criterion (too big `E_lim`). If the distance is increased to 8 Å, i.e., `E_lim` set to $1/8 = 0.125$, both Arg28 and Tyr42 do participate in the interaction interface, meaning there is an edge between them and some residue in the aptamer.

In this example one can see the shortcomings of the D-RIN approach. First, by looking at Fig. 9 one sees that the PPI interface is dominated by electrostatic interactions. The role of electrostatics on protein structure is a well-studied problem [126–129]. These

Table 3
Effect of water bridges on centrality C^{SV} shown by comparing protein–RNA (C^{SV}_{kl}) with protein–RNA–water C^{SV}_{klw} values for the tripartite graph $G_{k\cup l\cup w}$ for the TetR–RNA complex in water modeled as a D-RIN graph[a]

Subsustem	Residue[b]	C^{SV}_{kl}	C^{SV}_{klw}
1	GLU23**	0.0	0.416
1	GLY24**	0.0	0.418
1	LEU25*	0.0	0.242
1	THR26*	0.0	0.378
1	THR27*	0.326	0.443
1	GLN38	0.398	0.398
1	ASN47**	0.332	0.468
1	LYS48**	0.0	0.374
1	ARG49*	0.0	0.354
1	GLY138*	0.0	0.414
1	HIS139*	0.0	0.311
1	LEU142*	0.0	0.292
1	GLU147*	0.0	0.439
3	U13*	0.0	0.461
3	A15**	0.465	0.594
3	U16	0.398	0.398
3	G31*	0.0	0.341
3	A32*	0.0	0.359
4	HOH301	0.0	0.739
4	HOH302	0.0	0.439
4	HOH401	0.0	0.594
4	HOH405	0.0	0.461
4	HOH102	0.0	0.546
4	HOH106	0.0	0.592
4	HOH107	0.0	0.542

[a]Here, water bridges are reported between the protein chain A (k=1) and the aptamer (l=3). Water forms subsystem w=4.
[b]The number of asterisks is equal to the number of water-mediated contacts.

interactions can be, in contrast to dispersion interactions, quite long range. This is however no problem for the PIE-RIN model, as distance plays no role there and it is purely the pair interaction energy that counts. In D-RIN models one can easily omit residues

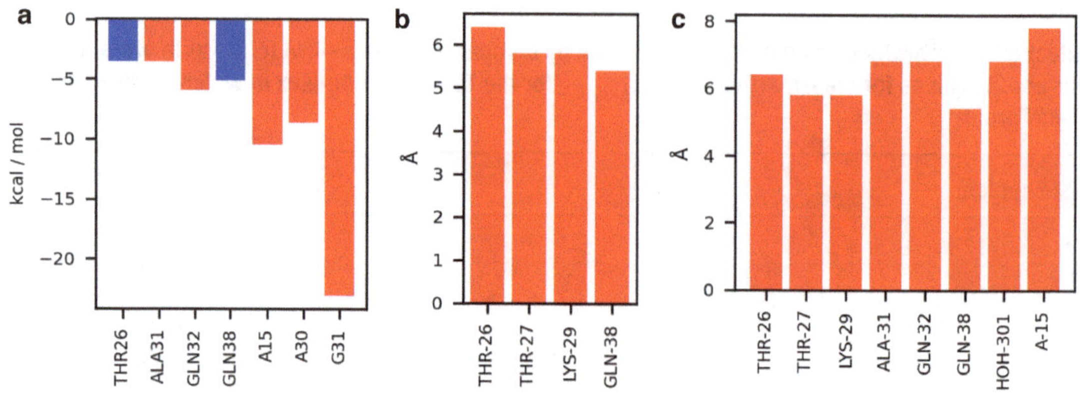

Fig. 10 Interactions of Arg28 captured in (**a**) the PIE–RIN model (`E_lim` = −3 kcal/mol), (**b**) `D-RIN` model with `E_lim` = 0.15 (dist. cut off ca. 6.7Å), (c) D-RIN model with E_lim = 0.125 (dist. cut off 8Å). Blue colour indicates a dispersion dominated interaction, whereas red means an electrostatic (Coulomb) type interaction. Visualization in pyProGA [82]

with potentially strong electrostatic interactions by imposing too short distance cut-off values. But how does one know? Importantly, just increasing the distance cut-off criterion does not automatically guarantee agreement with PIE–RIN models. Figure 10 can shed some light on the reasons why.

In the example of Arg28 the distance cut-off of 6.7 Å (i.e., `E_lim` = $1/6.7 \approx 0.15$) did not identify its interaction with any of the RNA residues nor any water. The PIE–RIN model does capture its interactions with A15, A30 and, G31, compare Fig. 10a, b. The D-RIN model with `E_lim` = $1/8 = 0.125$ does identify one contact to A15 in the RNA. Still, this contact is a "long distance" one, and hence its weight is quite small. This does not assign Arg28 a high rank in the C^{SV} score as it did in the PIE–RIN model. Erroneously, one water interaction is associated with Arg28, which is not found in the PIE–RIN model with `E_lim` = −3 kcal mol^{-1}. The data for the increased radius to 8 Å is in the directory step2/8A in the accompanying examples.

The concluding remark on D-RIN models is that they produce less consistent centrality ranking for various cut-off values than PIE–RIN models, which was noted also elsewhere [95]. They tend to include a lot of low-importance noise (energetically insignificant interactions) when long-range interactions are to be included. Also, several of the edges may represent frustrated contacts that do not contribute to the stability of the system. Moreover, there is no way to confirm if all significant potential long-range interactions are included without comparison to PIE–RIN models anyway.

5 Conclusions

The presented methodology reduces the complexity of protein–protein interaction interfaces governed by a myriad of residue–residue interactions to a ranking score more catered towards "human readable" format. In a PPI interface, the number of possible interactions grows as N^2, where N is the number of residues, whereas the number of residue scores in the singular value centrality is strictly linear, one score per residue. C^{SV} score have a good correlation with experiment on a qualitative level. Reasonable quantitative correlation can be observed, however one must bear in mind that the metric is not meant to reproduce $\Delta\Delta G$ values, but rater account for a residues "embeddedness" within the interaction interface (both number and strength of interactions affect this [116]). It is therefore reasonable to expect that a mutation of a residue with high C^{SV} score should have a more profound effect on the binding affinity between the interacting moieties. This was shown to hold true in the provided example.

It has been demonstrated that the presented technique of node importance ranking with respect to interaction interface stabilization can account for water-mediated interactions, which are considered non-negligible in PPIs. This is a novelty that cannot be routinely achieved by other techniques. However, conventionally, hotspots are considered "dry", meaning that they do not have water-mediated contacts as their dominant binding mode to the other molecule. To the best of our knowledge this is also the implicit assumption in several hotspot predictor software. It remains an open question whether residues interacting predominantly, if not exclusively, through a water bridge should be considered hotspots. Detailed experimental validation is due.

The huge advantage over molecular dynamics based hotspot predictions such as free energy perturbation or similar, is that one does not need to perform any in silico mutations. Instead one FMO calculation is performed on the wild-type system followed by the singular value centrality computation. The centrality evaluation takes a few seconds: 5 s for the protein–RNA example here (considering all interaction interfaces). The bottleneck is the FMO calculation, but with reasonable choice of the setup (as here PA/FMO2/DFTB/D3(BJ)/PCM) the calculation can take about 3–4 h on a 12 core CPU. FMO scales well with the number of CPU cores.

Lastly, it should be emphasized that the singular value centrality is not solely a hotspot predictor. Since it is directly connected to the PIE–RIN models of the interaction network within proteins (or similar molecules), it has also a strong interpretative value that can provide reasoning behind experimental observations. Coupled with the interaction visualization in the pyProGA PyMOL plugin one can quickly gain a very informative oversight of the interaction interface.

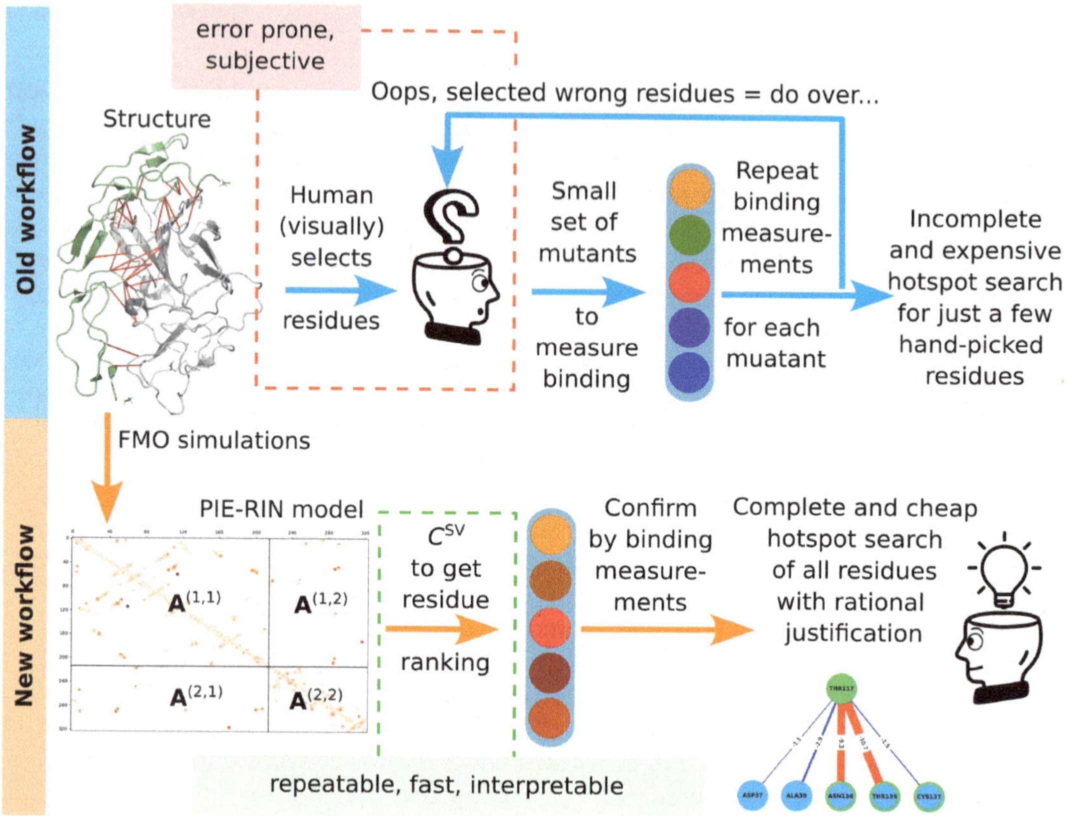

Fig. 11 PPI hotspot search protocol based on FMO coupled with singular value centrality calculation. The main advantage of the new integrative protocol is its reproducibility and interpretability

An overview of the described methods and comparison with old workflows is presented in Fig. 11.

An open challenge is the accounting for dynamic effects that certainly do play a role in real-life systems. Progress in this problem could provide even closer agreement with experimental findings and an even more revealing insight into residue roles in PPI temporal dynamics. Obviously, centrality evaluation has to be coupled with representative atomistic simulations of the structures.

Acknowledgments

This work was supported by the Scientific Grant Agency of the Ministry of Education of Slovak Republic and Slovak Academy of Sciences Grant VEGA 2/0061/23, and by the Slovak Research and Development Agency projects APVV-24-0351 and APVV-24-0594.

References

1. Anfinsen CB (1973) Principles that govern the folding of protein chains. Science 181:223–230
2. Finkelstein AV, Ptitsyn OB (2016) Protein physics, second, updated and extended edition. Elsevier, Amsterdam
3. Pauwels K, Van Molle I, Tommassen J, Van Gelder P (2007) Chaperoning Anfinsen: the steric foldases. Mol Microbiol 64:917–922
4. Kangueane, P, Nilofer, C, 2018. Protein–protein and domain–domain interactions. Springer, Singapore
5. Helms V, Kalinina OV (eds) (2023) Protein interactions. Wiley-VCH, Weinheim, Germany
6. Janin J (2007) Structural genomics: winning the second half of the game. Structure 15:1347–1349
7. Ponstingl H, Kabir T, Gorse D, Thornton JM (2005) Morphological aspects of oligomeric protein structures. Prog Biophys Mol Biol 89:9–35
8. Nooren IM (2003) Diversity of protein–protein interactions. EMBO J 22:3486–3492
9. Keskin O, Nussinov R (2007) Similar binding sites and different partners: implications to shared proteins in cellular pathways. Structure 15:341–354
10. Conte LL, Chothia C, Janin J (1999) The atomic structure of protein-protein recognition sites. J Mol Biol 285:2177–2198
11. Hwang H, Pierce B, Mintseris J, Janin J, Weng Z (2008) Protein–protein docking benchmark version 3.0. Proteins: Struct, Funct, Bioinf 73:705–709
12. Hwang H, Vreven T, Janin J, Weng Z (2010) Protein–protein docking benchmark version 4.0. Proteins: Struct, Funct, Bioinf 78:3111–3114
13. Janin J, Bahadur RP, Chakrabarti P (2008) Protein–protein interaction and quaternary structure. Q Rev Biophys 41:133–180
14. Chakrabarti P, Janin J (2002) Dissecting protein–protein recognition sites. Proteins: Struct, Funct, Bioinf 47:334–343
15. Rego NB, Xi E, Patel AJ (2021) Identifying hydrophobic protein patches to inform protein interaction interfaces. Proc Natl Acad Sci 118:e2018234118
16. Swapna LS, Bhaskara RM, Sharma J, Srinivasan N (2012) Roles of residues in the interface of transient protein-protein complexes before complexation. Sci Rep 2: 334
17. Yan C, Wu F, Jernigan RL, Dobbs D, Honavar V (2007) Characterization of Proteinâ€ "Protein interfaces. Protein J 27:59–70
18. Sowmya G, Anita S, Kangueane P (2011) Insights from the structural analysis of protein heterodimer interfaces. Bioinformation 6:137–143
19. Reichmann D, Phillip Y, Carmi A, Schreiber G (2007) On the contribution of water-mediated interactions to protein-complex stability. Biochemistry 47:1051–1060
20. Papoian GA, Ulander J, Wolynes PG (2003) Role of water mediated interactions in protein–protein recognition landscapes. J Am Chem Soc 125:9170–9178
21. Janin J (1999) Wet and dry interfaces: the role of solvent in proteinâ€"protein and proteinâ€ "DNA recognition. Structure 7:R277–R279
22. Martini S, Bonechi C, Foletti A, Rossi C (2013) Water–protein interactions: the secret of protein dynamics. Sci World J 2013:138916
23. Lisova O, Belkadi L, Bedouelle H (2014) Direct and indirect interactions in the recognition between a cross-neutralizing antibody and the four serotypes of dengue virus. J Mol Recognit 27:205–214
24. England P, Brégégère F, Bedouelle H (1997) Energetic and kinetic contributions of contact residues of antibody D1.3 in the interaction with lysozyme. Biochemistry 36:164–172
25. Barillari C, Taylor J, Viner R, Essex JW (2007) Classification of water molecules in protein binding sites. J Am Chem Soc 129:2577–2587
26. Levy Y, Onuchic JN (2004) Water and proteins: a love-hate relationship. Proc. Natl. Acad. Sci 101:3325–3326
27. Svedberg T (1927) The ultracentrifuge. Nobel Lecture
28. Sund H, Weber K (1966) The quaternary structure of proteins. Angew. Chem. - Int. Ed 5:231–245
29. Russell RB, Alber F, Aloy P, Davis FP, Korkin D, Pichaud M, Topf M, Sali A (2004) A structural perspective on protein–protein interactions. Curr Opin Struct Biol 14:313–324
30. Wodak SJ, Vajda S, Lensink MF, Kozakov D, Bates PA (2023) Critical assessment of methods for predicting the 3D structure of proteins and protein complexes. Annu Rev Biophys 52:183–206
31. Sedov IA, Zuev YF (2023) Recent advances in Protein–Protein interactions. Int J Mol Sci 24:1282
32. Katchalski-Katzir E, Shariv I, Eisenstein M, Friesem AA, Aflalo C, Vakser IA (1992) Molecular surface recognition: determination of geometric fit between proteins and their ligands by correlation techniques. Proc Natl Acad Sci 89:2195–2199
33. Abramson J et al (2024) Accurate structure prediction of biomolecular interactions with AlphaFold 3. Nature 630:493–500
34. Chen YC, Chen Y-H, Wright JD, Lim C (2022) PPI-HotspotDB: database of protein–protein interaction hot spots. J Chem Inf Model 62:1052–1060

35. Monteleone S, Fedorov DG, Townsend-Nicholson A, Southey M, Bodkin M, Heifetz A (2022) Hotspot identification and drug design of protein–protein interaction modulators using the fragment molecular orbital method. J Chem Inf Model 62:3784–3799
36. Di Micco P, Incarnato M, Pascarella G, Via A, Morea V (2025) Fast, comprehensive, and user customizable macromolecule interface analysis with FACE2FACE. J Chem Inf Model 65:9371–9377
37. Ibarra AA, Bartlett GJ, Hegedüs Z, Dutt S, Hobor F, Horner KA, Hetherington K, Spence K, Nelson A, Edwards TA, Woolfson DN, Sessions RB, Wilson AJ (2019) Predicting and experimentally validating hot-spot residues at Proteinâ€"Protein interfaces. ACS Chem Biol 14:2252–2263
38. Kortemme T, Baker D (2002) A simple physical model for binding energy hot spots in protein-protein complexes. Proc Natl Acad Sci 99:14116–14121
39. Liu Q, Chen P, Wang B, Zhang J, Li J (2018) Hot spot prediction in protein-protein interactions by an ensemble system. BMC Syst Biol 12: 132
40. Bajusz D et al (2021) Exploring protein hotspots by optimized fragment pharmacophores. Nat Commun 12: 3201
41. Chen YC, Sargsyan K, Wright JD, Chen Y-H, Huang Y-S, Lim C (2024) PPI-hotspotID for detecting protein–protein interaction hot spots from the free protein structure. eLife 13:RP96643
42. Jiang J, Wang N, Chen P, Zheng C, Wang B (2017) Prediction of protein hotspots from whole protein sequences by a random projection ensemble system. Int J Mol Sci 18:1543
43. Cukuroglu E, Engin HB, Gursoy A, Keskin O (2014) Hot spots in proteinâ€"protein interfaces: towards drug discovery. Prog Biophys Mol Biol 116:165–173
44. Moreira IS, Koukos PI, Melo R, Almeida JG, Preto AJ, Schaarschmidt J, Trellet M, Gümüş ZH, Costa J, Bonvin AMJJ (2017) SpotOn: high accuracy identification of protein–protein interface hot-spots. Sci Rep 7: 8007
45. Wang H, Liu C, Deng L (2018) Enhanced prediction of hot spots at protein–protein interfaces using extreme gradient boosting. Sci Rep 8: 14285
46. Morrow JK, Zhang S (2012) Computational prediction of protein hot spot residues. Curr Pharm Des 18:1255–1265
47. Schreiber, G, 2020. Protein–protein interaction regulators, pp 1. The Royal Society of Chemistry, p. 24
48. Morao I, Fedorov DG, Robinson R, Southey M, Townsend-Nicholson A, Bodkin MJ, Heifetz A (2017) Rapid and accurate assessment of GPCR-ligand interactions using the fragment molecular orbital-based density-functional. Tight-Binding Meth 38:1987–1990
49. Lu H, Zhou Q, He J, Jiang Z, Peng C, Tong R, Shi J (2020) Recent advances in the development of protein-protein interactions modulators: mechanisms and clinical trials. Sig Transduct Target Ther 5:213
50. Nada H, Choi Y, Kim S, Jeong KS, Meanwell NA, Lee K (2024) New insights into protein-protein interaction modulators in drug discovery and therapeutic advance. Sig Transduct Target Ther 9:341
51. Mabonga L, Kappo AP (2019) Protein–protein interaction modulators: advances, successes and remaining challenges. Biophys Rev 11:559–581
52. Lucero B, Francisco KR, Liu LJ, Caffrey CR, Ballatore C (2023) Protein–protein interactions: developing small-molecule inhibitors/ stabilizers through covalent strategies. Trends Pharmacol Sci 44:474–488
53. Wang J, Mao J, Li C, Xiang H, Wang X, Wang S, Wang Z, Chen Y, Li Y, No KT, Song T, Zeng X (2024) Interface-aware molecular generative framework for protein–protein interaction modulators. J Cheminf 16:142
54. Qiu Y, Li X, He X, Pu J, Zhang J, Lu S (2020) Computational methods-guided design of modulators targeting protein-protein interactions (PPIs). Eur J Med Chem 207:112764
55. Kannan N, Vishveshwara S (1999) Identification of side-chain clusters in protein structures by a graph spectral method. J Mol Biol 292:441–464
56. del Sol A, O'Meara P (2005) Small-world network approach to identify key residues in Proteinâ€"Protein Interaction. Proteins 58:672–682
57. Yeates TO, Beeby M (2006) Proteins in a small world. Science 314:1882–1883
58. Taylor NR (2013) Small world network strategies for studying protein structures and binding. Comput Struct Biotechnol J 5:e201302006
59. Viloria JS, Allega MF, Lambrughi M, Papaleo E (2017) An optimal distance cutoff for contact-based protein structure networks using side-chain centers of mass. Sci Rep 7:2838–2848
60. Pons C, Glaser F, Fernandez-Recio J (2011) Prediction of protein-binding areas by small-world residue networks and application to docking. BMC Bioinf 12:378–387
61. Yao X-Q, Momin M, Hamelberg D (2019) Establishing a framework of using residue–residue interactions in protein difference network analysis. J Chem Inf Model 59:3222–3228

62. Vendruscolo M, Dokholyan N, Paci E, Karplus M (2002) Small-world view of the amino acids that play a key role in protein folding. Phys Rev E Stat Nonlin Soft Matter Phys 65:061910
63. Greene LH, Higman VA (2003) Uncovering network systems within protein structures. J Mol Biol 334:781–791
64. Ribeiro AAST, Ortiz V (2014) Determination of signaling pathways in proteins through network theory: importance of the topology. J Chem Theory Comput 10:1762–1769
65. Yin Y, Maisuradze GG, Liwo A, Scheraga HA (2012) Hidden protein folding pathways in free-energy landscapes uncovered by network analysis. J Chem Theory Comput 8:1176–1189
66. Atilgan AR, Akan P, Baysal C (2004) Small-world communication of residues and significance for protein dynamics. Biophys J 86:85–91
67. VanWart AT, Eargle J, Luthey-Schulten Z, Amaro RE (2012) Exploring residue component contributions to dynamical network models of allostery. J Chem Theory Comput 8:2949–2961
68. Raimondi F, Felline A, Seeber M, Mariani S, Fanelli F (2013) A mixed protein structure network and elastic network model approach to predict the structural communication in biomolecular systems: the PDZ2 domain from tyrosine phosphatase 1E as a case study. J Chem Theory Comput 9:2504–2518
69. Dodd T, Yao X-Q, Hamelberg D, Ivanov I (2021) Subsets of adjacent nodes (SOAN): a fast method for computing suboptimal paths in protein dynamic networks. Mol Phys 119:e1893847
70. Doshi U, Holliday MJ, Eisenmesser EZ, Hamelberg D (2016) Dynamical network of residue-residue contacts reveals coupled allosteric effects in recognition, catalysis, and mutation. Proc. Natl. Acad. Sci 113:4735–4740
71. Yao X-Q, Hamelberg D (2022) From distinct to differential conformational dynamics to map allosteric communication pathways in proteins. J Phys Chem B 126:2612–2620
72. Yao X-Q, Hamelberg D (2022) Residue–residue contact changes during functional processes define allosteric communication pathways. J Chem Theory Comput 18:1173–1187
73. Ota K, Yamato T (2019) Energy exchange network model demonstrates protein allosteric transition: an application to an oxygen sensor protein. J Phys Chem B 123:768–775
74. Higman VA, Greene LH (2006) Elucidation of conserved long-range interaction networks in proteins and their significance in determining protein topology. Physica A 368:595–606
75. Del Conte A, Camagni GF, Clementel D, Minervini G, Monzon AM, Ferrari C, Piovesan D, Tosatto SE (2024) RING 4.0: faster residue interaction networks with novel interaction types across over 35,000 different chemical structures. Nucleic Acids Res 52:W306–W312
76. Contreras-Riquelme S, Garate J-A, Perez-Acle T, Martin AJM (2018) RIP-MD: a tool to study residue interaction networks in protein molecular dynamics. PeerJ 6:e5998
77. Guzzi PH, Di Paola L, Giuliani A, Veltri P (2022) PCN-Miner: an open-source extensible tool for the analysis of protein contact networks. Bioinformatics 38:4235–4237
78. Ser, cino˘glu O, Ozbek P (2018) gRINN: a tool for calculation of residue interaction energies and protein energy network analysis of molecular dynamics simulations. Nucleic Acids Res 46:W554–W562
79. Li Q, Luo R, Chen H-F (2019) Dynamical important residue network (DIRN): network inference via conformational change. Bioinformatics 35:4664–4670
80. Guarnera E, Tan ZW, Zheng Z, Berezovsky IN (2017) AlloSigMA: allosteric signaling and mutation analysis server. Bioinformatics 33:3996–3998
81. Woo H-M, Jeong H, Yoon B-J (2020) NAPAbench 2: a network synthesis algorithm for generating realistic protein–protein interaction (PPI) network families. PLoS One 15:e0227598
82. Sladek V, Yamamoto Y, Harada R, Shoji M, Shigeta Y, Sladek V (2021) pyProGA-A PyMOL plugin for protein residue network analysis. PLoS One 16:1–17
83. Estrada E (2010) Universality in protein residue networks. Biophys J 98:890–900
84. Estrada E (2000) Characterization of 3D molecular structure. Chem Phys Lett 319:713–718
85. Estrada E (2002) Characterization of the folding degree of proteins. Bioinformatics 18:697–704
86. Estrada E (2004) Characterization of the amino acid contribution to the folding degree of proteins. Proteins 54:727–737
87. del Sol A, Fujihashi H, O'Meara P (2005) Topology of small-world networks of protein-protein complex structures. Bioinformatics 21:1311–1315
88. Khor S (2015) Protein residue networks from a local search perspective. J. Complex Networks 4:245–278
89. Sladek V, Harada R, Shigeta Y (2020) Protein dynamics and the folding degree. J Chem Inf Model 60:1559–1567
90. Sladek V, Harada R, Shigeta Y (2021) Residue folding degree–relationship to secondary

structure categories and use as collective variable. Int J Mol Sci 22:13042
91. DiPaola L, Ruvo MD, Paci P, Santoni D, Giuliani A (2013) Protein contact networks: an emerging paradigm in chemistry. Chem Rev 113:1598–1613
92. Yehorova D, Di Geronimo B, Robinson M, Kasson PM, Kamerlin SC (2024) Using residue interaction networks to understand protein function and evolution and to engineer new proteins. Curr Opin Struct Biol 89:102922
93. Ferreiro DU, Komives EA, Wolynes PG (2014) Frustration in biomolecules. Q Rev Biophys 47:285–363
94. Vijayabaskar M, Vishveshwara S (2010) Interaction energy based protein structure networks. Biophys J 99:3704–3715
95. Sladek V, Tokiwa H, Shimano H, Shigeta Y (2018) Protein residue networks from energetic and geometric data: are they identical?. J Chem Theory Comput 14:6623–6631
96. Kitaura K, Ikeo E, Asada T, Nakano T, Uebayasi M (1999) Fragment molecular orbital method: an approximate computational method for large molecules. Chem Phys Lett 313:701–706
97. Fedorov DG (2020) In quantum mechanics in drug discovery. In: Heifetz A (Ed.)Eds. Methods in molecular biology. Humana, New York, NY
98. Fedorov DG, Kitaura K (2009) The fragnemt molecular orbital method - practical applications to large molecular systems. CRC Press, Taylor and Francis Group
99. Gordon MS, Fedorov DG, Pruitt SR, Slipchenko LV (2012) Fragmentation methods: a route to accurate calculations on large systems. Chem Rev 112:632–672
100. Fedorov DG (2017) The fragment molecular orbital method: theoretical development, implementation in GAMESS, and applications. WIREs Comput Mol Sci 7:e1322
101. Fukuzawa K, Tanaka S (2022) Fragment molecular orbital calculations for biomolecules. Curr Opin Struct Biol 72:127–134
102. Fukuzawa K, Watanabe C, Kurisaki I, Taguchi N, Mochizuki Y, Nakano T, Tanaka S, Komeiji Y (2014) Accuracy of the fragment molecular orbital (FMO) calculations for DNA: total energy, molecular orbital, and inter-fragment interaction energy. Comp. Theor. Chem 1034:7–16
103. Kato K et al (2020) High-precision atomic charge prediction for protein systems using fragment molecular orbital calculation and machine learning. J Chem Inf Model 60:3361–3368
104. Tanaka S, Mochizuki Y, Komeiji Y, Okiyama Y, Fukuzawa K (2014) Electron-correlated fragment-molecular-orbital calculations for biomolecular and nano systems. Phys Chem Chem Phys 16:10310–10344
105. Yamamoto Y, Nakano S, Shigeta Y (2022) Dynamical interaction analysis of proteins by a random forest-fragment molecular orbital (RF-FMO) method and application to Src tyrosine kinase. Bull. Chem. Soc. Jap 96:42–47
106. Heifetz A, Sladek V, Townsend-Nicholson A, Fedorov DG (2020) Quantum mechanics in drug discovery. In: Heifetz A (ed) Characterizing protein–protein interactions with the fragment molecular orbital method. Humana, New York, NY, pp 187–206
107. Sladek V, Fedorov DG (2022) The importance of charge transfer and solvent screening in the interactions of backbones and functional groups in amino acid residues and nucleotides. Int J Mol Sci 23:13514
108. Fedorov DG (2022) Parametrized quantum-mechanical approaches combined with the fragment molecular orbital method. J Chem Phys 157:231001
109. Estrada, E, 2011. The structure of complex networks. Oxford University Press
110. Newman, MEJ, 2010. Networks - anIntroduction. Oxford University Press
111. Borgatti SP (2005) Centrality and network flow. Soc Networks 27:55–71
112. Scardoni G, Laudanna C (2012) New frontiers in graph theory. In: Zhang Y (ed). InTech, pp 324–348
113. Wang S, Du Y, Deng Y (2017) A new measure of identifying influential nodes: efficiency centrality. Commun Nonlinear Sci Numer Simulat 47:151–163
114. Sladek V (2018) A note on the interpretation of the efficiency centrality. Commun Nonlinear Sci Numer Simulat 61:225–229
115. Latora V, Marchiori M (2001) Efficient behavior of small-world networks. Phys Rev Lett 87:198701
116. Sladek V, Artiushenko PV, Fedorov DG (2024) Effect of direct and water-mediated interactions on the identification of hotspots in biomolecular complexes with multiple subsystems. J Chem Inf Model 64:7602–7615
117. Maruyama K, Sheng Y, Watanabe H, Fukuzawa K, Tanaka S (2018) Application of singular value decomposition to the inter-fragment interaction energy analysis for ligand screening. Comput Theor Chem 1132:23–34
118. Tanaka S, Watanabe C, Honma T, Fukuzawa K, Ohishi K, Maruyama T (2020) Identification of correlated inter-residue interactions in protein complex based on the fragment molecular orbital method. J Mol Graphics Modell 100:107650
119. Grau FC, Jaeger J, Groher F, Suess B, Muller YA (2020) The complex formed between a

synthetic rna aptamer and the transcription repressor TetR is a structural and functional twin of the operator DNA–TetR regulator complex. Nucleic Acids Res 48: 3366–3378
120. Schmidt MW, Baldridge KK, Boatz JA, Elbert ST, Gordon MS, Jensen JH, Koseki S, Matsunaga N, Nguyen KA, Su S, Windus TL, Dupuis M, Montgomery JA (1993) General atomic and molecular electronic structure system. J Comput Chem 14:1347–1363
121. Barca GMJ et al (2020) Recent developments in the general atomic and molecular electronic structure system. J Chem Phys 152:154102
122. Gordon MS, Schmidt MW (2005) Theory and applications of computational chemistry. In: Dykstra CE, Frenking G, Kim KS, Scuseria GE (eds) Elsevier, pp 1167–1189
123. Fedorov DG, Kitaura K (2004) The importance of three-body terms in the fragment molecular orbital method 120:6832–6840
124. Suenaga M, Fedorov DG (2026) Quantum mechanics applications in drug design. Springer, chapter 1
125. Yuan S, Chan HS, Hu Z (2017) Using PyMOL as a platform for computational drug design. WIREs Comput Mol Sci 7:e1298
126. Zhou H-X, Pang X (2018) Electrostatic interactions in protein structure, folding, binding, and condensation. Chem Rev 118:1691–1741
127. Vascon F, Gasparotto M, Giacomello M, Cendron L, Bergantino E, Filippini F, Righetto I (2020) Protein electrostatics: from computational and structural analysis to discovery of functional fingerprints and biotechnological design. Comput Struct Biotechnol J 18:1774–1789
128. Fuxreiter M (2022) Electrostatics tunes protein interactions to context. Proc. Natl. Acad. Sci. 119:e2209201119
129. Zhang Z, Witham S, Alexov E (2011) On the role of electrostatics in protein–protein interactions. Phys Biol 8:035001

Chapter 10

Hidden Interactions: Quantum Perspectives in Drug Design

Taisiia Feoktistova and Tahsin Kellici

Abstract

Noncovalent intermolecular interactions play a key role in ligand recognition, affinity, and specificity. This chapter provides an overview, with literature examples, of underexplored interactions that contribute to molecular binding, including weak hydrogen bonding; anion–π and cation–π interactions; Bürgi–Dunitz and tetrel interactions; halogen and chalcogen bonding; and S···π, C···π, O···π, N···π, and amide···π interactions. These interactions are often overlooked, and in many of the examples described here, ligands were not initially designed to exploit them—even though reports show that, collectively, they can fine-tune ligand–protein recognition in drug design. IsoStar contour density surfaces are included to visualize the geometric preferences of each interaction. Additionally, where applicable, we include evaluation of the geometry and prevalence of these interactions through database mining and quantification of their energetics using quantum mechanical methods, like density functional theory.

Keywords Noncovalent interactions, Quantum mechanical calculations, DFT, Protein–ligand affinity, Amide–π stacking, Weak hydrogen bonding, Cation–π, Anion–π, Chalcogen bonding, Halogen bonding, Bürgi–Dunitz, Tetrel, Orthogonal multipolar interactions

1 Introduction

Unusual, underestimated, or otherwise hidden intermolecular interactions and their role in protein–ligand affinity and specificity have been repeatedly examined in the literature. Notable overviews include those by Bissantz et al. in 2010 [1], a Protein Data Bank (PDB)-based frequency analysis by Ferreira de Freitas et al. in 2017 [2], and a focused review on atypical interactions by Kuhn et al. in 2019 [3]. These interactions merit attention not only for their structural implications but also for their potential to significantly influence the enthalpic component of binding free energy when geometrically favorable. In certain cases, their contributions may rival that of conventional hydrogen bonds, thereby affecting both affinity and specificity. Their underappreciation reflects the challenges in quantifying them and the risk of conflating their enthalpic effects with other factors, including solvation and conformational or entropic

Alexander Heifetz (ed.), *Quantum Mechanics Applications in Drug Design*, Methods in Molecular Biology, vol. 3061, https://doi.org/10.1007/978-1-0716-5420-0_10,

changes (e.g., displacement of high-energy water molecules or shifts toward a bioactive conformation).

This chapter focuses on nine interactions that remain relatively underexplored in literature, such as weak H-bonding, halogen, chalcogen, and multipolar interactions. For each interaction, we outline key geometric and energetic parameters (typical distances, predominant angles, approximate strengths, and substitution effects) and present cases where structure–activity relationships (SARs) support the attribution of affinity gains to the interaction itself. Accordingly, most examples are drawn from recent medicinal chemistry efforts in which available data favor an enthalpic interpretation. Table 1 lists well-characterized, commonly encountered interactions that serve as critical determinants of binding. They have been extensively reviewed in prior literature and are included herein for completeness and comparative context.

Table 1
Summary of well-documented interactions with representative references

Interaction(s)	Description
Lipophilic interactions/hydrophobic effects	Estimated at ~30 kcal mol^{-1} Å^{-2}, the hydrophobic effect corresponds to ~0.7 kcal mol^{-1} per methyl-group-sized buried surface area, which can translate to an ~3.5-fold increase in the binding constant. By comparison, typical drug molecules contain only one to two hydrogen bond donors and three to four acceptors but around 16 hydrophobic atoms on average [1].
Hydrogen bonds	Interaction between a H-bond acceptor and a H-bond donor. Desolvation of both donor and acceptor must occur for the H bond to form; thus, the net free energy gain of the H-bond depends on the desolvation penalty → a solvent-exposed H-bond contributes substantially less to the overall interaction energy than the same H-bond within a buried hydrophobic pocket [4, 5].
π–π stacking	On average, 60% of aromatic sidechains (e.g., Phe, Tyr, and Trp) are involved in π–π interactions. The distribution may be uneven, depending on the aromatic amino acid pair, the distance between two aromatic rings, their locations in the protein, and whether clustering of aromatic amino acids is involved [6–8].
Salt bridges	A salt bridge is a noncovalent interaction between two ionized molecules, mixing two contributions, i.e., a hydrogen bond and an electrostatic interaction, and only substantially contributes to binding if not solvent-exposed. These interactions have been comprehensively explored computationally. Salt bridges are quite difficult to design because of exacting geometric requirements [9].

To visualize geometric preferences, we include IsoStar contour density surfaces derived from empirical contact counts around probe groups; red contours denote regions with twice the expected contact density, while green contours correspond to the expected density [10].

The reliability of the protein–ligand structure is critical for rationalizing these interactions. We focused on protein–ligand crystal structures with resolutions of 2.5 Å or better and verified the ligand and partner density before interpreting any putative interaction. In general, the quality of the structures should be assessed instead of over-relying on hidden bonds to elucidate SAR.

Each section reviews methods used to quantify these interactions. Quantum-mechanical (QM) approaches remain the most reliable for estimating interaction energies. Classical force fields often fail to accurately capture these noncovalent effects, whereas database-mining methods (Protein Data Bank [PDB]/Cambridge Structural Database [CSD]) are valuable for evaluating prevalence and geometry but are not suitable for estimating interaction strengths.

2 Weak H Bond Interactions with Polarized C–H Bonds

A C–H bond can serve as a hydrogen bond donor (HBD), with the hydrogen atom bearing a partial positive charge. This polarization arises from the influence of electron-withdrawing groups near the carbon atom or within ring structures (Fig. 1A) [11]. In macromolecules, a particularly prevalent example is the hydrogen bond formed between the Cα-H of amino acids and the oxygen atom of the peptide

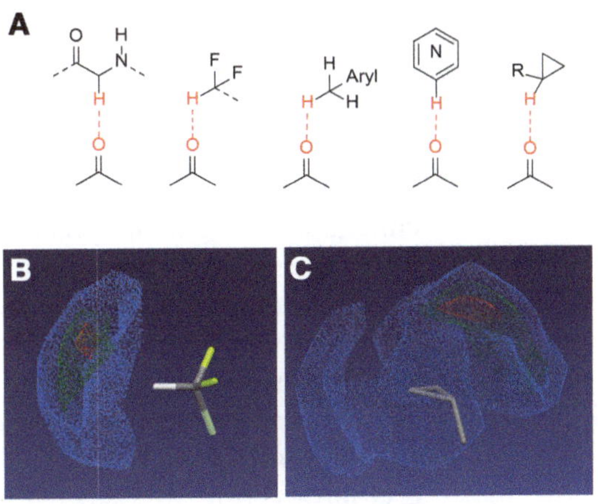

Fig. 1 (**A**) Interactions between a H bond acceptor (e.g., the oxygen atom in the carbonyl group of a peptide) and a polarized CH group, (**B**) IsoStar map of R–CF_2H and O=C interactions, and (**C**) IsoStar map of interactions between the hydrogen atom in the cyclopropyl group and O=C

backbone, which is estimated to be around half as strong as those mediated by NH donors. Such interactions are usually weak, with strengths ranging from 1.9 to 2.5 kcal mol^{-1} [11]. In the RCF_2H group, the hydrogen atom can act as a weak HBD, as revealed by the IsoStar contour map in Fig. 1B. Within rings, heteroaromatic and benzylic C–H bonds, as well as hydrogen atoms in cyclopropyl groups (Fig. 1C), are also weak donors. The latter interaction is notably frequent among kinase hinge binders [12, 13].

Of particular interest is the weak HBD capability of a methyl group attached to an aromatic system. In such cases, C–H bonds exhibit some degree of polarization due to hyperconjugation and inductive effects. Representative examples are shown in Fig. 2A, B. The first involves the crystal structure of an inhibitor bound to PI3Kα kinase (PDB ID: 6GVG) [14], where the methyl group CH attached to the pyrazolopyrimidine interacts with the backbone oxygen of Glu849. As suggested by Ouvry et al., this unconventional hydrogen bond substitutes for a traditional one formed between the NH_2 group of sapanisertib and the same residue [14]. The second example features a pan-RAF kinase inhibitor, where a polarized C–H bond of the *N*-methyl group is attached to pyridazinone (PDB ID: 6XLO), which interacts with the backbone carbonyl group of Cys532 [15].

The HBD capability of the CF_2H group is supported by theoretical studies corroborated with experimental data [16]. This group is usually considered a bioisostere of the methyl group, with moderately increased lipophilicity compared to that of the trifluoromethyl moiety [17]. In the example depicted in Fig. 2C, the difluoromethoxy moiety of the probe BAY-598 is oriented toward a hydrophobic pocket in the protein lysine methyltransferase SMYD2, with the polarized CH interacting with Thr105 [18].

As previously discussed, heteroaromatic C–H bonds have been extensively utilized as HBDs in nontraditional interactions with kinase hinge binders [12]. For IRAK4 (PDB ID: 6F3I), as shown in Fig. 2D, the highlighted hydrogen in pyrrolotriazine interacts with the hinge carbonyl group in Val263, substituting the conventional hydrogen bond in previous pyrrolopyrimidine hinge binders [19]. This substitution reduces the number of formal HBDs by one and enhances molecular lipophilicity, both being important factors affecting passive permeability [19].

Regarding cyclopropyl groups, the CH group on a tertiary carbon atom can become polarized because of ring strain and hybridization effects. Cyclopropylamides appear in numerous kinase hinge binders. For instance, for activin receptor-like kinase inhibitors (Fig. 2E), the tertiary carbon atom's highlighted hydrogen atom in the cyclopropyl ring interacts with the hinge carbonyl group in His283 [20]. The importance of this interaction is underscored by the distinct behavior of the cyclopropyl group compared to that of an isopropyl substituent [20].

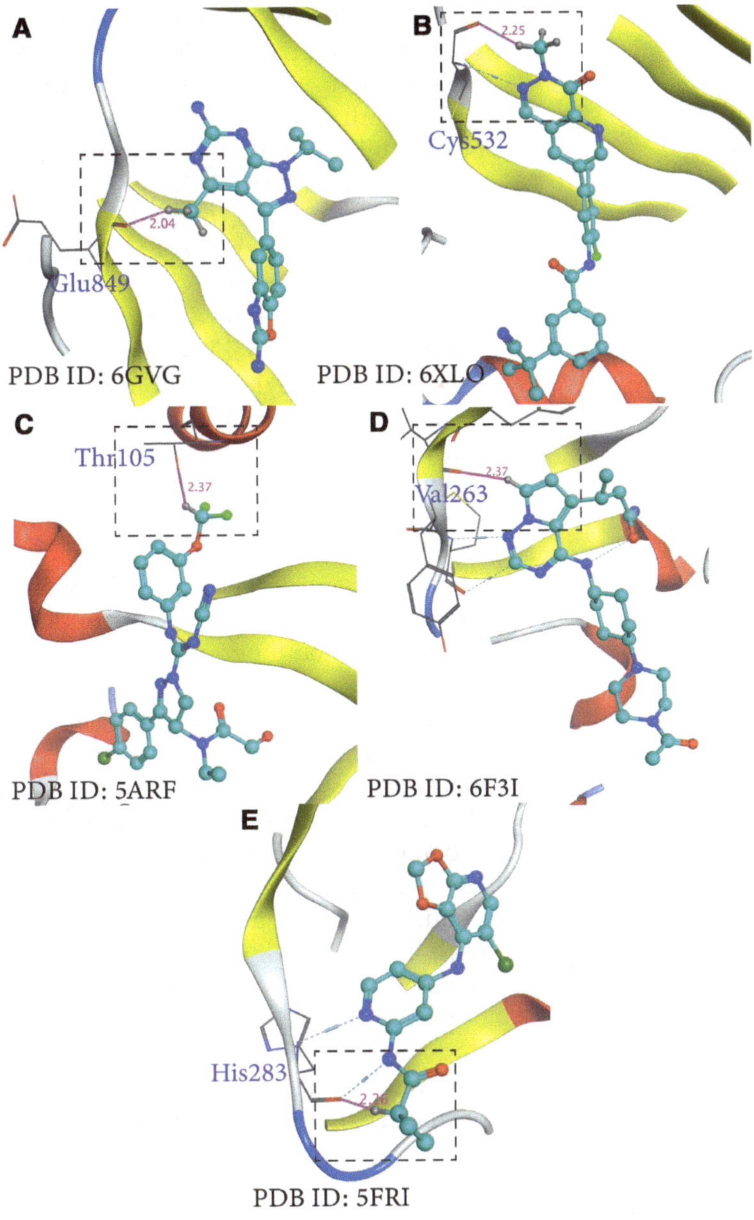

Fig. 2 The hydrogen bond formed between polarized CH groups and C=O is highlighted in magenta. Distances are reported in angstroms. (**A**) PI3Kα, (**B**) bRAF, (**C**) SMYD2, (**D**) IRAK4, and (**E**) ALK5

3 Halogen Bonding

Halogen bonding (XB) is a directional noncovalent interaction enabled by the presence of a σ-hole (a region of positive electrostatic potential) along the covalent bond axis. It can also be described as a

Lewis acid–base interaction, wherein the halogen atom acts as a Lewis acid, and a lone pair-bearing atom in a N- or O-containing molecule, a π-system, or an anion, acts as a Lewis base (LB) (Fig. 3A) [21–23]. XB strength follows the trend I > Br > Cl»F, reflecting the increasing polarizability of the halogen atom and, consequently, the size of the σ-hole (Fig. 3B–D) [21, 24]. Fluorine is a notable exception; owing to its high electronegativity, it is considered σ-hole deficient leaving negligible positive electrostatic potential along the extension of its covalent bond axis [23, 24]. Substituents that contribute to the deepening of the σ-hole through electron withdrawal can also enhance XB strength [21, 25]. XB is highly directional, with a preference for linearity, where angle deviations contribute to energy loss [23, 26]. Its strength is maximized when the distance is slightly shorter than the sum of the van der Waals radii, typically between 3.0 and 3.5 Å, with the angle between the LB and σ-hole being between 150° and 180° [23, 24, 26, 27]. XB has found applications in catalysis, materials, and particularly medicinal chemistry owing to its tunability as well as the frequent use of halogen substitution in rational drug design for improving absorption, distribution, metabolism, and excretion properties, including metabolic stability of molecules during lead optimization [21, 23, 26, 28, 29].

This prevalence of halogens across various systems has generated interest and prompted exploration using various computational methods like quantum mechanics (e.g., density functional theory (DFT) and the quantum theory of atoms in molecules [QTAIM]) and informatics. Historically, XB has often been compared to hydrogen bonding (HB) and is sometimes described as "weak hydrogen bond." This analogy stems from the similarities between the two: both interactions involve an LB; are noncovalent, directional, and tunable; and are widely applicable, with contributions from both dispersion and electrostatics [21, 24, 25, 30, 31]. A computational study that directly compared XB and HB in protein–ligand systems found that their fundamental bonding mechanisms are similar. Using bromobenzene as a model for a halogenated ligand, the authors

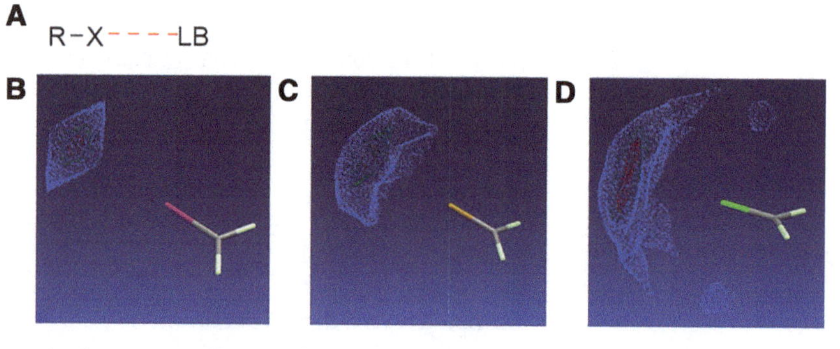

Fig. 3 (**A**) Interaction between an LB and a halogen atom (X); IsoStar maps of (**B**) Ar–I, (**C**) Ar–Br, and (**D**) Ar–Cl interactions with O=C

optimized geometries using DFT at the BLYP-D3(BJ)/TZ2P level of theory and performed Kohn–Sham molecular orbital and energy decomposition analyses. These calculations demonstrated that, in addition to electrostatics and dispersion, orbital-level interactions, specifically donor–acceptor interactions and steric repulsion between highest occupied molecular orbitals (HOMO–HOMO), are important for XB [31].

Recently, an extensive distance-based PDB survey revealed that in XB, the most common LB partner is the backbone carbonyl oxygen atom, followed by the tyrosine sidechain [23]. Furthermore, a study on nonclassical interactions using protein–ligand structures in the PDB reconfirmed that halogens form strong and favorable interactions with carbonyl group oxygen atoms at a diverse set of binding sites [32]. Additionally, a PDB survey focused on iodinated ligands similarly revealed that the backbone carbonyl oxygen atom is the most common binding partner, followed by O, S, and Se atoms or π-systems of amino acid sidechains. Additionally, the energetic evaluation of four selected structures using the RI-MP2/def2-TZVP level of theory, followed by QTAIM analysis and noncovalent interaction plots, enabled the authors to confirm the variable attractiveness of halogen bonding, with the variability being partially attributed to the differences between the properties of the Lewis base [30]. Finally, a QM study on halogen–water interactions at protein binding sites using the MP2/TZVPP level of theory showed that although these interactions are energetically weaker than halogen–protein interactions, they can be a key component to improve protein–ligand affinities by lowering the desolvation penalty. As an example, the authors reported the structure of 2-cyclopentyl-7-iodo-1*H*-indole-3carbonitrile bound to dual-specificity tyrosine-phosphorylation-regulated kinase 1A (DYRK1a kinase), where iodine–water XB was observed [25].

In medicinal chemistry, because of the prominence of halogens as substituents, a few examples of structures have been reported in the literature to date, highlighting XB as one of the key interactions driving selectivity. Factor Xa (fXa), a serine protease, is one of the key targets for preventing and treating thromboembolic disorders, such as acute myocardial infarction. The inhibition of fXa has an antithrombotic effect. An X-ray structure of phenylenediamine analog (Fig. 4A, PDB ID: 3Q3K) shows a halogen bond between the indoline 4-chloro substituent and Tyr228 in the S1 pocket, a key interaction contributing to its potent anti-fXa activity ($IC_{50} = 7.4$ nM) [33]. In another example, a halogen bond formed between one of the iodo substituents of erythrosine extra bluish and the oxygen atom in Thr77 in the human biliverdin IX beta reductase (BLVRB) complex (Fig. 4B, PDB Id: 5OOH). The parent, xanthene, showed low inhibitory activity against BLVRB, whereas the halogenated dyes, erythrosine extra bluish and phloxine B, displayed moderate inhibition, with IC_{50}

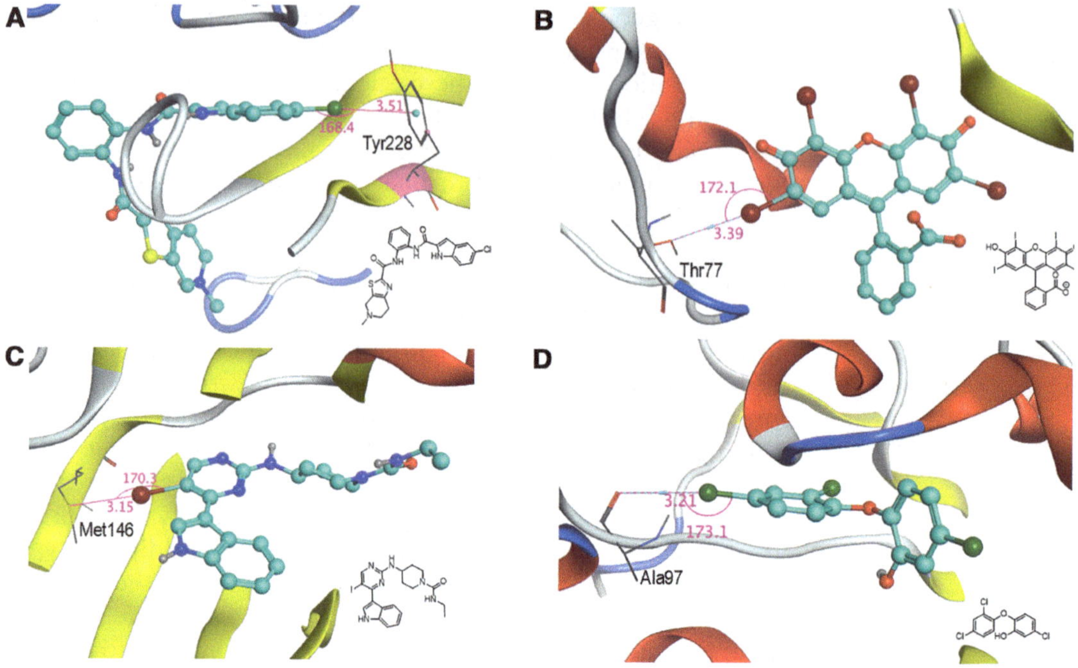

Fig. 4 Protein–ligand interactions involving halogen bonds are highlighted in magenta. Distances are reported in Å. (**A**) The chloro substituent in the phenylenediamine analog and the tyrosine 228 residue in factor Xa (PDB ID: 3Q3K), (**B**) one of the iodo substituents in erythrosine extra bluish and the sidechain's oxygen atom in the threonine 77 residue in human biliverdin IX beta reductase (PDB ID: 5OOH), (**C**) an iodo-substituted aminopyrimidine inhibitor and methionine 146 c-Jun *N*-terminal kinase 3 (PDB ID: 4×21), and (**D**) the chlorine atom at position four and the oxygen atom in the alanine 97 residue in *S. aureus* enoyl–acyl carrier protein reductase (PDB ID: 4ALI)

values of 0.41 and 0.70 μM, respectively [34]. A recent example of XB to the sulfur atom of Met146 is the c-Jun *N*-terminal kinase 3 (JNK3) complex with an iodo-substituted aminopyrimidine inhibitor (Fig. 4C, PDB ID: 4×21). Although all the halogenated ligands bound JNK3 ~5-fold tighter than the nonhalogenated analog, the affinity difference among the iodo, bromo, and chloro analogs was unclear, underscoring that ligand–protein binding is multifactorial and can be modulated by factors such as residue conformational strain and enthalpy–entropy compensation [35]. In the final example, a halogen bond between triclosan's 4-Cl and the carbonyl group's oxygen atom in Ala97 in *S. aureus*, enoyl–ACP reductase (FabI) reportedly accounted for the improved K_i (inhibition constant) versus that of 5-chloro-2-phenoxyphenol, an inhibitor that lacks chlorine atoms at positions 2 and 4 (Fig. 4D, PDB ID: 4ALI) [36].

4 Orthogonal Multipolar Interactions

4.1 Fluorine Interactions

As the most electronegative halogen in the periodic table, fluorine might be expected to readily participate in both HB and XB interactions (Fig. 5A). However, experimental evidence indicates that its bonding behavior is more complex.

In the context of HB, the fluorine atom in C–F bonds is a poor HB acceptor. This was first systematically documented by Dunitz in his seminal 1997 study on fluorine interactions [37]. Subsequent crystallographic analyses by Taylor using CSD confirmed that even the strongest HBs involving fluorine are weaker than those formed with many common acceptors [38]. This weak performance is reflected in the low frequency of X–H···F–C HBs observed in small-molecule crystal structures. Similar findings from PDB analyses suggest that when fluorine does participate in HB within biological systems, such interactions are typically forced by favorable neighboring ligand–receptor contacts rather than by the intrinsic acceptor capability of fluorine [39].

The limitations of fluorine also extend to XB. XB strength correlates with atomic polarizability, which determines the positive characteristics of σ-holes. Polarizability increases in the order F < Cl < Br < I, and accordingly, fluorine exhibits the weakest σ-hole potential and thus the poorest XB capability. Although isolated cases of fluorine-mediated XB have been reported, they are exceptional and rarely observed in protein–ligand interactions [40, 41].

While fluorine is not typically involved in hydrogen bonding, the abundance of contacts in the CSD and PDB suggests that it is favorable for the C–F dipole to undergo multipolar interactions with other dipoles [42]. Dipolar interactions between C–F and N–H bonds are particularly common [38]. These interactions typically occur at distances ranging from 2.5 to 3.5 Å, while their angle (C–F···H) varies between 110° and 140°. A recent interaction of interest is the one

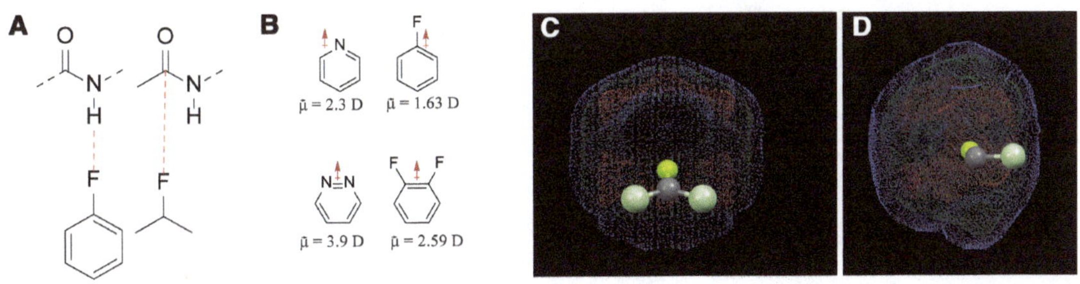

Fig. 5 (**A**) Common interactions of fluorine with NH and the carbon atom of the amide bond based on PDB data, (**B**) dipolar moments of azines compared to those of fluorine-substituted phenyl rings, IsoStar contours of the interactions between (**C**) aromatic and (**D**) aliphatic fluorine atoms with carbonyl groups in amides according to the PDB

observed in the sphingosine 1-phosphate receptor bound to CBP-307 [43]. The fluorine atom in the ligand interacts with the NH in Asn101, a crucial residue comprising orthosteric binding pockets and mediating receptor activation (Fig. 6A). As further evidence of C–F···NH multipolar interactions, fluoroaryls have been described as azine bioisosteres, –not because of hydrogen bond acceptor capability, but because of similar dipoles (Fig. 5B) [44].

Another favorable multipolar interaction is the one between a C–F bond and the electrophilic carbon atom of a carbonyl group [45]. These interactions typically yield a free energy gain of 0.3–0.6 kcal mol^{-1}, with distances between 3.3 and 3.6 Å and angles (F···C=O) ranging from 70° to 110° (Fig. 5C, D). One of the most interesting examples where this interaction becomes important is a series of compounds used as TYK2 and JAK1 dual inhibitors. The difluorocyclopropyl *S*-enantiomer in Fig. 6B is a eutomer possessing a balanced TYK2 and JAK1 inhibition profile (TYK2 and JAK1 IC_{50} values of 23 and 17 nM, respectively) [46]. As the interactions that this enantiomer develops with the protein (more specifically, the carbon atom of the carbonyl groups in Glu883 and Gly882) cannot be replicated by the *R*-enantiomer (Fig. 6B), the *S*-enantiomer shows better activity. Although an extensive matched molecular pair-based

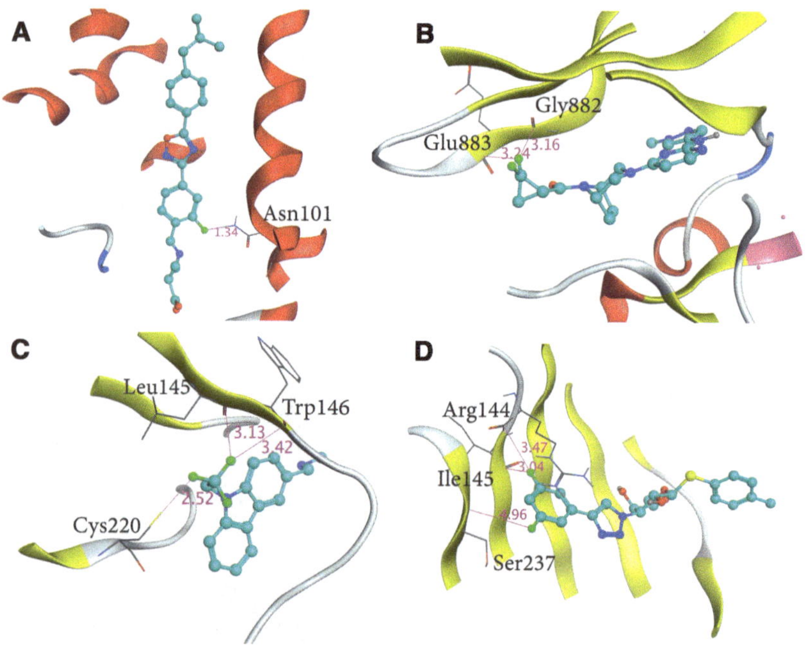

Fig. 6 Multipolar interactions of fluorine (highlighted in magenta). Distances are reported in angstroms. (**A**) Sphingosine 1-phosphate receptor bound to CBP-307 (PDB ID:7VIG [43]), (**B**) Jak1 with difluorocyclopropyl inhibitor (PDB ID: 6DBN [46]), (**C**) P53 Y220C mutant complexed with the PhiKan083 inhibitor (docking pose in PDB ID: 2VUK [50]), and (**D**) galectin-3 complexed with a fluorinated inhibitor (PDB ID: 6QLR [49])

study revealed a weak benefit, it was often counterbalanced by the R-enantiomer's impact on logD [47].

Interactions between fluorine atoms and carbonyl groups in peptide bonds influence the binding of the Y220C mutant of tumor suppressor p53 to the small molecule PhiKan083. Incorporation of fluorine atoms resulted in a fivefold increase in binding affinity [48]. The docking of the fluorinated compound on a similar crystal structure showed that the orientation of the C–F vector toward the backbone carbonyl groups in Leu145 and Trp146 residues yielded the most favorable interaction energy (Fig. 6C).

Finally, the significance of fluorine interactions with carbonyl groups in amide bonds has been very well explored in a series of compounds targeting galectin-3 [49]. Within the series, the SAR analysis highlighted the importance of orthogonal multipolar fluorine–amide interactions with Arg144, Ile145, and Ser237. QM calculations based on two single-point BP-D3 calculations using the TZVP basis set, as required by this method, revealed that *meta* substitution offers optimal interaction directionality, while *para* substitution with the Ser237 backbone further enhances the binding affinity (Fig. 6D).

4.2 Chlorine Interactions

In drug design, the strategic incorporation of chlorine substituents has attracted considerable attention. A 2023 perspective highlighted the so-called *magic chloro* effect [51], which arises from multiple complementary mechanisms that enhance molecular interactions and pharmacological properties. Chlorine substitution increases dipole moments, reduces the basicity of heteroaromatic systems, and renders benzene rings less electron-rich, thereby modulating electronic properties. Additionally, chlorine can alter scaffold conformations and enhance lipophilic binding interactions with target proteins [51].

Beyond these effects, chlorine can engage in XB, as previously discussed herein. However, a less-characterized interaction involves orthogonal multipolar interactions formed by chlorine atoms (Fig. 7A), presenting an interesting contrast to XB: although halogen bonds exploit positive electrostatic potential regions at chlorine atoms, orthogonal multipolar interactions leverage the nucleophilic character of chlorine (arising from its partial negative charge) to interact with electrophilic centers, such as the carbon atoms of amide carbonyl groups.

A standard review on multipolar interactions by Paulini et al. noted that the PDB contains numerous examples of chlorine interactions [52], spanning distances of 2.65–4.15 Å and almost orthogonal angles ((Cl···C=O) = 78° and (C–Cl···C) = 154°) (Fig. 7B). Although both the PDB and CSD document chlorine interactions, to the best of our knowledge, no study to date has yet examined their effects on SARs or on bond strength. One example suggesting the importance of chlorine interactions is between A-196, a potent and selective inhibitor of SUV420H1 and SUV420H2, two protein lysine

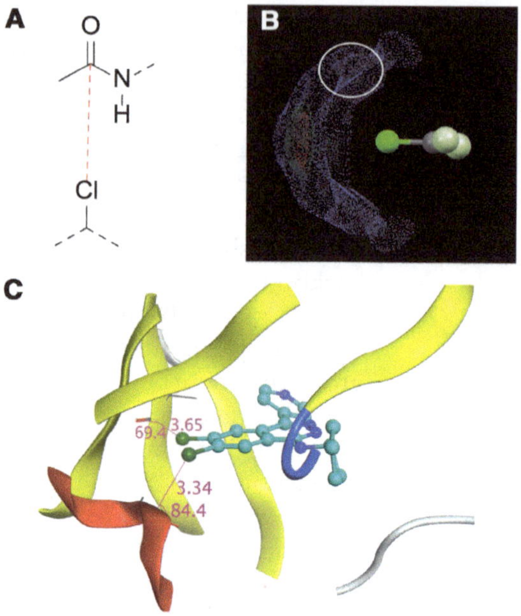

Fig. 7 (**A**) Interaction between a chlorine atom and the carbon atom of an amide functional group. (**B**) IsoStar contour map showing interactions between aromatic chlorine atoms and amide groups. The circled region shows amide groups and carbon atoms oriented perpendicular to the chlorine atom, eliminating the possibility of an XB interaction. (**C**) Interaction between a chlorine atom and the protein lysine methyltransferase (highlighted in magenta). Distances are reported in angstrom (PDB ID: 5CPR)

methyltransferases [53]. Both chlorine atoms in the inhibitor interact with the Ala289 and Ala268 backbones (distances of 3.65 and 3.34 Å and angles of 69.4 and 84.4°, respectively) (Fig. 7C). The limited SAR shows that transforming from the monochloro to the dichloro analog enhances potency tenfold (for the monochloro and dichloro analogs of SUV420H1, $IC_{50} = 5.9$ and 0.59 μM, respectively) [53].

5 Putative Directional Interactions: Bürgi–Dunitz and Tetrel

One of the commonly underappreciated protein–ligand interactions is the Bürgi–Dunitz interaction (trajectory), also sometimes referred to as the orthogonal multipolar or $n \rightarrow \pi^*$ interaction. Depending on the origins of the π^* orbital, the $n \rightarrow \pi^*$ interaction can be categorized as either amide or aromatic [54, 55]. The $n \rightarrow \pi^*$ (amide) stereoelectronic interaction occurs when the electrophilic carbon atom in a carbonyl group is in proximity to a nucleophilic atom, such as oxygen, nitrogen, or sulfur (Fig. 8A). Key features that distinguish Bürgi–Dunitz interactions from simple dipole–dipole interactions

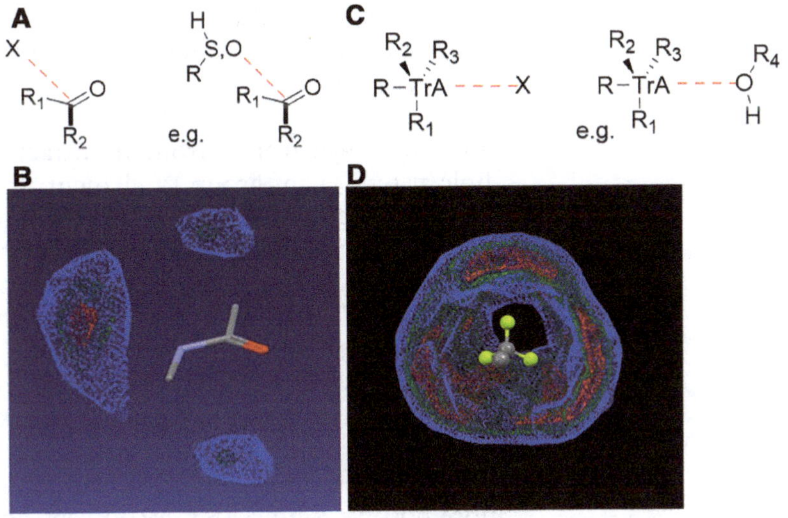

Fig. 8 (**A**) Interaction between a nucleophilic atom (X) and a carbonyl carbon atom, a representative example of RS(O)R_1, (**B**) IsoStar map of the backbone amide interaction with phenylalanine, (**C**) the interaction between a nucleophilic atom (X) and a tetrel atom (TrA), a representative example of OHR_4, and (**D**) IsoStar contour analysis of a trifluoromethyl moiety interacting with carbonyl groups

include the distance between the key atoms, approximately 2.6–3.2 Å, which is between the typical distances of van der Waals interactions and covalent bonds, and an interaction angle of approximately 109 ± 10°, enabling proper orbital overlap (Fig. 8B) [52, 54, 56]. This key angular requirement directionalizes Bürgi–Dunitz interactions. Depending on the stage of the transition between Bürgi–Dunitz and dipole–dipole interactions, the carbonyl group's carbon atom is sometimes displaced [56]. For the $n \rightarrow \pi^*$ (aromatic) interaction, the lone pair of electrons of the oxygen atom in a carbonyl group is delocalized to the π^* orbital of an aromatic ring [54]. Although the Bürgi–Dunitz interaction is relatively weak (<1 kcal mol^{-1}), it plays a critical role in stabilizing protein secondary structures, well exemplified in collagen. Furthermore, because of the ubiquity of the carbonyl functionality, the Bürgi–Dunitz interaction substantially contributes to both protein–ligand recognition specificity and selectivity [52, 54–56]. For example, Yeung et al. observed a Bürgi–Dunitz interaction between one of the sulfonamide oxygen atoms and a side chain amide carbonyl group in the Gln414 residue in a hepatitis C virus NS5B/inhibitor complex co-crystal structure. As Gln414 is a key residue that differentiates the gt-2 NS5B protein from that of gt-1, this Bürgi–Dunitz interaction was hypothesized to be responsible for the improved coverage of gt-2a in the C6-sulfonamide and C6-N-ethylamino dessulfonamide analogs. Both the distance between the interacting nucleophilic oxygen and amide carbon atoms (3.5 Å) and the O···C=O angle (108.0°) are consistent with the Bürgi–Dunitz interaction requirements (Fig. 9A, PDB ID: 5QJ0) [57].

Another important directional interaction to consider is the tetrel interaction, which arises from the attraction between an LB and a σ -hole at tetravalent group IV element atoms (TrAs), the namesake of this interaction [58, 59]. The tetrel interaction represents a specific subset within the broader family of σ-hole interactions, which also includes halogen and chalcogen bonding, also discussed in this chapter. Tetrel atoms, such as carbon and silicon, have four σ -holes; the requirement for the interaction angle to be approximately 180°, relative to the TrA–R bond axis, highly directionalizes the tetrel interaction, optimizing alignment with the σ-hole (Fig. 8C). Although the tetrel interaction is primarily electrostatic, polarization and dispersion can also affect it, e.g., more polarizable atoms have more positive σ-holes [58, 59]. Even though biologically relevant small-molecule ligands are rarely designed to introduce tetrel interactions, it can still be quite relevant for fine-tuning the binding, especially with functional groups such as CF_3, which is very common in medicinal chemistry [60]. A PDB survey of protein–drug structures containing both a CF_3 group and electron-rich atoms revealed a total of 419 unique interactions, with C···O and C···N interactions being the most prevalent. Following this survey, a theoretical study on the located structures using the RI-MP2/def2-TZVPD level of theory revealed that the interaction energy for a tetrel interaction was up to $-2\,\text{kcal mol}^{-1}$ ($-8.4\,\text{kJ mol}^{-1}$) [61]. An example of a tetrel interaction is shown in Fig. 9B, where the X-ray structure of 2-pyridone (compound 2) complexed with bovine trypsin revealed that

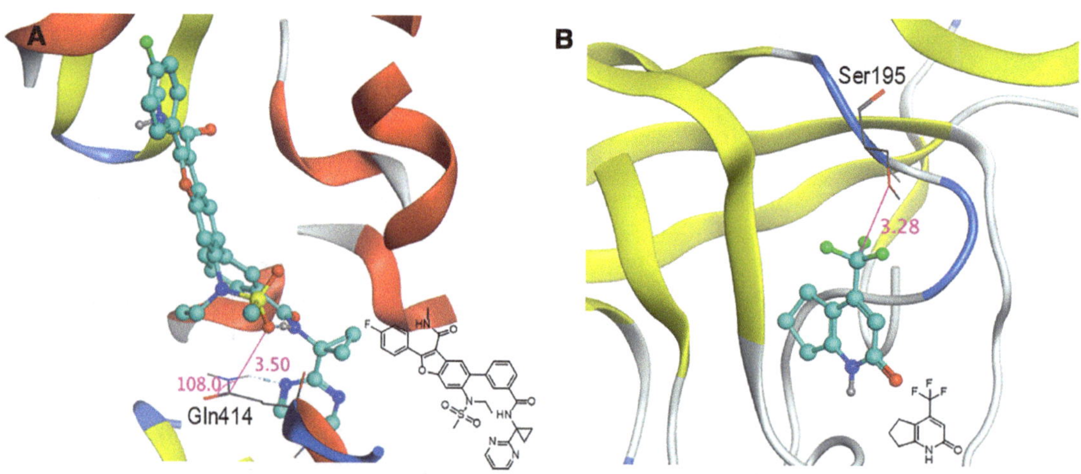

Fig. 9 Bürgi–Dunitz interactions (highlighted in magenta). Distances are reported in angstroms. (**A**) Bürgi–Dunitz interaction between the amide carbonyl group of Glc414 in hepatitis C virus NS5B and a sulfonamide oxygen atom in the inhibitor (PDB ID: 5QJ0), and (**B**) CF_3···O tetrel interaction between the compound and the sidechain oxygen atom of the Ser195 residue in bovine trypsin (PDB ID: 3NK8)

the trifluoromethyl substituent is positioned in a manner enabling a tetrel interaction with the sidechain oxygen atom in Ser195 (PDB ID: 3NK8) [60, 62].

6 Interactions Involving Sulfur: Chalcogen Bonds

In the context of protein–ligand interactions, sulfur participates in the following noncovalent interactions:

- HB: Because of its low electronegativity and a moderate HBD, sulfur is considered a weak hydrogen bond acceptor [63]. Notably, HB interactions have been exploited in replacing the kinase hinge binder with patterns using sulfur as the hydrogen bond acceptor [64, 65];
- S⋯π interactions: as the sulfur atom is approximately 3–4 Å directly above the aromatic ring, its lone electron pairs interact with the aromatic system. The *S*–π interaction is of interest in both the folding and structural stabilization of macromolecules such as proteins. Although there are examples of pharmaceutically relevant compounds developing these interactions (as shown in a detailed review by Meyer et al. [66]), their direct impact on binding affinity is generally modest;
- Chalcogen bonds: Divalent sulfur atoms possess low-lying σ* (antibonding) orbitals, which are the foundation of the positive potential (σ-hole) formations that can interact with electron-donating LBs, such as oxygen, nitrogen, and sulfur atoms. (see the section on Halogen Bonding). IsoStar analysis (Fig. 10B) showed that the C–S⋯O/N/S angle was between 150° and 180°, which is considered an optimal chalcogen bond angle range [67]. Comparable patterns are observed in the interactions between amide carbonyl groups and sulfur atoms in proteins (Fig. 10C). Interaction energies for the S–O interaction have been calculated using the M06-2X-D3/6-31++

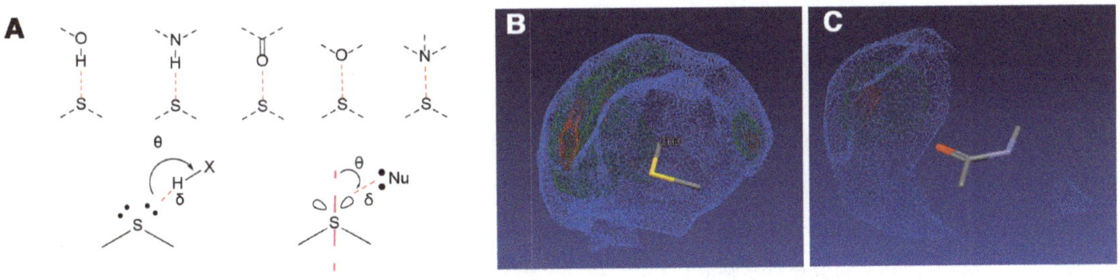

Fig. 10 **(A)** Interactions between sulfur and polarized hydrogen atoms, and the chalcogen bonds formed between sulfur and O or N atoms, **(B)** IsoStar map depicting R–S–R and O=C interactions, and **(C)** IsoStar map showing interactions between the oxygen atom of the amide group and the sulfur atoms in proteins

G** method [68]. The interaction energies associated with aromatic sulfur atoms engaging with oxygen atoms are comparable with those characteristic of weaker hydrogen bonds. Data indicate that these interaction energies are enhanced by the presence of additional electronegative atoms within the aromatic framework. In particular, the interaction energies increase in the order thiophene < thiazole < thiadiazole, implying that the S···O interaction involving aromatic divalent sulfur atoms exhibits strength comparable to that of HB [68, 69]. The frequency of S···N interactions is substantially lower than that of S···O interactions (e.g., 31 vs. 474, for ether and tertiary amine, respectively), likely due to the conjugation of nitrogen atoms in most PDB structures, which reduces the availability of lone pair electrons necessary for chalcogen bond formation [70].

In drug discovery, the importance of chalcogen bonds can be easily assessed by analyzing the SARs of recent cases in drug design. The sulfur atom in the thiazole moiety in the compound shown in Fig. 11A (an inhibitor of human liver glycogen phosphorylase) interacts with His377 O and Asn284 OD1 at distances of 3.2 and 3.4 Å, respectively [71]. In the same series, another analog where the sulfur atom in the thiazole moiety is shifted by one position, can only interact with Asn284 at 3.62 Å (Fig. 11B). The inhibition constants for the first and second compounds are $K_i = 43.6$ and $78.0\,\mu$ M, respectively. The higher potency of the first compound compared with that of the second compound is attributable to the ability of the former to develop a second chalcogen bond interaction [71].

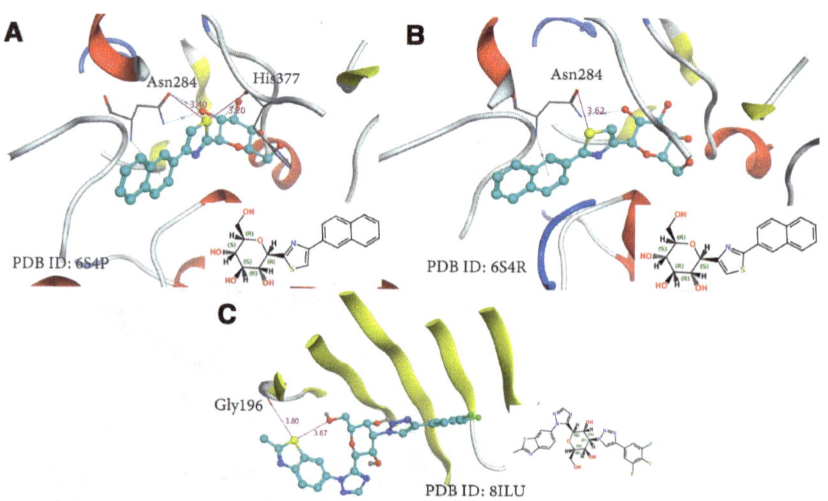

Fig. 11 Chalcogen bonds formed between sulfur atoms and C=O groups (highlighted in magenta). Distances are reported in angstroms. (**A**) Glycogen phosphorylase complexed with a 4-naphthalen-2-yl-1,3-thiazol-2-yl inhibitor, (**B**) Glycogen phosphorylase complexed with a 2-naphthalen-2-yl-1,3-thiazol-4-yl inhibitor,and (**C**) galectin-3 complexed with a benzothiazole-containing inhibitor

Additionally, the authors pursued galectin-3 inhibitors by exploring the potential of replacing a chlorine atom in a previous chemical series by a sulfur atom to generate a similar interaction [72]. The compound shown in Fig. 11C contains a benzothiazole moiety instead of an m-chlorophenyl moiety, its matched molecular pair [72]. The sulfur atom interacts with the backbone carbonyl of Gly196 and forms an intramolecular bond with the hydroxyl group in the pyranose ring. The effect of the substitution becomes apparent when comparing the IC_{50} values of both analogs. The m-Cl phenyl analog displays human and mouse galectin-3 IC_{50} values of 111 and 5670 nM, respectively, while the benzothiazole moiety shows corresponding values of 37 and 939 nM [72]. The sulfur-mediated chalcogen bonds, both intermolecularly and intramolecularly, help increase binding affinity and lock the hydroxyl group into its bioactive conformation, which, in turn, is involved in additional interactions with the protein. This intramolecular bond exemplifies the underappreciated role of chalcogen interactions in locking the bioactive conformations of inhibitors [73]. For sulfur specifically, an intramolecular interaction is energetically more favorable than the intermolecular interactions more frequently associated with XB [74].

7 Cation–π and Anion–π Interactions

Cation–π and anion–π interactions have similar characteristics, with two main terms contributing to the interaction energy: electrostatic and ion-induced polarization. The electrostatic term originates from the electric field generated by the cation (or anion), which polarizes the π–electron cloud of the aromatic sidechain [75]. This polarization term is always favorable and independent of the charge sign of the ion. The ion-induced polarization term relates to the electrostatically induced dipole moment of the aromatic ring, which interacts with the polarizing positive (or negative) charge [76]. For anion–π interactions, the negative quadrupole moment of the ring can be turned into a positive one by attaching electron-withdrawing substituents, transforming an unfavorable anion–π interaction into a favorable one [75].

Cation–π interactions are frequently observed on protein surfaces (particularly in membrane proteins), suggesting that they are more resilient to solvent-screening effects than salt bridges, which are typically buried or located at the interface between two proteins or domains of a protein [76]. Compared to anions, cations are smaller and have greater polarizing power, enabling them to approach the π-cloud of the aromatic ring more closely. Therefore, cation–π interactions occur over shorter distances than anion–π interactions (Fig. 12A) [75, 76]. This observation is reinforced by calculations revealing that when the ion moves away from the ring centroid, the binding energy reduction is more pronounced for the cation–π complex than for the anion–π complex [77, 78]. IsoStar contours in Fig. 12D–F illustrate this behavior. In Fig. 12D, the aromatic rings and basic amine are in close proximity and very well defined (red dots represent high-density regions for a specific interaction). The same scenario applies to the interaction between a phenyl ring and a guanidinium cation (Arg). For interactions between a carboxylate group and a phenyl ring, the distance is much longer (Fig. 12F). The distance and angular dependence of cation–π interactions have been analyzed using DLPNO-CCSD(T) with ORCA calculations and interpreted using symmetry-adapted perturbation theory (SAPT) [79]. Interaction distances typically range from 3 to 5 Å (for arginine–arene complexes, the distance is longer) [79]. Interplanar angles between the positively charged (e.g., guanidium) moiety and aromatic group are predominantly clustered below 30°. To determine the strength of these bonds, binding energies were calculated using the diffuse basis set def2-qzvppd (excluding hydrogen atoms) for revDSD-PBE86-D4 after briefly validating the calculated binding energies at the LNO-CCSD(T) level of calculations [80]. At best,

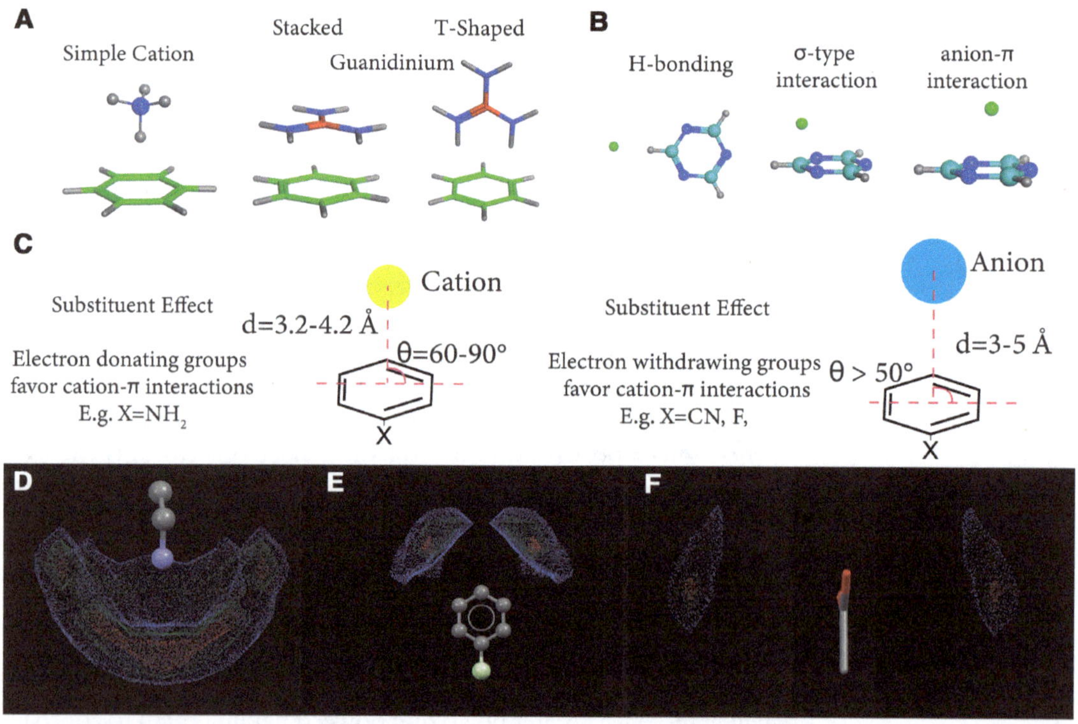

Fig. 12 (**A**) Different directionalities and approaches of cations to a π-system. (**B**) Depending on the approach of an anion toward a π-system, the interaction can be considered as H-bonding, σ-type, or anion–π interactions; (**C**) distances, angles, and substituent effects in cation–π and anion–π interactions; IsoStar contour for (**D**) charged amine group and aromatic residues, (**E**) phenyl ring and guanidinium cation, and (**F**) carboxylic acid and any aromatic residue

the binding energy reached −4.7 kcal mol^{-1} for the T-shaped interaction of an arginine moiety with an aromatic ring [80]. Regarding the directionality of the interaction, *ab initio* studies have suggested that cation–π interactions are stronger above the center of the ring, whereas anion–π interactions exhibit relatively uniform strength over the entire π-system (Fig. 12B, C).

The substituent effect plays important roles in both cation–π and anion–π interactions [81]. The effects of substituents on cation–π interactions have been analyzed in depth by Dougherty et al. [82], Wheeler and Houk [83], and other groups [84, 85]. Although the exact mechanism of the substituent effect remains controversial, researchers agree that the substituent effect can be explained by inductive, resonance, and through-space effects [82–86]. A good rule of thumb for all three physical explanations is that electron-withdrawing groups weaken cation–π interactions while strengthening anion–π interactions [75].

In drug design, cation–π interactions are one of the most ubiquitous interactions because aromatic rings are a universal feature. An analysis of the published output from the medicinal chemistry departments of large pharmaceutical companies revealed that more than 90% of as-synthesized biologically active compounds contain at least one aromatic ring [87]. In 2019, a similar analysis for US Food and Drug Administration (US FDA)-approved drugs showed that >66% of them contained at least one aromatic ring [88]. The importance of cation–π interactions is also evident in the databases developed for their in-depth study and for guiding molecular design in drug discovery [89]. One of the foremost examples of the use of cation–π interactions in medicinal chemistry is the structural mechanism for inhibiting hydroxy-3-methylglutaryl coenzyme A (HMG-CoA) reductase by statins, such as fluvastatin, simvastatin, and cerivastatin. In these drugs, the aromatic ring (fluorophenyl moiety) interacts with the guanidinium group in the Arg590 residue, enhancing the drugs' binding affinities (Fig. 13A) [90, 91]. More recent examples of interactions between a ligand's π-system and a positively charged residue in a protein include a nanomolar inhibitor of the galectin-8 *N*-terminal domain, where the cation–π interaction is non-canonical because it forms between an arginine residue (Arg45 in the protein) and an alkyne group in the ligand (Fig. 13B) [92].

Additionally, at physiological pH, some ligands, such as small-molecule neurotransmitters, including acetylcholine (ACh), gamma aminobutyric acid (GABA), 5-hydroxytryptamine (5-HT), glycine, and glutamate [93, 94], contain cationic groups. Figure 13C, D shows the binding modes of GABA and L-dopamine with their receptors. The positively charged moiety interacts with multiple aromatic residues. This basic amine remains unchanged in multiple pharmaceuticals targeting the 5-HT3 receptor, such as ondansetron, granisetron, and palonosetron (Fig. 13E) [93, 95, 96]. Usually, the

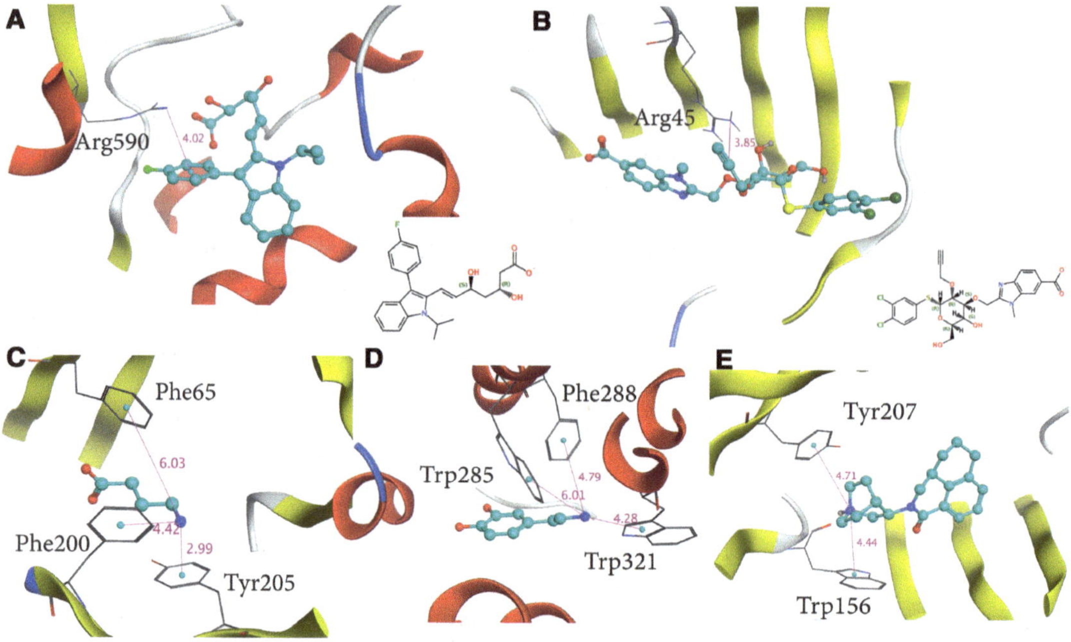

Fig. 13 Cation–π interactions are highlighted in magenta. Distances are measured in Å. Binding modes of (**A**) fluvastatin in human HMG-CoA (PDB ID: [90]), (**B**) inhibitor of the galectin-8 *N*-terminal domain (PDB ID: [92]), (**C**) GABA complexed with GABA-A receptor (PDB ID: [98]), (**D**) L-dopamine in dopamine receptor 1 (PDB ID: [99]), and (**E**) Palonosetron in the mouse serotonin 5HT3 receptor (PDB ID: [96])

charged motif is in an "aromatic box," where the binding site contains multiple aromatic amino acids [97].

More recently, compared with cation–π interactions, anion–π interactions have been attracting increased attention in drug design. Anion–π bonds are counterintuitive because π-systems, such as aromatic rings, are notably electron-rich [100]. Therefore, if any interactions occurred, anions and π-systems would repel one other. However, the analysis of ribonucleic acid (RNA) molecules, and especially the interactions between the ribophosphate backbone and nucleobases, has shown that anion–π interactions are important in both nucleic acids and drug-like molecules [77]. Noncovalent bonds between phosphates and nucleobases are a common feature in various RNA structural motifs, especially those involving guanine or uracil bases (Fig. 14A). In drug design, anion–π interactions have been especially important in elucidating the inhibition of malate synthase (GlcB) by phenyl-diketo acids (PDKAs) [101]. Notably, all the crystal structures of GlcB inhibitor complexes with PDKA analogs maintain close contact between the Asp633 residue's sidechain and inhibitor's aryl ring. Recently, this observation was further investigated [102]. The authors used the M06-2X density functional and 6-31+G(d) basis set and added a fine DFT integration grid to account for the sensitivity of the M06-2X functional to the integration grid's density

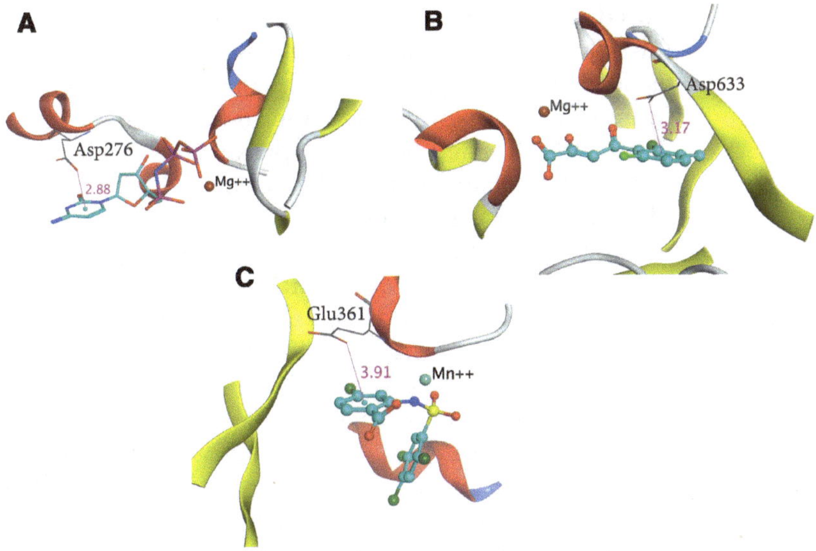

Fig. 14 Anion–π interactions (highlighted in magenta). Distances are reported in angstroms. Binding modes of (**A**) cytidine triphosphate in human DNA polymerase beta (PDB ID: 4MFC [105]); (**B**) potent inhibitor at the malate synthase site in *Mycobacterium tuberculosis* (PDB ID: 3SB0 [101]); and (**C**) inhibitor complexed with non-nucleoside thumb pocket 2 HCV NS5B polymerase (PDB ID: 4J04 [103])

[102]. By assigning the right importance to the anion–π interaction, the authors built a predictive model that accurately replicated the orientation of the Asp633 residue while handling the cooperativity and substituent effects of noncovalent interactions (Fig. 14B).

Another interesting example is a series of non-nucleoside thumb pocket 2 hepatitis C virus (HCV) NS5B polymerase inhibitors. A series of 5-bromo- and 4-chloroanthranilic acid sulfonamides was designed, where the central phenyl ring was substituted with different electron-withdrawing groups. Replacement of the phenyl group with a difluoro-bromo-phenyl ring resulted in a tenfold enhancement in the activity ($IC_{50} = 43$ versus 4.5 μM) [103]. In the crystal structure, this aromatic moiety interacted with the Glu361 sidechain at 3.9 Å, revealing the importance of the aromatic ring and the impact of electron-withdrawing groups (Fig. 14C). When bonding external ligands to macromolecules, cation-induced electron density relocation has a special importance [104].

8 X–H···π Interactions

X–H···π interactions, where X = O, N, or C (Fig. 15A), are sometimes referred to as weak hydrogen bonds and can be characterized as a noncovalent interaction between a hydrogen-containing group (hydrogen donor) and an aromatic system (hydrogen acceptor).

Residues that can participate in this interaction include tryptophan, histidine, phenylalanine, and tyrosine (as acceptors) and leucine, isoleucine, valine, alanine, and proline (as donors). These interactions mainly arise from dispersion and electrostatic attraction and are frequently classified as weak (approximately 1–4 kcal mol^{-1}, depending on the polarizability of the donor and acceptor groups and the donor–acceptor geometry) [1]. Despite the C–H···π interaction being the weakest of the three, the CSD indicates that the C–H···π interaction is the most prevalent [106–108]. Although weaker than classical hydrogen bonds, X–H···π interactions should not be overlooked because of their contributions to molecular recognition, binding affinity, and protein folding [106, 107, 109]. Additionally, the cooperativity of these individual weak noncovalent interactions can result in a greater overall contribution, making them cumulatively important. Reportedly, this cooperativity can be either positive or negative, depending on the directionality [106, 109–111]. Because of the highlighted importance and prevalence of C,N,O–H···π interactions, numerous theoretical (DFT, Møller–Plesset perturbation theory (MP2), Hartree–Fock (HF), and molecular dynamics (MD) and experimental (nuclear magnetic resonance (NMR) spectroscopy and structural biology) studies dedicated to the characterization of these interactions have been published [106, 107, 109–114]. Often, model systems, such as benzene–water/ammonia/methylated compounds to represent O,N,C–H···π interactions, respectively, are used to study these interactions in isolation [108].

In protein–ligand complexes, C–H···π interactions have been described as beneficial and stabilizing. A C–H···π interaction can occur when a CH group is orthogonal to a π-system less than 3.5 Å away (Fig. 15B) [107, 112–115]. C–H···π interactions are particularly ubiquitous in protein–carbohydrate binding sites, as highlighted

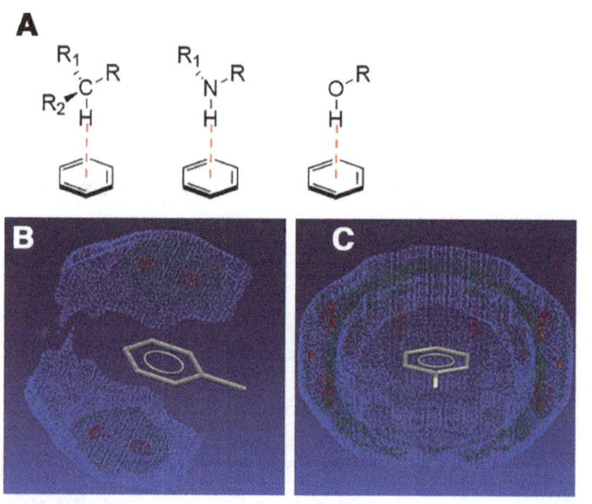

Fig. 15 (**A**) Interaction between C,N,O–H and a π-system represented by a phenyl ring; IsoStar maps of (**B**) R–CH_2–Phe interaction with R–CH and (**C**) Phe interaction with N/O–H

by some recent examples shown in Fig. 16. In the first example, the binding of lactose to human galectin-3 strongly depends on the presence of several C–H···π interactions with the indole moiety of the Trp181 (Fig. 16A, PDB ID: 3ZSJ) [110]. In the second example, the methyl group in α-methyl mannose generates a C–H··· π interaction with the sidechain in the Tyr729 in C-type carbohydrate recognition domain 4 (CRD4) (Fig. 16B, PDB ID: 7JUD) [116].

Similar to the C–H···π interaction, the N–H···π interaction occurs in several systems where the hydrogen atom in an amino or an amide group is above or below a π-system less than 3.5 Å away (Fig. 15C) [115, 117]. Figure 16C shows one of the very few examples of an N–H···π interaction in a protein–ligand complex exemplified in literature. In this example, an N–H···π interaction occurs between the benzimidazole ligand and the backbone of the Asp148 in human Chk1 kinase [118]. Although beyond the scope of this chapter, several examples of intramolecular amide N–H···π interactions are known, such as the N–H···π interaction between the sidechain amide in the Asn37 residue and the phenol ring in the Tyr33 in the third

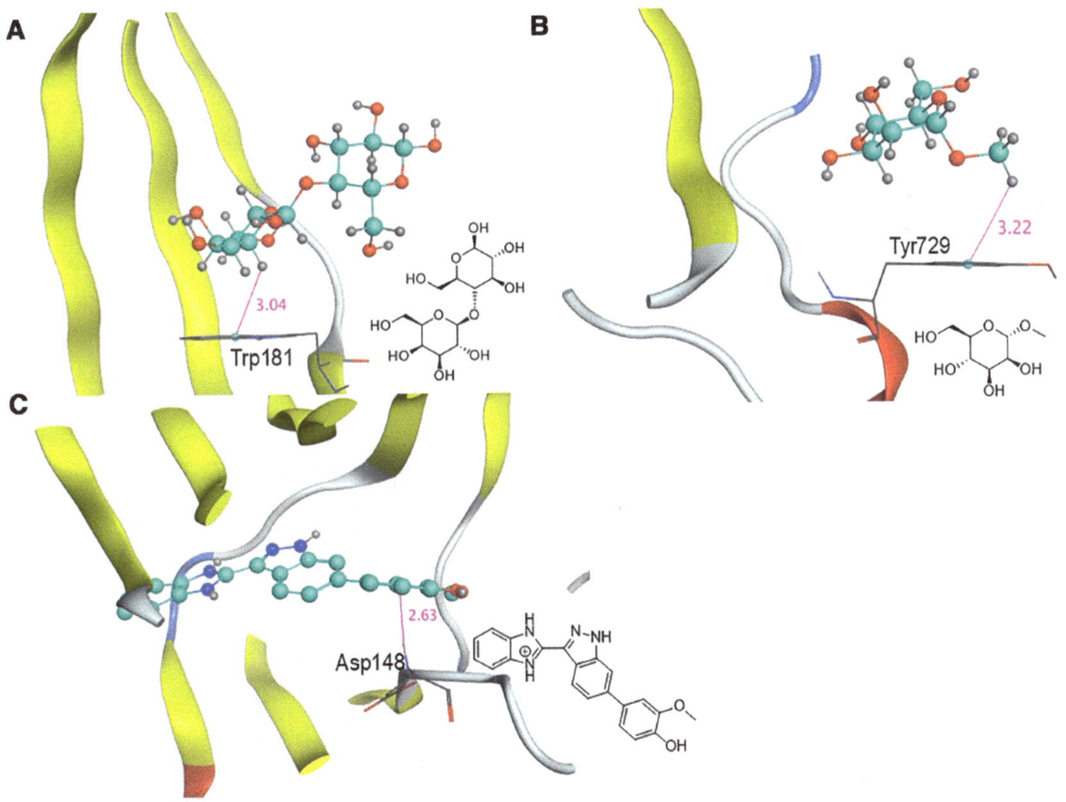

Fig. 16 X–H··· π interactions are highlighted in magenta. Distances are measured in angstroms. (**A**) C–H···π interaction between lactose and human galectin-3 (PDB ID: 3ZSJ), (**B**) C–H··· π interaction between α-methyl mannose and tyrosine in C-type carbohydrate recognition domain 4 (PDB ID: 7JUD), and (**C**) N–H··· π interaction between a ligand and human Chk1 kinase (PDB ID: 2C3K)

IgG-binding domain of Protein G (GB3) (PDB ID: 2OED) [119]. In another example, an apparent N–H···π interaction occurs in a collagen-like peptide containing glutamine–phenylalanine pairs (PDB ID: 7VEG) [117]. In both examples, although both the interaction distances and strengths vary, the axial geometry is quite apparent, despite the CSD statistics (Fig. 15C).

Perhaps the most elusive of the three, the O–H···π interaction has rarely been reported, which could be partially attributed to the fact that many O–H bonds are partially deprotonated in physiological conditions. Although both O–H and N–H can theoretically be stronger than the CH···π interaction because the C–H bond is more polarizable, the energy gain is potentially overcompensated by a substantially higher desolvation penalty [1, 108, 115]. Additionally, the lack of a clearly predominant geometry between the O–H bond and π-orbital contributes to the lack of O–H···π interaction examples in protein–ligand systems (Fig. 15C) [120].

9 Amide···π Stacking Interactions or Nontraditional Stacking

Amide···π stacking, a variation of π···π stacking, is the sixth most prevalent noncovalent interaction in the PDB (Fig. 17A) [2]. As its name indicates, amide···π stacking occurs between an amide and the π-electron cloud of an aromatic ring, such as amino acid sidechains in tryptophan, phenylalanine, or tyrosine (Fig. 17B). Because amide functionalities are abundant in proteins and peptides, as part of backbones and sidechains of glutamine and asparagine, and are one of the most common functional groups in many biologically active molecules and materials, this interaction cannot be overlooked [121, 122]. Diederich et al. mined the PDB and used quantum chemical calculations to study the energetics and predominant geometries

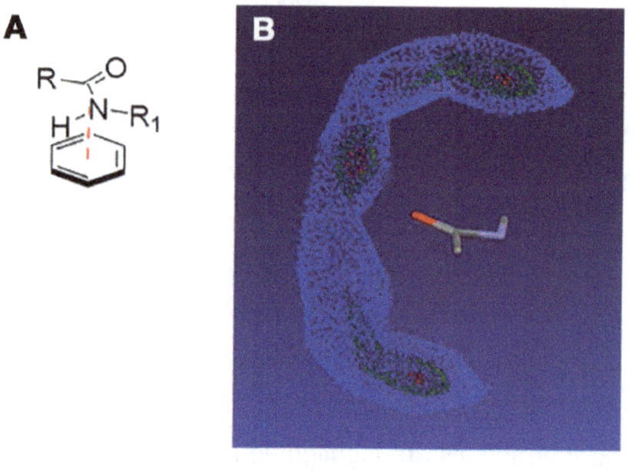

Fig. 17 (**A**) Interaction between an amide and a π-system represented by a phenyl ring, and (**B**) IsoStar map of the backbone amide's interaction with phenylalanine

of heteroarenes relevant in medicinal chemistry and a model amide (*N*-methylacetamide). In their extensive computational study using the spin-component-scaled second-order Møller–Plesset (SCS-MP2) ab initio method, the authors demonstrated that (1) offset (antiparallel) stacking, where dipole moments of an amide and an arene (represented by a pyridine moiety) are approximately opposite, is thermodynamically favorable by ~2 kcal mol^{-1}. This trend was also confirmed in the PDB analysis; (2) the optimal distance between *N*-methylacetamide and pyridine is within 3.2–3.8 Å, with the global minimum being at 3.4 Å; (3) strengthening the dipole moment is linearly correlated with the interaction-derived energy gain, with energies ranging from −4.1 to −1.3 kcal mol^{-1}; (4) decreasing the π-electron density of the arene is energetically favorable [123]. In a follow-up study by Wheeler et al., the DLPNO-CCSD(T)/cc-pVQZ//ωB97X-D/def2-TZVP level of theory was used for *N*-methylacetamide dimers containing 28 heterocycles common in pharmaceuticals, mostly confirming the findings previously reported by Diederich et al., with the caveat that some interactions, such as N···H_3C, can override the molecular dipole effect [124]. These principles were further confirmed in an experimental study using reversible inhibitors bound to one of the two bacterial serine hydrolase CTX-M-14/27 isoforms. In this study, one phenyl and eighteen heteroarene analogs of a known CTX-M inhibitor were synthesized. It is notable that the measured K_i values revealed an approximately three to sixfold improved activity for heteroarene, confirming that enhancing dipole–dipole interactions produces strong stacking. The X-ray structures of select complexes confirmed the predicted binding mode. In the structures, amide···π stacking interactions between the protein backbone (Asp240–Gly238 bond) and heteroarene were one of the key interactions, with distances ranging between 3.7 and 3.9 Å (Fig. 18A, PDB ID: 6OOK) [125]. A similar example was reported for heteroarene analogs of triazine ligands bound to cysteine protease human cathepsin L (hCatL) at the entrance of

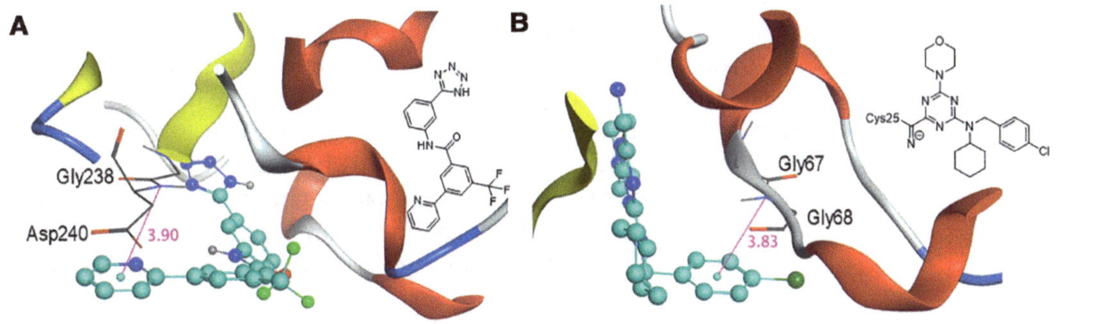

Fig. 18 Amide···π interactions are highlighted in magenta. Distances are measured in angstroms. (**A**) A 2-pyridine analog bound to bacterial serine hydrolase CTX-M-14 (PDB ID: 6OOK),and (**B**) a triazine nitrile ligand bound to cysteine protease human cathepsin L (PDB ID: 4AXM)

the S3 pocket. In this study, heteroarene analogs showed improved potencies compared to those of phenyl analogs because of improved amide⋯π stacking with the Gly67–Gly68 peptide amide bond (Fig. 18B, PDB ID: 4AXM) [126].

10 Conclusions

Unusual noncovalent interactions, such as those discussed in this chapter, play essential roles in molecular recognition, protein folding, and supramolecular assembly [127]. These interactions can often be overlooked by methods such as docking, which is widely utilized in structure-based drug design and virtual screening because of its ability to quickly and cost-effectively identify and score binding poses with common types of scoring functions, including physics-based functions (e.g., GoldScore), knowledge-based statistical potentials (e.g., DrugScore), empirical functions (e.g., GlideScore), or machine-learned interatomic potentials (e.g., AtomNet) [128–130]. Although most scoring functions can reliably detect stronger noncovalent interactions, such as classical hydrogen bonding, they insufficiently account for contributions from polarization and thus inaccurately score predicted binding poses. Consequently, the key binding pose and interactions may be misidentified, compromising SAR interpretation and structure-based drug design and therefore prolonging design–synthesize–test–analyze cycles [131].

By contrast, QM methods offer a more accurate yet more computationally demanding way of evaluating these interactions and their contributions. Because of the higher computational cost, QM methods are not suitable for high-throughput workflows and typically require simplified model systems. Both wavefunction theory (WFT), such as Møller–Plesset perturbation theory (MP2) or coupled-cluster singles and doubles CCSD(T), and dispersion-corrected DFT are suitable for estimating the energetics of weak noncovalent interactions [132–134]. Although CCSD(T) is often referred to as the gold standard in computational chemistry, particularly for describing weak noncovalent interactions, it is best suited for benchmarking studies of smaller systems, given the much higher computational cost compared to that of DFT: even a small increase in the number of atoms drastically increases the CPU time [133, 135, 136]. Although DFT is more practical than other methods because of its lower scaling degree, its performance predominantly depends on the choice of the functional, system type, and inclusion of dispersion corrections. Some DFT functionals exhibiting performances comparable to that of WFT, such as MP2, are Head–Gordon functionals, such as ω B97X-D [137]. Additionally, the choice of a basis is critical for accurately describing weak interactions. The use of a basis with diffuse and polarization functions and correlation-consistent sets for WFT is especially important as, compared to DFT, it is more sensitive to the size of the basis set. Some excellent options are correlation-consistent polarized double-zeta (cc-pVDZ) for WFT and second-generation Ahlrichs' triple-zeta with valence and polarization (def2-TZVP) for DFT [133, 136, 138–140]. These computational chemistry methods can be well complemented by informatics analyses, such as surveys and analyses of relevant structures in PDB or CSD [131].

Collectively, these less-common interactions affect both biological and chemical systems and can be used in rational drug design to fine-tune ligand–protein binding and affinity. Although the examples showcased in this chapter often improved potency, most ligands were not originally designed to form these interactions. More high-throughput methods, such as docking, might overlook these interactions, resulting in the misinterpretation of SARs. Although computationally intensive physics-based methods might reveal these interactions and elucidate their energetics, systems may have to be simplified to reduce the computational cost. Improved characterization of these atypical and underappreciated interactions will also benefit computational structure prediction.

Finally, as deep-learning-based co-folding methods, such as Boltz-2 [141], continue to model protein–ligand and protein–protein complexes, the incorporation of more accurate representations of weak, geometrically constrained noncovalent contacts, and their enthalpic contributions, should increase the accuracy and predictive power of generated complexes.

References

1. Bissantz C, Kuhn B, Stahl M (2010) A medicinal chemist's guide to molecular interactions. J Med Chem 53(14):5061–5084. https://doi.org/10.1021/jm100112j
2. Ferreira de Freitas R, Schapira M (2017) A systematic analysis of atomic protein-ligand interactions in the PDB. MedChemComm 8(10):1970–1981. https://doi.org/10.1039/c7md00381a
3. Kuhn B, Gilberg E, Taylor R, Cole J, Korb O (2019) How significant are unusual protein-ligand interactions? Insights from database mining. J Med Chem 62(22):10441–10455. https://doi.org/10.1021/acs.jmedchem.9b01545
4. Kenny PW, Montanari CA, Prokopczyk IM, Ribeiro JFR, Sartori GR (2016) Hydrogen bond basicity prediction for medicinal chemistry design. J Med Chem 59(9):4278–4288. https://doi.org/10.1021/acs.jmedchem.5b01946
5. Laurence C, Brameld KA, Graton J, Le Questel J-Y, Renault E (2009) The pKBHX database: toward a better understanding of hydrogen-bond basicity for medicinal chemists. J Med Chem 52(14):4073–4086. https://doi.org/10.1021/jm801331y
6. Lee EC, Hong BH, Lee JY, Kim JC, Kim D, Kim Y, Tarakeshwar P, Kim KS (2005) Substituent effects on the edge-to-face aromatic interactions. J Am Chem Soc 127(12):4530–4537. https://doi.org/10.1021/ja037454r
7. Meyer EA, Castellano RK, Diederich F (2003) Interactions with aromatic rings in chemical and biological recognition. Angew Chem Int Ed 42(11):1210–1250. https://doi.org/10.1002/anie.200390319
8. Carter-Fenk K, Liu M, Pujal L, Loipersberger M, Tsanai M, Vernon RM, Forman-Kay JD, Head-Gordon M, Heidar-Zadeh F, Head-Gordon T (2023) The energetic origins of Pi–Pi contacts in proteins. J Am Chem Soc 145(45):24836–24851. https://doi.org/10.1021/jacs.3c09198
9. Kurczab R, Śliwa P, Rataj K, Kafel R, Bojarski AJ (2018) Salt bridge in ligand–protein complexes—systematic theoretical and statistical investigations. J Chem Inf

Model 58(11):2224–2238. https://doi.org/10.1021/acs.jcim.8b00266

10. Bruno IJ, Cole JC, Lommerse JP, Rowland RS, Taylor R, Verdonk ML (1997) IsoStar: a library of information about nonbonded interactions. J Comput Aided Mol Des 11(6):525–537. https://doi.org/10.1023/a:1007934413448
11. Derewenda ZS (2023) C-H groups as donors in hydrogen bonds: a historical overview and occurrence in proteins and nucleic acids. Int J Mol Sci 24(17):13165. https://doi.org/10.3390/ijms241713165
12. Derewenda ZS, Hawro I, Derewenda U (2020) C-H…O hydrogen bonds in kinase-inhibitor interfaces. IUBMB Life 72(6):1233–1242. https://doi.org/10.1002/iub.2282
13. Pierce AC, Sandretto KL, Bemis GW (2002) Kinase inhibitors and the case for CH…O hydrogen bonds in protein-ligand binding. Proteins 49(4):567–576. https://doi.org/10.1002/prot.10259
14. Ouvry G, Clary L, Tomas L, Aurelly M, Bonnary L, Borde E, Bouix-Peter C, Chantalat L, Defoin-Platel C, Deret S, Forissier M, Harris CS, Isabet T, Lamy L, Luzy AP, Pascau J, Soulet C, Taddei A, Taquet N, Thoreau E, Varvier E, Vial E, Hennequin LF (2019) Impact of minor structural modifications on properties of a series of mTOR inhibitors. ACS Med Chem Lett 10(11):1561–1567. https://doi.org/10.1021/acsmedchemlett.9b00401
15. Huestis MP, Durk MR, Eigenbrot C, Gibbons P, Hunsaker TL, La H, Leung DH, Liu W, Malek S, Merchant M, Moffat JG, Muli CS, Orr CJ, Parr BT, Shanahan F, Sneeringer CJ, Wang W, Yen I, Yin J, Rudolph J, Siu M (2021) Targeting KRAS mutant cancers via combination treatment: discovery of a pyridopyridazinone pan-RAF kinase inhibitor. ACS Med Chem Lett 12(5):791–797. https://doi.org/10.1021/acsmedchemlett.1c00063
16. Sessler CD, Rahm M, Becker S, Goldberg JM, Wang F, Lippard SJ (2017) CF_2H, a hydrogen bond donor. J Am Chem Soc 139(27):9325–9332. https://doi.org/10.1021/jacs.7b04457
17. Zafrani Y, Sod-Moriah G, Yeffet D, Berliner A, Amir D, Marciano D, Elias S, Katalan S, Ashkenazi N, Madmon M, Gershonov E, Saphier S (2019) CF_2H, a functional group-dependent hydrogen-bond donor: is it a more or less lipophilic bioisostere of OH, SH, and CH_3?. J Med Chem 62(11):5628–5637. https://doi.org/10.1021/acs.jmedchem.9b00604
18. Eggert E, Hillig RC, Koehr S, Stockigt D, Weiske J, Barak N, Mowat J, Brumby T, Christ CD, Ter Laak A, Lang T, Fernandez-Montalvan AE, Badock V, Weinmann H, Hartung IV, Barsyte-Lovejoy D, Szewczyk M, Kennedy S, Li F, Vedadi M, Brown PJ, Santhakumar V, Arrowsmith CH, Stellfeld T, Stresemann C (2016) Discovery and characterization of a highly potent and selective aminopyrazoline-based in vivo probe (BAY-598) for the protein lysine methyltransferase SMYD2. J Med Chem 59(10):4578–4600. https://doi.org/10.1021/acs.jmedchem.5b01890
19. Degorce SL, Anjum R, Dillman KS, Drew L, Groombridge SD, Halsall CT, Lenz EM, Lindsay NA, Mayo MF, Pink JH, Robb GR, Scott JS, Stokes S, Xue Y (2018) Optimization of permeability in a series of pyrrolotriazine inhibitors of IRAK4. Bioorg Med Chem 26(4):913–924. https://doi.org/10.1016/j.bmc.2018.01.008
20. Goldberg FW, Daunt P, Pearson SE, Greenwood R, Grist M, Debreczeni JÉ (2016) Identification and optimisation of a series of N-(4-anilino-2-pyridyl)acetamide activin receptor-like kinase 1 (ALK1) inhibitors. MedChemComm 7(6):1204–1208. https://doi.org/10.1039/c6md00039h
21. Sivickyte O, Costa PJ (2023) Intrinsic bond strength index as a halogen bond interaction energy predictor. Phys Chem Chem Phys 25(26):17535–17546. https://doi.org/10.1039/d2cp04786a
22. Nunes RS, Vila-Vicosa D, Costa PJ (2021) Halogen bonding: an underestimated player in membrane-ligand interactions. J Am Chem Soc 143(11):4253–4267. https://doi.org/10.1021/jacs.0c12470
23. Shinada NK, de Brevern AG, Schmidtke P (2019) Halogens in protein-ligand binding mechanism: a structural perspective. J Med Chem 62(21):9341–9356. https://doi.org/10.1021/acs.jmedchem.8b01453
24. Lu Y, Wang Y, Xu Z, Yan X, Luo X, Jiang H, Zhu W (2009) C–X···H contacts in biomolecular systems: how they contribute to protein–ligand binding affinity. J Phys Chem B 113(37):12615–12621. https://doi.org/10.1021/jp906352e
25. Engelhardt MU, Zimmermann MO, Dammann M, Stahlecker J, Poso A, Kronenberger T, Kunick C, Stehle T, Boeckler FM (2024) Halogen bonding on water horizontal line a drop in the ocean?. J Chem Theory Comput 20(23):10716–10730. https://doi.org/10.1021/acs.jctc.4c00834
26. Wilcken R, Zimmermann MO, Lange A, Joerger AC, Boeckler FM (2013) Principles and applications of halogen bonding in medicinal chemistry and chemical biology. J Med Chem 56(4):1363–1388. https://doi.org/10.1021/jm3012068

27. Zhang Q, Xu Z, Shi J, Zhu W (2017) Underestimated halogen bonds forming with protein backbone in protein data bank. J Chem Inf Model 57(7):1529–1534. https://doi.org/10.1021/acs.jcim.7b00235
28. Yang Z, Liu Y, Chen Z, Xu Z, Shi J, Chen K, Zhu W (2015) A quantum mechanics-based halogen bonding scoring function for protein-ligand interactions. J Mol Model 21(6):138. https://doi.org/10.1007/s00894-015-2681-6
29. Ali S, Tian X, Meccia SA, Zhou J (2025) Highlights on U.S. FDA-approved halogen-containing drugs in 2024. Eur J Med Chem 287:11380. https://doi.org/10.1016/j.ejmech.2025.117380
30. Frontera A, Bauza A (2021) Biological halogen bonds in protein-ligand complexes: a combined QTAIM and NCIPlot study in four representative cases. Org Biomol Chem 19(31):6858–6864. https://doi.org/10.1039/d1ob01212f
31. Margiotta E, van der Lubbe SCC, de Azevedo Santos L, Paragi G, Moro S, Bickelhaupt FM, Fonseca Guerra C (2020) Halogen bonds in ligand-protein systems: molecular orbital theory for drug design. J Chem Inf Model 60(3):1317–1328. https://doi.org/10.1021/acs.jcim.9b00946
32. Kuhn B, Gilberg E, Taylor R, Cole J, Korb O (2019) How significant are unusual protein–ligand interactions? Insights from database mining. J Med Chem 62:10441–10455. https://doi.org/10.1021/acs.jmedchem.9b01545
33. Yoshikawa K, Yoshino T, Yokomizo Y, Uoto K, Naito H, Kawakami K, Mochizuki A, Nagata T, Suzuki M, Kanno H, Takemura M, Ohta T (2011) Design, synthesis and SAR of novel ethylenediamine and phenylenediamine derivatives as factor Xa inhibitors. Bioorg Med Chem Lett 21:2133–2140. https://doi.org/10.1016/j.bmcl.2011.01.132
34. Nesbitt NM, Zheng X, Li Z, Manso JA, Yen WY, Malone LE, Ripoll-Rozada J, Pereira PJB, Mantle TJ, Wang J, Bahou WF (2018) In silico and crystallographic studies identify key structural features of biliverdin IX beta reductase inhibitors having nanomolar potency. J Biol Chem 293:5431–5446. https://doi.org/10.1074/jbc.RA118.001803
35. Lange A, Günther M, Büttner FM, Zimmermann MO, Heidrich J, Hennig S, Zahn S, Schall C, Sievers-Engler A, Ansideri F, Koch P, Laemmerhofer M, Stehle T, Laufer SA, Boeckler FM (2015) Targeting the gatekeeper MET146 of C-Jun N-terminal kinase 3 induces a bivalent halogen/chalcogen bond. J Am Chem Soc 137:14640–14652. https://doi.org/10.1021/jacs.5b07090
36. Schiebel J, Chang A, Lu H, Baxter MV, Tonge PJ, Kisker C (2012) *Staphylococcus aureus* FabI: inhibition, substrate recognition, and potential implications for in vivo essentiality. Structure 20:802–813. https://doi.org/10.1016/j.str.2012.03.013
37. Dunitz JD, Taylor R (1997) Organic fluorine hardly ever accepts hydrogen bonds. Chem Eur J 3(1):89–98. https://doi.org/10.1002/chem.19970030115
38. Taylor R (2017) The hydrogen bond between N-H or O-H and organic fluorine: favourable yes, competitive no. Acta Crystallogr B Struct Sci Cryst Eng Mater 73(Pt 3): 474–488. https://doi.org/10.1107/S2052520617005923
39. Pietrus W, Kafel R, Bojarski AJ, Kurczab R (2022) Hydrogen bonds with fluorine in ligand-protein complexesthe PDB analysis and energy calculations. Molecules 27(3): 1005. https://doi.org/10.3390/molecules27031005
40. Metrangolo P, Murray JS, Pilati T, Politzer P, Resnati G, Terraneo G (2011) Fluorine-centered halogen bonding: a factor in recognition phenomena and reactivity. Cryst Growth Des 11(9):4238–4246. https://doi.org/10.1021/cg200888n
41. Eskandari K, Lesani M (2015) Does fluorine participate in halogen bonding. Chemistry 21(12):4739–4746. https://doi.org/10.1002/chem.201405054
42. Wehrhan L, Keller BG (2024) Fluorinated protein-ligand complexes: a computational perspective. J Phys Chem B 128(25):5925–5934. https://doi.org/10.1021/acs.jpcb.4c01493
43. Yu L, He L, Gan B, Ti R, Xiao Q, Hu H, Zhu L, Wang S, Ren R (2022) Structural insights into sphingosine-1-phosphate receptor activation. Proc Natl Acad Sci U S A 119(16):e2117716119. https://doi.org/10.1073/pnas.2117716119
44. Meanwell NA (2018) Fluorine and fluorinated motifs in the design and application of bioisosteres for drug design. J Med Chem 61(14):5822–5880. https://doi.org/10.1021/acs.jmedchem.7b01788
45. Olsen JA, Banner DW, Seiler P, Obst Sander U, D'Arcy A, Stihle M, Muller K, Diederich F (2003) A fluorine scan of thrombin inhibitors to map the fluorophilicity/fluorophobicity of an enzyme active site: evidence for C-F…C=O interactions. Angew Chem Int Ed Engl 42(22):2507–2511. https://doi.org/10.1002/anie.200351268

46. Fensome A, Ambler CM, Arnold E, Banker ME, Brown MF, Chrencik J, Clark JD, Dowty ME, Efremov IV, Flick A, Gerstenberger BS, Gopalsamy A, Hayward MM, Hegen M, Hollingshead BD, Jussif J, Knafels JD, Limburg DC, Lin D, Lin TH, Pierce BS, Saiah E, Sharma R, Symanowicz PT, Telliez JB, Trujillo JI, Vajdos FF, Vincent F, Wan ZK, Xing L, Yang X, Yang X, Zhang L (2018) Dual inhibition of TYK2 and JAK1 for the treatment of autoimmune diseases: discovery of ((S)-2,2-difluorocyclopropyl)((1 R,5 S)-3-(2-((1-methyl-1 H-pyrazol-4-yl)amino) pyrimidin-4-yl)-3,8-diazabicyclo[3.2.1] octan-8-yl)methanone (PF-06700841). J Med Chem 61(19):8597–8612. https://doi.org/10.1021/acs.jmedchem.8b00917
47. Xing L, Keefer C, Brown MF (2017) Fluorine multipolar interaction: toward elucidating its energetics in binding recognition. J Fluorine Chem 198:47–53. https://doi.org/10.1016/j.jfluchem.2016.12.013
48. Bauer MR, Jones RN, Baud MG, Wilcken R, Boeckler FM, Fersht AR, Joerger AC, Spencer J (2016) Harnessing fluorine-sulfur contacts and multipolar interactions for the design of p53 mutant Y220C rescue drugs. ACS Chem Biol 11(8):2265–2274. https://doi.org/10.1021/acschembio.6b00315
49. Kumar R, Ignjatovic MM, Peterson K, Olsson M, Leffler H, Ryde U, Nilsson UJ, Logan DT (2019) Structure and energetics of ligand-fluorine interactions with galectin-3 backbone and side-chain amides: insight into solvation effects and multipolar interactions. ChemMedChem 14(16):1528–1536. https://doi.org/10.1002/cmdc.201900293
50. Boeckler FM, Joerger AC, Jaggi G, Rutherford TJ, Veprintsev DB, Fersht AR (2008) Targeted rescue of a destabilized mutant of p53 by an *in silico* screened drug. Proc Natl Acad Sci 105(30):10360–10365. https://doi.org/10.1073/pnas.0805326105
51. Chiodi D, Ishihara Y (2023) "Magic Chloro": profound effects of the chlorine atom in drug discovery. J Med Chem 66(8):5305–5331. https://doi.org/10.1021/acs.jmedchem.2c02015
52. Paulini R, Muller K, Diederich F (2005) Orthogonal multipolar interactions in structural chemistry and biology. Angew Chem Int Ed Engl 44(12):1788–1805. https://doi.org/10.1002/anie.200462213
53. Bromberg KD, Mitchell TR, Upadhyay AK, Jakob CG, Jhala MA, Comess KM, Lasko LM, Li C, Tuzon CT, Dai Y, Li F, Eram MS, Nuber A, Soni NB, Manaves V, Algire MA, Sweis RF, Torrent M, Schotta G, Sun C, Michaelides MR, Shoemaker AR, Arrowsmith CH, Brown PJ, Santhakumar V, Martin A, Rice JC, Chiang GG, Vedadi M, Barsyte-Lovejoy D, Pappano WN (2017) The SUV4-20 inhibitor A-196 verifies a role for epigenetics in genomic integrity. Nat Chem Biol 13(3):317–324. https://doi.org/10.1038/nchembio.2282
54. Singh SK, Das A (2015) The n →> pi* interaction: a rapidly emerging non-covalent interaction. Phys Chem Chem Phys 17(15):9596–9612. https://doi.org/10.1039/c4cp05536e
55. Robert W, Newberry RTR (2017) The n→π* interaction. Acc Chem Res 50:1838–1846. https://doi.org/10.1021/acs.accounts.7b00121
56. Fufezan C (2010) The role of Buergi-Dunitz interactions in the structural stability of proteins. Proteins 78(13):2831–2838. https://doi.org/10.1002/prot.22800
57. Yeung KS, Beno BR, Mosure K, Zhu J, Grant-Young KA, Parcella K, Anjanappa P, Bora RO, Selvakumar K, Wang YK, Fang H, Krause R, Rigat K, Liu M, Lemm J, Sheriff S, Witmer M, Tredup J, Jardel A, Kish K, Parker D, Haskell R, Santone K, Meanwell NA, Soars MG, Roberts SB, Kadow JF (2018) Structure-property basis for solving transporter-mediated efflux and pan-genotypic inhibition in HCV NS5B inhibitors. ACS Med Chem Lett 9(12):1217–1222. https://doi.org/10.1021/acsmedchemlett.8b00379
58. Murray JS, Riley KE, Politzer P, Clark T (2010) Directional weak intermolecular interactions: r-hole bonding. Aust J Chem 63:1598–1607. https://doi.org/10.1071/CH10259
59. Clark T (2013) σ-Holes. Wiley Interdiscip Rev Comput Mol Sci 3:13–20. https://doi.org/10.1002/wcms.1113
60. Garcia LX, Bauza A, Seth SK, Frontera A (2017) Importance of R-CF(3)...O tetrel bonding interactions in biological systems. J Phys Chem A 121(28):5371–5376. https://doi.org/10.1021/acs.jpca.7b06052
61. de las Nieves Piña M, Sahu AK, Frontera A, Biswal HS, Bauzá A (2023) Tetrel bonds involving a CF3 group participate in protein–drug recognition: a combined crystallographic and computational study. Phys Chem Chem Phys 25:12409–12419. https://doi.org/10.1039/D3CP00839H
62. Vulpetti A, Schiering N, Dalvit C (2010) Combined use of computational chemistry, NMR screening, and X-ray crystallography for identification and characterization of fluorophilic protein environments. Proteins 78:3281–3291. https://doi.org/10.1002/prot.22836
63. Chand A, Sahoo DK, Rana A, Jena S, Biswal HS (2020) The prodigious hydrogen bonds with

sulfur and selenium in molecular assemblies, structural biology, and functional materials. Acc Chem Res 53(8):1580–1592. https://doi.org/10.1021/acs.accounts.0c00289
64. Eduful BJ, O'Byrne SN, Temme L, Asquith CRM, Liang Y, Picado A, Pilotte JR, Hossain MA, Wells CI, Zuercher WJ, Catta-Preta CMC, Zonzini Ramos P, Santiago AS, Counago RM, Langendorf CG, Nay K, Oakhill JS, Pulliam TL, Lin C, Awad D, Willson TM, Frigo DE, Scott JW, Drewry DH (2021) Hinge binder scaffold hopping identifies potent calcium/calmodulin-dependent protein kinase kinase 2 (CAMKK2) inhibitor chemotypes. J Med Chem 64(15):10849–10877. https://doi.org/10.1021/acs.jmedchem.0c02274
65. Hua Z, Bregman H, Buchanan JL, Chakka N, Guzman-Perez A, Gunaydin H, Huang X, Gu Y, Berry V, Liu J, Teffera Y, Huang L, Egge B, Emkey R, Mullady EL, Schneider S, Andrews PS, Acquaviva L, Dovey J, Mishra A, Newcomb J, Saffran D, Serafino R, Strathdee CA, Turci SM, Stanton M, Wilson C, Dimauro EF (2013) Development of novel dual binders as potent, selective, and orally bioavailable tankyrase inhibitors. J Med Chem 56(24):10003–10015. https://doi.org/10.1021/jm401317z
66. Meyer EA, Castellano RK, Diederich F (2003) Interactions with aromatic rings in chemical and biological recognition. Angew Chem 42(11):1210–1250. https://doi.org/10.1002/anie.200390319
67. Kriz K, Fanfrlik J, Lepsik M (2018) Chalcogen bonding in protein-ligand complexes: PDB survey and quantum mechanical calculations. ChemPhysChem 19(19):2540–2548. https://doi.org/10.1002/cphc.201800409
68. Koebel MR, Cooper A, Schmadeke G, Jeon S, Narayan M, Sirimulla S (2016) S...O and S...N sulfur bonding interactions in protein-ligand complexes: empirical considerations and scoring function. J Chem Inf Model 56(12):2298–2309. https://doi.org/10.1021/acs.jcim.6b00236
69. Zhang X, Gong Z, Li J, Lu T (2015) Intermolecular sulfur···oxygen interactions: theoretical and statistical investigations. J Chem Inf Model 55(10):2138–2153. https://doi.org/10.1021/acs.jcim.5b00177
70. Adhav VA, Shelke SS, Balanarayan P, Saikrishnan K (2023) Sulfur-mediated chalcogen versus hydrogen bonds in proteins: a see-saw effect in the conformational space. QRB Discov 4:1–10. https://doi.org/10.1017/qrd.2023.3
71. Kyriakis E, Karra AG, Papaioannou O, Solovou T, Skamnaki VT, Liggri PGV, Zographos SE, Szennyes E, Bokor É, Kun S, Psarra A-MG, Somsák L, Leonidas DD (2020) The architecture of hydrogen and sulfur σ-hole interactions explain differences in the inhibitory potency of C-β-d-glucopyranosyl thiazoles, imidazoles and an N-β-d glucopyranosyl tetrazole for human liver glycogen phosphorylase and offer new insights to structure-based design. Bioorg Med Chem 28(1):115196. https://doi.org/10.1016/j.bmc.2019.115196
72. Liu C, Wang W, Feng J, Beno B, Raja T, Swidorski J, Manepalli RKVLP, Vetrichelvan M, Rao Jalagam P, Nair SK, Gupta A, Panda M, Ghosh K, Kaushikkumar Shukla J, Sale H, Shah D, Singh Gautam S, Patel D, Mathur A, Ellsworth BA, Cheng D, Regueiro-Ren A (2024) Identification of benzothiazole derived monosaccharides as potent, selective, and orally bioavailable inhibitors of human and mouse galectin-3; a rare example of using a S···O binding interaction for drug design. Bioorg Med Chem 101:117638. https://doi.org/10.1016/j.bmc.2024.117638
73. Hudson BM, Nguyen E, Tantillo DJ (2016) The influence of intramolecular sulfur-lone pair interactions on small-molecule drug design and receptor binding. Org Biomol Chem 14(16):3975–3980. https://doi.org/10.1039/c6ob00254d
74. Beno BR, Yeung KS, Bartberger MD, Pennington LD, Meanwell NA (2015) A survey of the role of noncovalent sulfur interactions in drug design. J Med Chem 58(11):4383–4438. https://doi.org/10.1021/jm501853m
75. Frontera A, Quiñonero D, Deyà PM (2011) Cation–π and anion–π interactions. WIREs Comput Mol Sci 1(3):440–459. https://doi.org/10.1002/wcms.14
76. Infield DT, Rasouli A, Galles GD, Chipot C, Tajkhorshid E, Ahern CA (2021) Cation-pi Interactions and their functional roles in membrane proteins. J Mol Biol 433(17):167035. https://doi.org/10.1016/j.jmb.2021.167035
77. Lucas X, Bauza A, Frontera A, Quinonero D (2016) A thorough anion-pi interaction study in biomolecules: on the importance of cooperativity effects. Chem Sci 7(2):1038–1050. https://doi.org/10.1039/c5sc01386k
78. Gamez P (2014) The anion–π interaction: naissance and establishment of a peculiar supramolecular bond. Inorg Chem Front 1(1):35–43. https://doi.org/10.1039/c3qi00055a
79. Kumar K, Woo SM, Siu T, Cortopassi WA, Duarte F, Paton RS (2018) Cation-pi interactions in protein-ligand binding: theory and

data-mining reveal different roles for lysine and arginine. Chem Sci 9(10):2655–2665. https://doi.org/10.1039/c7sc04905f

80. Calinsky R, Levy Y (2024) Aromatic residues in proteins: re-evaluating the geometry and energetics of pi-pi, Cation-pi, and CH-pi Interactions. J Phys Chem B 128(36):8687–8700. https://doi.org/10.1021/acs.jpcb.4c04774
81. Mahadevi AS, Sastry GN (2013) Cation-pi interaction: its role and relevance in chemistry, biology, and material science. Chem Rev 113(3):2100–2138. https://doi.org/10.1021/cr300222d
82. Mecozzi S, West AP, Dougherty DA (1996) Cation−π interactions in simple aromatics: electrostatics provide a predictive tool. J Am Chem Soc 118(9):2307–2308. https://doi.org/10.1021/ja9539608 Cation−π interactions in simple aromatics: electrostatics provide a predictive tool
83. Wheeler SE, Houk KN (2009) Substituent effects in cation/π interactions and electrostatic potentials above the centers of substituted benzenes are due primarily to through-space effects of the substituents. J Am Chem Soc 131(9):126–3127. https://doi.org/10.1080/00268-970802537614
84. Bauza A, Deya PM, Frontera A, Quinonero D (2014) Substituent effects in cation-pi interactions revisited: a general approach based on intrinsic properties of the arenes. Phys Chem Chem Phys 16(4):1322–1326. https://doi.org/10.1039/c3cp54147a
85. Yourdkhani S, Chojecki M, Korona T (2019) Substituent effects in the so-called cationpi interaction of benzene and its boron-nitrogen doped analogues: overlooked role of sigma-skeleton. Phys Chem Chem Phys 21(12):6453–6466. https://doi.org/10.1039/c8cp04962a
86. Sayyed FB, Suresh CH (2011) Substituent effects in cation-pi interactions: a unified view from inductive, resonance, and through-space effects. J Phys Chem A 115(22):5660–5664. https://doi.org/10.1021/jp202441x
87. Roughley SD, Jordan AM (2011) The medicinal chemist's toolbox: an analysis of reactions used in the pursuit of drug candidates. J Med Chem 54(10):3451–3479. https://doi.org/10.1021/jm200187y
88. Das P, Delost MD, Qureshi MH, Smith DT, Njardarson JT (2019) A survey of the structures of US FDA approved combination drugs. J Med Chem 62(9):4265–4311. https://doi.org/10.1021/acs.jmedchem.8b01610
89. Yang JF, Wang F, Wang MY, Wang D, Zhou ZS, Hao GF, Li QX, Yang GF (2023) CIPDB: a biological structure databank for studying cation and pi interactions. Drug Discov Today 28(5):103546. https://doi.org/10.1016/j.drudis.2023.103546
90. Istvan ES, Deisenhofer J (2001) Structural mechanism for statin inhibition of HMG-CoA reductase. Science 292(5519):1160–1164. https://doi.org/10.1126/science.1059344
91. Ahmadi A, Ayoubi-Chianeh M, Kassaee MZ, Fattahi A (2023) Revealing a notable interaction in the binding of type II statins to HMG-CoA reductase: stacked cation–π interaction. J Phys Org Chem 36(9), e4517. https://doi.org/10.1002/poc.4517
92. Puric E, Hassan M, Sjovall F, Tomasic T, Pevec M, Lah J, Forteza JA, Sundin A, Leffler H, Nilsson UJ, Logan DT, Anderluh M (2025) Nanomolar inhibitor of the galectin-8 N-terminal domain binds via a non-canonical cation-pi interaction. Commun Chem 8(1):59. https://doi.org/10.1038/s42004-025-01458-6
93. Zacharias N, Dougherty DA (2002) Cation–π interactions in ligand recognition and catalysis. Trends Pharmacol Sci 23(6):281–287. https://doi.org/10.1016/S0165-6147(02)02027-8
94. Pless SA, Hanek AP, Price KL, Lynch JW, Lester HA, Dougherty DA, Lummis SC (2011) A cation-pi interaction at a phenylalanine residue in the glycine receptor binding site is conserved for different agonists. Mol Pharmacol 79(4):742–748. https://doi.org/10.1124/mol.110.069583
95. Dougherty DA (2025) The cation-pi interaction in chemistry and biology. Chem Rev 125(5):2793–2808. https://doi.org/10.1021/acs.chemrev.4c00707
96. Zarkadas E, Zhang H, Cai W, Effantin G, Perot J, Neyton J, Chipot C, Schoehn G, Dehez F, Nury H (2020) The binding of palonosetron and other antiemetic drugs to the serotonin 5-HT3 receptor. Structure 28(10):1131–1140 e1134. https://doi.org/10.1016/j.str.2020.07.004
97. Davis MR, Dougherty DA (2015) Cation-pi interactions: computational analyses of the aromatic box motif and the fluorination strategy for experimental evaluation. Phys Chem Chem Phys 17(43):29262–29270. https://doi.org/10.1039/c5cp04668h
98. Zhu S, Noviello CM, Teng J, Walsh Jr.RM, Kim JJ, Hibbs RE (2018) Structure of a human synaptic GABA(A) receptor. Nature 559(7712):67–72. https://doi.org/10.1038/s41586-018-0255-3
99. Zhuang Y, Krumm B, Zhang H, Zhou XE, Wang Y, Huang XP, Liu Y, Cheng X, Jiang Y, Jiang H, Zhang C, Yi W, Roth BL, Zhang Y, Xu HE (2021) Mechanism of

dopamine binding and allosteric modulation of the human D1 dopamine receptor. Cell Res 31(5):593–596. https://doi.org/10.1038/s41422-021-00482-0

100. Wang DX, Wang MX (2020) Exploring anion-pi interactions and their applications in supramolecular chemistry. Acc Chem Res 53(7):1364–1380. https://doi.org/10.1021/acs.accounts.0c00243

101. Krieger IV, Freundlich JS, Gawandi VB, Roberts JP, Gawandi VB, Sun Q, Owen JL, Fraile MT, Huss SI, Lavandera JL, Ioerger TR, Sacchettini JC (2012) Structure-guided discovery of phenyl-diketo acids as potent inhibitors of *M. tuberculosis* malate synthase. Chem Biol 19(12):1556–1567. https://doi.org/10.1016/j.chembiol.2012.09.018

102. Ellenbarger JF, Krieger IV, Huang HL, Gomez-Coca S, Ioerger TR, Sacchettini JC, Wheeler SE, Dunbar KR (2018) Anion-pi Interactions in computer-aided drug design: modeling the inhibition of malate synthase by phenyl-diketo acids. J Chem Inf Model 58(10):2085–2091. https://doi.org/10.1021/acs.jcim.8b00417

103. Stammers TA, Coulombe R, Rancourt J, Thavonekham B, Fazal G, Goulet S, Jakalian A, Wernic D, Tsantrizos Y, Poupart MA, Bos M, McKercher G, Thauvette L, Kukolj G, Beaulieu PL (2013) Discovery of a novel series of non-nucleoside thumb pocket 2 HCV NS5B polymerase inhibitors. Bioorg Med Chem Lett 23(9):2585–2589. https://doi.org/10.1016/j.bmcl.2013.02.110

104. Kuzniak-Glanowska E, Glanowski M, Kurczab R, Bojarski AJ, Podgajny R (2022) Mining anion-aromatic interactions in the Protein Data Bank. Chem Sci 13(14):3984–3998. https://doi.org/10.1039/d2sc00763k

105. Koag MC, Lee S (2014) Metal-dependent conformational activation explains highly promutagenic replication across O6-methylguanine by human DNA polymerase beta. J Am Chem Soc 136(15):5709–5721. https://doi.org/10.1021/ja500172d

106. Wang J, Yao L (2019) Dissecting C-H•••pi and N-H•••pi interactions in two proteins using a combined experimental and computational approach. Sci Rep 9(1):20149. https://doi.org/10.1038/s41598-019-56607-4

107. Plevin MJ, Bryce DL, Boisbouvier J (2010) Direct detection of CH/π interactions in proteins. Nat Chem 2:466–471.

108. Suresh CH, Mohan N, Vijayalakshmi KP, George R, Mathew JM (2009) Typical aromatic noncovalent interactions in proteins: a theoretical study using phenylalanine. J Comput Chem 30:1392–1404.

109. Di Mino C, Seel AG, Clancy AJ, Headen TF, Földes T, Rosta E, Sella A, Skipper NT (2023) Strong structuring arising from weak cooperative O-H···π and C-H···O hydrogen bonding in benzene-methanol solution. Nat Commun 14:5900.

110. Diehl RC, Chorghade RS, Keys AM, Alam MM, Early SA, Dugan AE, Krupkin M, Ribbeck K, Kulik HJ, Kiessling LL (2024) CH-pi interactions are required for human galectin-3 function. JACS Au 4(8):3028–3037. https://doi.org/10.1021/jacsau.4c00357

111. Keys AM, Kastner DW, Kiessling LL, Kulik HJ (2025) CH-pi interactions confer orientational flexibility in protein-carbohydrate binding sites. Chemrxiv. Peprint at https://doi.org/10.26434/chemrxiv-2025-5n9ml-v2

112. Hayakawa D, Gouda H (2025) Three-dimensional CH/pi and CH/N interactions from quantum-mechanical and database analyses. J Chem Inf Model 65(8):4116–4127. https://doi.org/10.1021/acs.jcim.5c00124

113. Platzer G, Mayer M, Beier A, Bruschweiler S, Fuchs JE, Engelhardt H, Geist L, Bader G, Schorghuber J, Lichtenecker R, Wolkerstorfer B, Kessler D, McConnell DB, Konrat R (2020) PI by NMR: probing CH-pi interactions in protein-ligand complexes by NMR spectroscopy. Angew Chem Int Ed Engl 59(35):14861–14868. https://doi.org/10.1002/anie.202003732

114. Xiao Y, Woods RJ (2023) Protein–ligand CH–π interactions: structural informatics, energy function development, and docking implementation. J Chem Theor Comput 19:5503–5515.

115. Tsuzuki S, Honda K, Uchimaru T, Mikami M, Tanabe K (2000) Origin of the attraction and directionality of the NH/π interaction: comparison with OH/π and CH/π interactions. J Am Chem Soc 122(46):11450–11458. https://doi.org/10.1021/ja001901a

116. Feinberg H, Jegouza SAF, Lasanajak Y, Smith DF, Drickamer K, Weis WI, Taylor ME (2021) Structural analysis of carbohydrate binding by the macrophage mannose receptor CD206. JBC 296:100368. https://doi.org/10.1016/j.jbc.2021.100368

117. Zhang R, Xu Y, Lan J, Fan S, Huang J, Xu F (2022) Structural achievability of an NH–π Interaction between Gln and Phe in a crystal structure of a collagen-like peptide. Biomolecules 12:1433.

118. Foloppe N, Fisher LM, Francis G, Howes R, Kierstan P, Potter A (2006) Identification of a buried pocket for potent and selective inhibition of Chk1: prediction and verification. Bioorg

Med Chem 14(6):1792–1804. https://doi.org/10.1016/j.bmc.2005.10.022

119. Chen X, Zhang X, Chen J, Wang M, Yang Y, An L, Liu Z, Song X, Yao L (2024) Quantification of CH and NH/π-stacking interactions in cells using nuclear magnetic resonance spectroscopy. Anal Chem 96(36):14354–14362. https://doi.org/10.1021/acs.analchem.4c00688

120. Di Mino C, Seel AG, Clancy AJ, Headen TF, Földes T, Rosta E, Sella A, Skipper NT (2023) Strong structuring arising from weak cooperative O-H···π and C-H···O hydrogen bonding in benzene-methanol solution. Nat Commun 14:5900. https://doi.org/10.1038/s41467-023-41451-y

121. de Figueiredo RM, Suppo J-S, Campagne J-M (2016) Nonclassical routes for amide bond formation. Chem Rev 116(19):12029–12122. https://doi.org/10.1021/acs.chemrev.6b00237

122. Kumari S, Carmona AV, Tiwari AK, Trippier PC (2020) Amide bond bioisosteres: strategies, synthesis, and successes. J Med Chem 63(21):12290–12358. https://doi.org/10.1021/acs.jmedchem.0c00530

123. Harder M, Kuhn B, Diederich F (2013) Efficient stacking on protein amide fragments. Chem Med Chem 8(3):397–404. https://doi.org/10.1002/cmdc.201200512

124. Bootsma AN, Wheeler SE (2018) Stacking interactions of heterocyclic drug fragments with protein amide backbones. ChemMedChem 13(8):835–841. https://doi.org/10.1002/cmdc.201700721

125. DeFrees K, Kemp MT, ElHilali-Pollard X, Zhang X, Mohamed A, Chen Y, RA R (2019) An empirical study of amide–heteroarene π-stacking interactions using reversible inhibitors of a bacterial serine hydrolase. Org Chem Front 6(11):1749–1756. https://doi.org/10.1039/c9qo00342h

126. Giroud M, Ivkovic J, Martignoni M, Fleuti M, Trapp N, Haap W, Kuglstatter A, Benz J, Kuhn B, Schirmeister T, Diederich F (2017) Inhibition of the cysteine protease human cathepsin L by triazine nitriles: amide···heteroarene π-stacking interactions and chalcogen bonding in the S3 pocket. Chem Med Chem 12(3):257–270. https://doi.org/10.1002/cmdc.201600563

127. Hobza P, Muller-Dethlefs K (2009) Non-covalent interactions: theory and experiment. 1st ed. R Soc Chem 1–20. https://doi.org/10.1039/9781847559906

128. Pantsar T, Poso A (2018) Binding affinity via docking: fact and fiction. Molecules 23(8):1899. https://doi.org/10.3390/molecules23081899

129. Mohanty M, Mohanty PS (2023) Molecular docking in organic, inorganic, and hybrid systems: a tutorial review. Vol.154. Monatsh Chem 1–25. https://doi.org/10.1007/s00706-023-03076-1

130. Yang C, Chen EA, Zhang Y (2022) Protein-ligand docking in the machine-learning era. Molecules 27(14):4568. https://doi.org/10.3390/molecules27144568

131. Tosstorff A, Kuhn B (2025) Comparison of molecular recognition in docking versus experimental CSD and PDB data. J Chem Inf Model 65:10390–10399. https://doi.org/10.1021/acs.jcim.5c00893

132. Noga J, Bartlett RJ (1987) The full CCSDT model for molecular electronic structure. J Chem Phys 86(12):7041–7050. https://doi.org/10.1063/1.452353

133. Hobza P, Muller-Dethlefs K (2009) Non-covalent interactions: theory and experiment. R Soc Chem 21–69. https://doi.org/10.1039/9781847559906

134. Møller C, Plesset MS (1934) Note on an approximation treatment for many-electron systems. Phys Rev 46(7):618–622. https://doi.org/10.1103/PhysRev.46.618

135. Matczak P, Wojtulewski S (2015) Performance of Moller–Plesset second-order perturbation theory and density functional theory in predicting the interaction between stannylenes and aromatic molecules. J Mol Model 21(3):41. https://doi.org/10.1007/s00894-015-2589-1

136. Gaussian 16, Revision C.01, Frisch MJ, Trucks GW, Schlegel HB, Scuseria GE, Robb MA, Cheeseman JR, Scalmani G, Barone V, Petersson GA, Nakatsuji H, Li X, Caricato M, Marenich AV, Bloino J, Janesko BG, Gomperts R, Mennucci B, Hratchian HP, Ortiz JV, Izmaylov AF, Sonnenberg JL, Williams-Young D, Ding F, Lipparini F, Egidi F, Goings J, Peng B, Petrone A, Henderson T, Ranasinghe D, Zakrzewski VG, Gao J, Rega N, Zheng G, Liang W, Hada M, Ehara M, Toyota K, Fukuda R, Hasegawa J, Ishida M, Nakajima T, Honda Y, Kitao O, Nakai H, Vreven T, Throssell K, Montgomery JA, Peralta JE, Ogliaro F, Bearpark MJ, Heyd JJ, Brothers EN, Kudin KN, Staroverov VN, Keith TA, Kobayashi R, Normand J, Raghavachari K, Rendell AP, Burant JC, Iyengar SS, Tomasi J, Cossi M, Millam JM, Klene M, Adamo C, Cammi R, Ochterski JW, Martin RL, Morokuma K, Farkas O, Foresman JB, Fox DJ Gaussian, Inc. https://gaussian.com

137. Chai JD, Head-Gordon M (2008) Long-range corrected hybrid density functionals with damped atom-atom dispersion corrections.

Phys Chem Chem Phys 10(44):6615–6620. https://doi.org/10.1039/b810189b

138. Bursch M, Mewes JM, Hansen A, Grimme S (2022) Best-practice DFT protocols for basic molecular computational chemistry. Angew Chem Int Ed Engl 61(42):e202205735. https://doi.org/10.1002/anie.202205735
139. Weigend F, Ahlrichs R (2005) Balanced basis sets of split valence, triple zeta valence and quadruple zeta valence quality for H to Rn: design and assessment of accuracy. Phys Chem Chem Phys 7(18):3297–3305. https://doi.org/10.1039/b508541a
140. Dunning Jr. TH (1989) Gaussian basis sets for use in correlated molecular calculations. I. The atoms boron through neon and hydrogen. J Chem Phys 90(2):1007–1023. https://doi.org/10.1063/1.456153
141. Passaro S, Corso G, Wohlwend J, Reveiz M, Thaler S, Somnath VR, Getz N, Portnoi T, Roy J, Stark H, Kwabi-Addo D, Beaini D, Jaakkola T, Barzilay R (2025) Boltz-2: towards accurate and efficient binding affinity prediction. bioRxiv: 2025.2006.2014.659707. https://doi.org/10.1101/2025.06.14.659707

Chapter 11

Quantum Chemistry in Drug Design: Applications, Challenges, and the Need for Speed

Christofer S. Tautermann, Matthias Degroote and Benjamin Ries

Abstract

Quantum mechanics (QM) provides a rigorous framework for modeling molecular interactions, making it increasingly valuable in drug design. However, its broader adoption is limited by computational cost and scalability. This manuscript reviews key QM applications—torsional profiles, spectra prediction, reactivity analysis, and modeling of non-covalent interactions—highlighting their impact and limitations. A detailed in-house study compares multiple methods across 3000 drug-like fragments, illustrating the trade-off between speed and accuracy. While machine learning and hybrid approaches offer faster alternatives, they often fall short of density functional theory (DFT) precision. The manuscript also explores scaling strategies and the emerging role of quantum computing. To fully leverage QM in drug discovery, faster and more scalable methods are essential.

Keywords Quantum chemistry, Drug design, Quantum chemistry in drug design, Torsion barriers

1 Introduction

Quantum mechanics is the theoretical foundation of chemistry, offering a rigorous framework for understanding molecular structure, dynamics, and interactions. In drug design, where precision and predictive power are essential, QM methods have become increasingly relevant [1]. They enable accurate modeling of torsional barriers, spectra, reactivity, and binding energies—areas where classical approaches often fall short.

Despite their utility, QM methods are still underutilized in pharmaceutical research. The reasons are clear: high computational cost, long runtimes, and limited scalability. This chapter explores some aspects of the practical applications of QM in drug design, highlighting specific use cases and their impact. As a showcase, we discuss the important use case of the calculation of torsional barriers in drug-like molecules in detail to compare several approaches and to demonstrate the difficulty that QM calculations introduce

Alexander Heifetz (ed.), *Quantum Mechanics Applications in Drug Design*, Methods in Molecular Biology, vol. 3061, https://doi.org/10.1007/978-1-0716-5420-0_11,

in interactive design approaches. At last, we also briefly discuss the emerging role of quantum computing (QC), before concluding with a call to action: *much faster, scalable QM is essential to unlock its full potential in drug discovery.*

2 Quantum Mechanics in Drug Design: Practical Applications

Computational approaches to the solution of Schrödinger's equation have existed for decades, and with the rise of density functional methods, tailor-made applications to specific use cases have become feasible [2]. Still, in most cases, quantum mechanical treatment of systems relevant to drug design is restricted either to model systems resembling the macromolecule, systems that neglect the actual surrounding (such as solvent) or neglect the fact that biological systems exist as thermal ensembles. As a result, most ligand-centric approaches (such as conformational ensembles, spectra and reactivities) are confined to the calculations of electronic energies of molecule geometries (nuclei fixed) in an implicit solvent. The computation of intramolecular interactions is usually either done for model systems (hydrogen bonds in probes, π-interactions) or, if it comes to QM/MM approaches, where the level of theory of the QM region has to be severely reduced to allow for sufficient sampling of ensembles. The following applications are prototypical QM use cases in pharma—this list is not meant to be comprehensive, but it should give an idea which problems are tackled. The use case of torsion barriers is discussed in detail, as it is symptomatic of the dilemma that fast calculations are inaccurate and long, accurate calculations don't allow for interactive modeling.

2.1 Torsional Barriers and Conformational Analysis

Torsional strain influences molecular stability, conformational preferences, and bioactivity. The bioactive conformation of a drug binding to a target is usually a low-energy state [3]. Therefore, the accurate description of dihedral angles is a prerequisite for small-molecule drug design. Most commonly, target-based drug design is performed based on force fields like OFF2.0 [4] or MMFF94 [5] or very quick scoring functions like TorsionNet [6]. Large efforts are undertaken to generate fast and accurate descriptions of molecules, recently often hybridizing traditional methods with AI methods in models like ANI-2x [7] or MACE-OFF23 [8]. However, speed comes at the cost of accuracy, and the gold standard to calculate torsion profiles is still QM methods, particularly density functional theory (DFT). In an optimal setting for drug design, all methods should be at the fingertips of computational scientists. Boehringer Ingelheim's TorsionProfiler [9] (available as a GitHub repository) exemplifies this and shows how QM can be integrated into drug design workflows, enabling rapid profiling across diverse molecular datasets.

In the following, we will present our in-house study using several methods at different theory levels. Each method was employed to calculate around 3000 torsion profiles of drug-like fragments [6, 10, 11]. After calculating the profiles, we compared each method to our reference method at the DFT level using the ωB97XD functional.

But which differences do we expect to capture in those comparisons? Generally, two torsion profiles using different methods for the same bond in a molecule can vary in the height or depth of a profile region, which can be interpreted as a scaling issue in comparison. Another difference between the two profiles could be a shift of one profile, such that minima or maxima don't align anymore. A very big difference could be derived from a combination of scaling and shifting issues leading to an alternative functional form, like changing the multiplicity of the profile (see Fig. 1).

Understanding the possible differences in the profiles helped us to automate the torsion profile comparisons using several metrics. Those metrics compare two profiles capturing different aspects, like RMSD, correlations or earth mover distance. The collection of metrics leads to a feature vector, which is next interpreted by an on-augmented data pre-trained random forest model classifying into similar, uncertain and dissimilar (see Fig. 2).

The results of the in-house study demonstrate the scalability of QM-based torsion analysis. Comparisons between MM, ML, and QM methods reveal that while ML models like ANI2X [7] and MMFF94 [5] offer speed, they often lack the accuracy needed for nuanced conformational predictions. Hybrid approaches using semiempirical methods (e.g., GFN2-xTB [12]) strike a balance but still fall short of full DFT accuracy (see Fig. 3).

Beyond being useful for modeling purposes, torsion profiles are essential for identifying atropisomerism—where rotational barriers exceed ~20 kcal/mol—leading to stable stereoisomers. Such stereoisomers can exhibit distinct pharmacokinetic and pharmacodynamic properties. We used a two-stage process in order to efficiently calculate the relevant profiles. First screening with an automatic fragmentation algorithm [13] and the MACE-OFF23 [8] ML-trained force field for potential dihedral angles of interest. Next calculating the torsion profile for the identified dihedral angles with our gold standard DFT-Method ωB97XD to get the most accurate

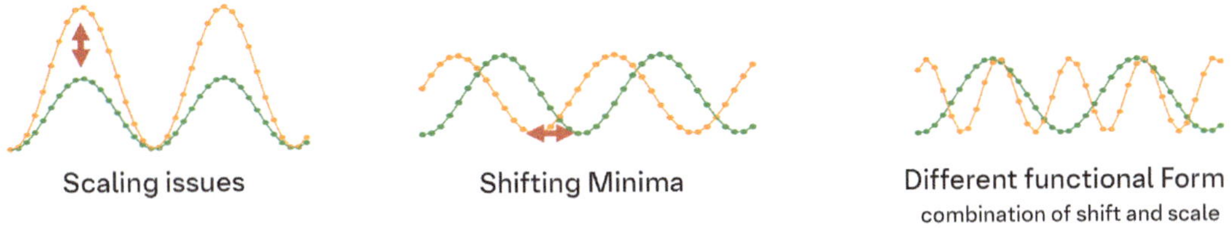

Fig. 1 Changes that can be observed between multiple torsion profiles

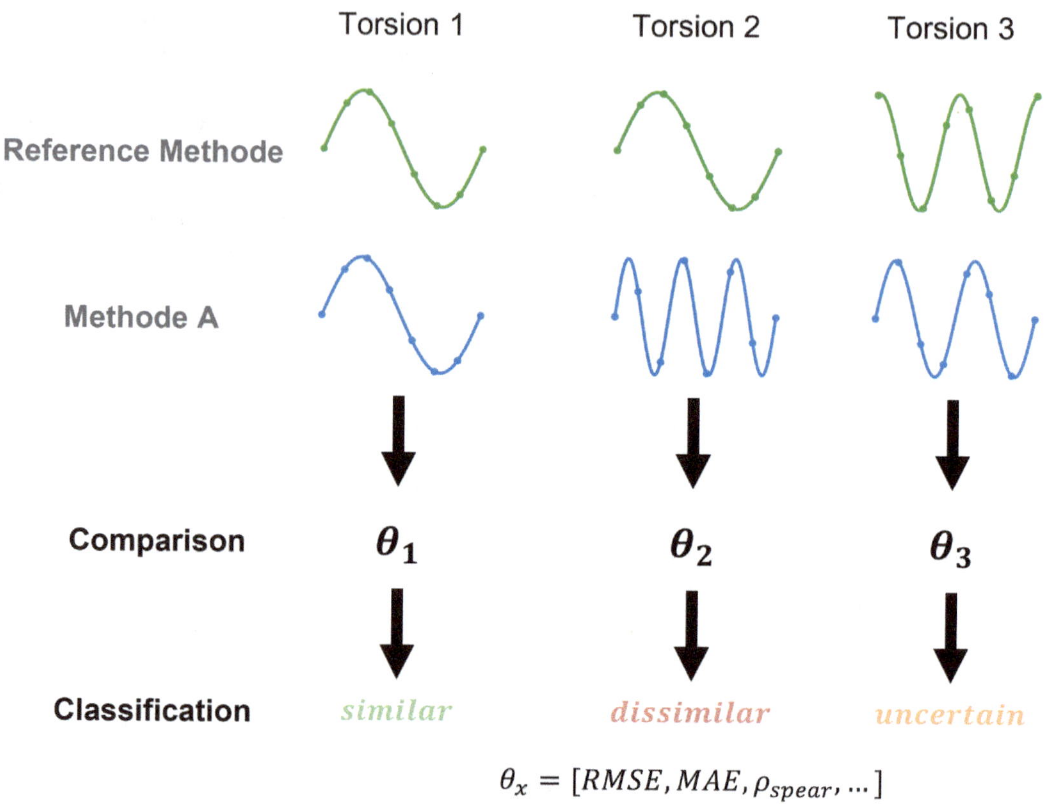

Fig. 2 The comparison process used in our in-house study. Two profiles were compared using the described metrics, which were afterward classified by our random forest model.

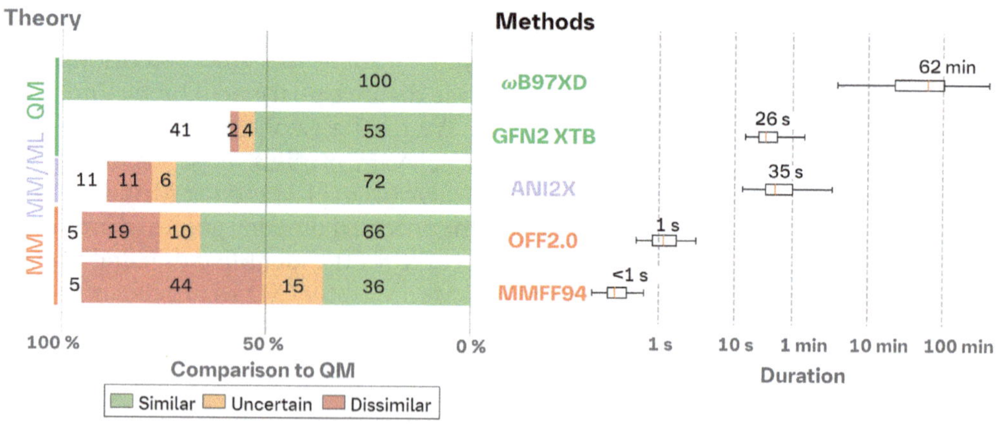

Fig. 3 Method comparison. This figure shows a classification comparison on the left, showing the similarity of torsion profiles of different methods to ωB97XD/cc-pVDZ. On the right, the distributions of calculation durations are shown in boxplots

profile possible. The QM calculations help predict these barriers most reliably, guiding synthetic strategies and informing crystallization behavior, as demonstrated in a very recent study on Protacs [14].

2.2 Spectral Predictions and Structural Validation

QM methods are indispensable for predicting molecular spectra—IR, NMR, VCD, and UV. These theoretical spectra serve as fingerprints for structural validation, stereochemical assignments, and comparison with experimental data. For example, NMR chemical shifts and coupling constants derived from QM calculations can distinguish between closely related isomers or conformers [15].

In drug design, such predictions aid in confirming the identity of synthetic intermediates, validating final products, and exploring stereoisomerism by vibrational dichroism spectra. QM-based spectral analysis also supports regulatory submissions, corroborating analytical data.

2.3 Reactivity and Quantum Descriptors

Understanding chemical reactivity is central to lead optimization of covalent ligands, where the non-covalent affinity needs to be balanced with the reactivity of the warhead (K_i and k_{inact}). QM methods offer descriptors such as HOMO-LUMO gaps, electrostatic potentials, and Fukui functions that correlate with reactivity trends. For more accurate descriptions of reactions, the activation barriers can be calculated by determining the transition states of the reaction. These parameters are increasingly integrated into machine learning (ML) models to enhance the speed and usability of predictions. Activation barriers have been shown to be very predictive for the experimental reactivity of covalent warheads (such as acrylamides) towards glutathione, which serves as a mimic of target cysteines [16]. However, determination of transition states is cumbersome, and confirmation by calculation of the Hessian is computationally demanding. Therefore, the pragmatic approach is to train ML models on a large number of QM-barriers to have fast reactivity prediction in drug design [16].

QM-derived reactivity assessments, such as Fukui indices, are less demanding to calculate than activation barriers, because they are based on the geometry minima of the educts. With these estimates, the regioselectivity of reactions can be predicted [17].

2.4 Specific Interactions: π-Stacking, Hydrogen Bonds, and Halogen-π Effects

Non-covalent interactions—such as π–π stacking, hydrogen bonding, halogen–π interactions—play a pivotal role in molecular recognition and binding [18]. QM methods allow for precise quantification of these interactions, surpassing the limitations of molecular mechanics.

Studies on heteroaromatic π-stacking reveal complex energy landscapes that depend on geometry, substituents, and electronic effects [19, 20]. Accurate, QM-derived potential energy surfaces (PES) of model systems, such as benzene interacting with drug fragments, help elucidate preferred binding modes [20]. π–interactions are known to be dominated by dispersion interactions [21] and therefore the inclusion of implicit or explicit dispersion corrections, such as Grimme's D3 [22], has significantly enhanced the performance of DFT methods for such systems.

Hydrogen bond strength is heavily investigated because directed hydrogen bonds are crucial anchoring motifs in drug recognition by the target [18]. The strength of hydrogen bonds is usually determined by QM calculations, where a series of donors is combined with a standard H-bond acceptor (and vice versa) and the dimerization energy is calculated by the supermolecule approach [23]. This is also very time-consuming, and ML-based approaches, based on QM data, provide a pragmatic access to reasonable estimates of H-bond strengths [24].

3 Scaling QM to Larger Systems

Most of the above-mentioned applications use model systems of small size in non-natural surroundings (such as implicit solvent) to describe certain aspects of the drug/target interaction. This approach has to be chosen due to the size limitations and scaling behaviors most QM methods bring with them. However, there are some very interesting approaches that aim towards the description of the full system, and they have to be mentioned here.

3.1 Fragmentation and Linear Scaling

One approach to applying QM to large biomolecular systems is fragmentation. By dividing a protein–ligand complex into smaller, manageable fragments, QM calculations can be performed on each part individually and recombined to approximate the whole. This method has been used successfully in hydration-free energy calculations and binding affinitypredictions. One prominent approach in this context is the FMO (fragment molecular orbital) method [25], where the large drug/target complex is divided into smaller fragments of interacting partners. For all fragments, the molecular orbitals are calculated and then combined to describe the full system again. Although this is not a free energy method, it has been demonstrated that binding affinities for certain target/drug combinations can be well explained by FMO [26].

A complementary approach, where the full system is described by QM methods, is linear-scaling DFT. These methods enable QM calculations on large systems by exploiting sparsity in the electronic structure. Runtime linearly scales in these methods with the number of electrons (at a given basis set/plane wave cutoff), and this allows for the description of very large systems. Codes such as ONETEP [27] have demonstrated the feasibility of QM on full protein-ligand complexes, though computational time remains a challenge [28].

3.2 Electrostatic Embedding and QM/MM

Electrostatic embedding techniques allow QM regions to be influenced by the surrounding MM environment, improving accuracy without incurring full QM cost. QM/MM methods are widely used in enzymatic reaction modeling and ligand binding studies, offering a practical compromise between accuracy and efficiency [29, 30].

3.3 Binding Free Energy Calculations

Predicting binding affinity is a cornerstone of drug discovery as it is often directly related to the predicted human dose. Traditional molecular mechanics [MM] with Poisson–Boltzmann [PB] and surface area solvation (MM-PBSA) methods offer speed but lack accuracy, and if based on force fields, they lack electronic detail. QM-PBSA attempts to refine these estimates by incorporating electronic effects, but the computational burden remains high, and the gain in accuracy may not justify the large effort [31, 32]. Direct sampling of the phase space of larger systems with QM accuracy is currently not feasible due to the need for thousands of independent snapshots. Hybrid Monte Carlo methods [33] and semi-empirical approximations (e.g., GFN2-xTB [12]) offer partial solutions but often compromise accuracy or system size. Approaches such as free energy calculations on QM/MM potentials are significant improvements [34, 35], but sampling times still limit the applications to systems that show very quick convergence—that is unfortunately rarely the case for drug-target systems.

4 Quantum Computing: A Brief Outlook

Quantum computing (QC) applications in ground state calculations, binding affinitypredictions, and coherent molecular dynamics [36].

To drive and demonstrate progress in the theoretical foundations of electronic structure calculations for pharma-relevant systems, the active site of cytochrome P450 (i.e., the coordinated heme-group) has been chosen as a model system [37]. Recent advances in Hamiltonian compression and initial state preparation have reduced runtime estimates for cytochrome P450 model calculations from 3 days (2021) [37] to under 1 h (2024) [38]. However, practical QC applications in drug design currently remain limited. Most algorithms are not tailored for pharma use cases, and no significant speedup over DFT has been demonstrated for routine tasks.

The potential is enormous, but foundational research is needed to adapt QC methods to the unique demands of drug discovery. For now, the focus should remain on accelerating classical QM methods [36].

5 Conclusion: The Need for Speed

Quantum mechanics has become a standard tool in drug design, enabling detailed insights into molecular behavior, interactions, and reactivity. Its applications—from torsional profiling to binding energy calculations—are well-established and impactful. However, the reliance on approximations and the slow computational speed limit its broader adoption. We demonstrated this problem of accuracy vs. speed on the example of torsion barriers: methods that allow interactive modeling with response times in the seconds range don't yield the required accuracy, and DFT methods are still not quick enough to allow modeling without longer breaks.

Quantum computing offers a glimpse into a future where these limitations may be overcome. Yet, the field is still in its infancy, and practical integration into drug design workflows remains a challenge.

To truly unlock the potential of QM in drug discovery, the community must prioritize speed. A maybe controversial statement in this context may be: *No new functionals needed—just faster calculations.* Achieving sub-second DFT evaluations on a reasonable level of theory (hybrid functional and triple-zeta basis set) for drug-sized molecules could revolutionize interactive modeling and simulation, accelerating the path from molecule to medicine.

The call to action is clear: invest in faster, scalable QM methods—whether through algorithmic innovation, hardware acceleration, or quantum computing. Only then can we realize the full impact of quantum chemistry in drug design.

References

1. Tautermann CS (2020) Quantum mechanics in drug discovery. Methods Mol Biology Clifton NJ 2114:1–17. https://doi.org/10.1007/978-1-0716-0282-9_1
2. Bursch M, Mewes J, Hansen A, Grimme S (2022) Best-practice DFT protocols for basic molecular computational chemistry. Angew Chem Int Ed 61:e202205735. https://doi.org/10.1002/anie.202205735
3. Tong J, Zhao S (2021) Large-scale analysis of bioactive ligand conformational strain energy by ab initio calculation. J Chem Inf Model 61:1180–1192. https://doi.org/10.1021/acs.jcim.0c01197
4. Boothroyd S, Behara PK, Madin O et al (2023) Development and benchmarking of open force field 2.0.0: the sage small molecule force field. J Chem Theory Comput 19:3251–3257. https://doi.org/10.26434/chemrxiv-2022-n2z1c-v2
5. Halgren TA, Nachbar RB (1996) Merck molecular force field. IV. conformational energies and geometries for MMFF94. J Comput Chem 17:587–615
6. Rai BK, Sresht V, Yang Q, et al (2022) TorsionNet: a deep neural network to rapidly predict small-molecule torsional energy profiles with the accuracy of quantum mechanics. J Chem Inf Model 62:785–800. https://doi.org/10.1021/acs.jcim.1c01346
7. Devereux C, Smith JS, Davis KK et al (2020) Extending the applicability of the ANI deep learning molecular potential to sulfur and halogens. J Chem Theory Comput 16:4192–4202. https://doi.org/10.1021/acs.jctc.0c00121
8. Kovács DP, Moore JH, Browning NJ, et al (2025) MACE-OFF: Short-range transferable machine learning force fields for organic molecules. J Am Chem Soc 147(21):17598–17611. https://doi.org/10.1021/jacs.4c07099
9. Ries B, Gordiy I, Hermann M, Fox T (2025) Torsion profile calculation and analysis made easy! https://github.com/Boehringer-Ingelheim/TorsionProfiler. Accessed 8 Oct 2025

10. Dahlgren MK, Schyman P, Tirado-Rives J, Jorgensen WL (2013) Characterization of biaryl torsional energetics and its treatment in OPLS all-atom force fields. J Chem Inf Model 53:1191–1199. https://doi.org/10.1021/ci4001597
11. D'Amore L, Hahn DF, Dotson DL et al (2022) Collaborative assessment of molecular geometries and energies from the open force field. J Chem Inf Model 62:6094–6104. https://doi.org/10.1021/acs.jcim.2c01185
12. Bannwarth C, Ehlert S, Grimme S (2019) GFN2-xTB—an accurate and broadly parametrized self-consistent tight-binding quantum chemical method with multipole electrostatics and density-dependent dispersion contributions. J Chem Theory Comput 15:1652–1671. https://doi.org/10.1021/acs.jctc.8b01176
13. Stern CD, Bayly CI, Smith DGA et al (2022) Capturing non-local through-bond effects in molecular mechanics force fields I: fragmenting molecules for quantum chemical torsion scans. Preprint at bioRxiv https://doi.org/10.1101/2020.08.27.270934
14. Stamos NA, Ries B, Schneider R et al (2025) Beyond barriers, big crystallization hurdles: atropisomerism in beyond rule of five compounds explored by computational and NMR studies. Mol Pharm 22:3268–3285. https://doi.org/10.1021/acs.molpharmaceut.5c00204
15. Xin D, Sader CA, Choudhary O et al (2017) Development of a 13C NMR chemical shift prediction procedure using B3LYP/cc-pVDZ and empirically derived systematic error correction terms: a computational small molecule structure elucidation method. J Org Chem 82:5135. https://doi.org/10.1021/acs.joc.7b00321
16. Palazzesi F, Hermann MR, Grundl MA et al (2020) BIreactive: a machine-learning model to estimate covalent warhead reactivity. J Chem Inf Model 60:2915–2923. https://doi.org/10.1021/acs.jcim.9b01058
17. Kuttruff CA, Haile M, Kraml J, Tautermann CS (2018) Late-stage functionalization of drug-like molecules using diversinates. ChemMedChem 13:983–987. https://doi.org/10.1002/cmdc.201800151
18. Bissantz C, Kuhn B, Stahl M (2010) A medicinal chemist's guide to molecular interactions. J Med Chem 53:5061–5084. https://doi.org/10.1021/jm100112j
19. Bootsma AN, Doney AC, Wheeler SE (2019) Predicting the strength of stacking interactions between heterocycles and aromatic amino acid side chains. J Am Chem Soc 141:11027–11035. https://doi.org/10.1021/jacs.9b00936
20. Huber RG, Margreiter MA, Fuchs JE et al (2014) Heteroaromatic π-stacking energy landscapes. J Chem Inf Model 54:1371–1379. https://doi.org/10.1021/ci500183u
21. Phipps MJS, Fox T, Tautermann CS, Skylaris C-K (2015) Energy decomposition analysis approaches and their evaluation on prototypical protein–drug interaction patterns. Chem Soc Rev 44:3177–3211. https://doi.org/10.1039/c4cs00375f
22. Grimme S, Antony J, Ehrlich S, Krieg H (2010) A consistent and accurate ab initio parametrization of density functional dispersion correction (DFT-D) for the 94 elements H-Pu. J Chem Phys 132:154104. https://doi.org/10.1063/1.3382344
23. Kerdawy AE, Tautermann CS, Clark T, Fox T (2013) Economical and accurate protocol for calculating hydrogen-bond-acceptor strengths. J Chem Inf Model 53:3262–3272. https://doi.org/10.1021/ci4006222
24. Melo DU, Carneiro LM, Coutinho-Neto MD et al (2025) Developing a machine learning model for hydrogen bond acceptance based on natural bond orbital descriptors. J Org Chem 90:9776–9788. https://doi.org/10.1021/acs.joc.5c00724
25. Fedorov DG, Nagata T, Kitaura K (2012) Exploring chemistry with the fragment molecular orbital method. Phys Chem Chem Phys 14:7562–7577. https://doi.org/10.1039/c2cp23784a
26. Heifetz A, Chudyk EI, Gleave L et al (2016) The fragment molecular orbital method reveals new insight into the chemical nature of GPCR–ligand interactions. J Chem Inf Model 56:159–172. https://doi.org/10.1021/acs.jcim.5b00644
27. Prentice JCA, Aarons J, Womack JC et al (2020) The ONETEP linear-scaling density functional theory program. J Chem Phys 152:174111. https://doi.org/10.1063/5.0004445
28. Dziedzic J, Fox SJ, Fox T et al (2013) Large-scale DFT calculations in implicit solvent—a case study on the T4 lysozyme L99A/M102Q protein. Int J Quantum Chem 113:771–785. https://doi.org/10.1002/qua.24075
29. Cui Q, Pal T, Xie L (2021) Biomolecular QM/MM simulations: what are some of the "Burning Issues"?. J Phys Chem B 125:689–702. https://doi.org/10.1021/acs.jpcb.0c09898
30. Woods CJ, Shaw KE, Mulholland AJ (2015) Combined quantum mechanics/molecular mechanics (QM/MM) simulations for protein–ligand complexes: free energies of binding of water molecules in influenza neuraminidase. J Phys Chem B 119:997–1001. https://doi.org/10.1021/jp506413j
31. Gundelach L, Fox T, Tautermann CS, Skylaris C-K (2022) BRD4: quantum mechanical protein–ligand binding free energies using the

full-protein DFT-based QM-PBSA method. Phys Chem Chem Phys 24:25240–25249. https://doi.org/10.1039/d2cp03705j
32. Fox S, Wallnoefer HG, Fox T et al (2011) First principles-based calculations of free energy of binding: application to ligand binding in a self-assembling superstructure. J Chem Theory Comput 7:1102–1108. https://doi.org/10.1021/ct100706u
33. Sampson C, Fox T, Tautermann CS et al (2015) A "Stepping Stone" approach for obtaining quantum free energies of hydration. J Phys Chem B 119:7030–7040. https://doi.org/10.1021/acs.jpcb.5b01625
34. Lu X, Fang D, Ito S et al (2016) QM/MM free energy simulations: recent progress and challenges. Mol Simul 42:1056–1078. https://doi.org/10.1080/08927022.2015.1132317
35. Pregeljc D, Hügli RJR, Riniker S (2025) Efficient multistate free-energy calculations with QM/MM accuracy using replica-exchange enveloping distribution sampling. J Phys Chem B 129:5948–5960. https://doi.org/10.1021/acs.jpcb.5c02086
36. Santagati R, Aspuru-Guzik A, Babbush R et al (2024) Drug design on quantum computers. Nat Phys 20:549–557. https://doi.org/10.1038/s41567-024-02411-5
37. Goings JJ, White A, Lee J et al (2022) Reliably assessing the electronic structure of cytochrome P450 on today's classical computers and tomorrow's quantum computers. Proc Natl Acad Sci U S A 119:e2203533119. https://doi.org/10.1073/pnas.2203533119
38. Caesura A, Cortes CL, Pol W et al (2025) Faster quantum chemistry simulations on a quantum computer with improved tensor factorization and active volume compilation. PRX Quantum 6:030337. https://doi.org/10.1103/yngp-5fpm

Check for updates

Chapter 12

Drug Censorship at Scale: The Practicalities of Automation and Parallelism for Fragment Molecular Orbital (FMO) in *GAMESS*

Jintao Long

Abstract

Quantum mechanics (QM) methods are increasingly applied in drug discovery, with the fragment molecular orbital (FMO) approach in GAMESS enabling large-scale calculations on biomolecular systems. This chapter outlines key lessons from integrating FMO-based QM workflows into high-throughput pipelines, focusing on challenges such as input preparation (protonation, charge assignment, minimization, and truncation), error classification, and output parsing. It also highlights limitations in scaling GAMESS on modern HPC resources and offers practical guidance on designing user-friendly interfaces. The aim is to provide non-proprietary insights for researchers seeking to adopt QM methods in drug discovery.

Keywords GAMESS, Automation, Fragment molecular orbital (FMO)

1 Introduction

When I was invited to work on the FMO-HT (fragment molecular orbital-high throughput) project at Evotec, I had little more than the notion that its methodology was rooted in quantum physics. Looking back after two years of intensive development on this tool, I see the journey as a unique blend of scientific methodology, research and software development.

GAMESS (general atomic and molecular electronic structure system) is one of the most widely used quantum chemistry software packages. It provides a rich set of quantum-mechanical methods, including Hartree–Fock (HF), density functional theory (DFT), second-order Møller–Plesset perturbation (MP2), coupled-cluster (CC) methods, and the fragment molecular orbital (FMO) approach [1]. This chapter focuses on using GAMESS as the primary software for quantum-related calculations in customized FMO applications, with an emphasis on my experience automating its workflows for high-throughput screening. A complete processing flow of using

Alexander Heifetz (ed.), *Quantum Mechanics Applications in Drug Design*, Methods in Molecular Biology, vol. 3061,
https://doi.org/10.1007/978-1-0716-5420-0_12,

FMO can usually be split into several steps—input preparation, GAMESS execution, output analysis and visualization (see Fig. 1).

2 Input Preparation Is Key to the Success of Running FMO

Input preparation is crucial to ensure successful calculations for any scientific software, but even more so for running FMO with GAMESS due to its peculiarities [2, 3].

First, while GAMESS supports a broad range of *quantum methods* (HF, DFT, MP2, CC, and FMO), it does **not** provide classical force-field or molecular dynamics calculations. Instead, it is designed as an ab initio quantum-chemistry engine. The flexibility of methods and keywords means the configuration space for each execution is vast, and small nuances in input can lead to dramatic differences in results.

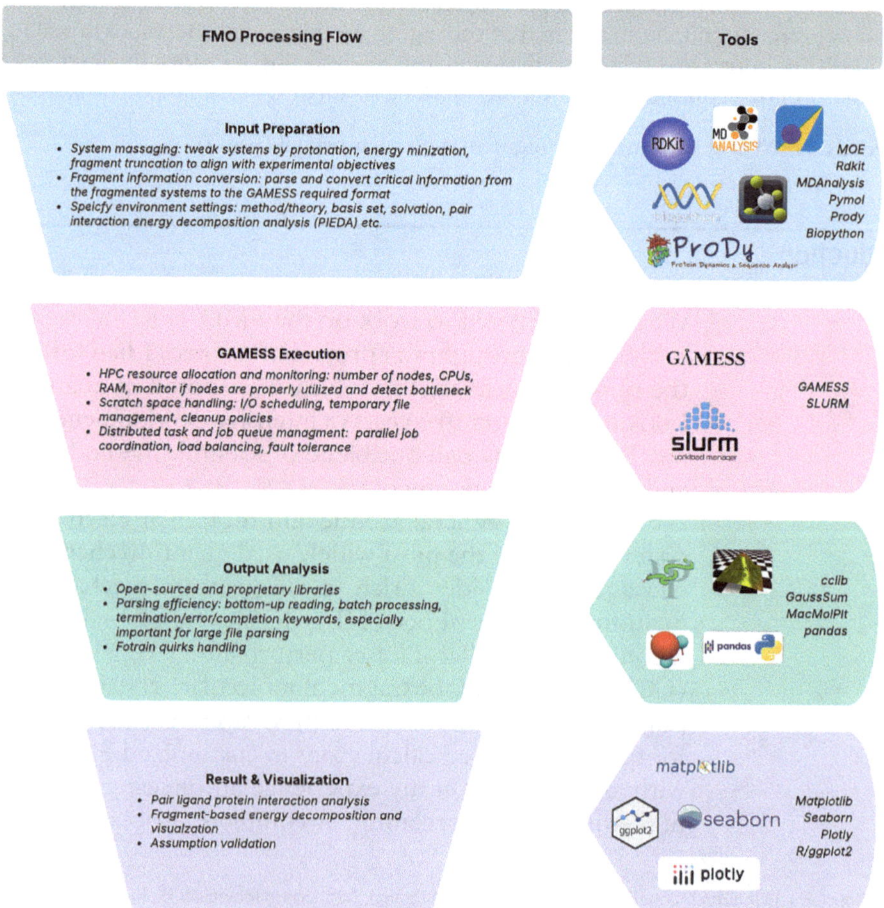

Fig. 1 Three types of correctness for evaluating FMOautomation software

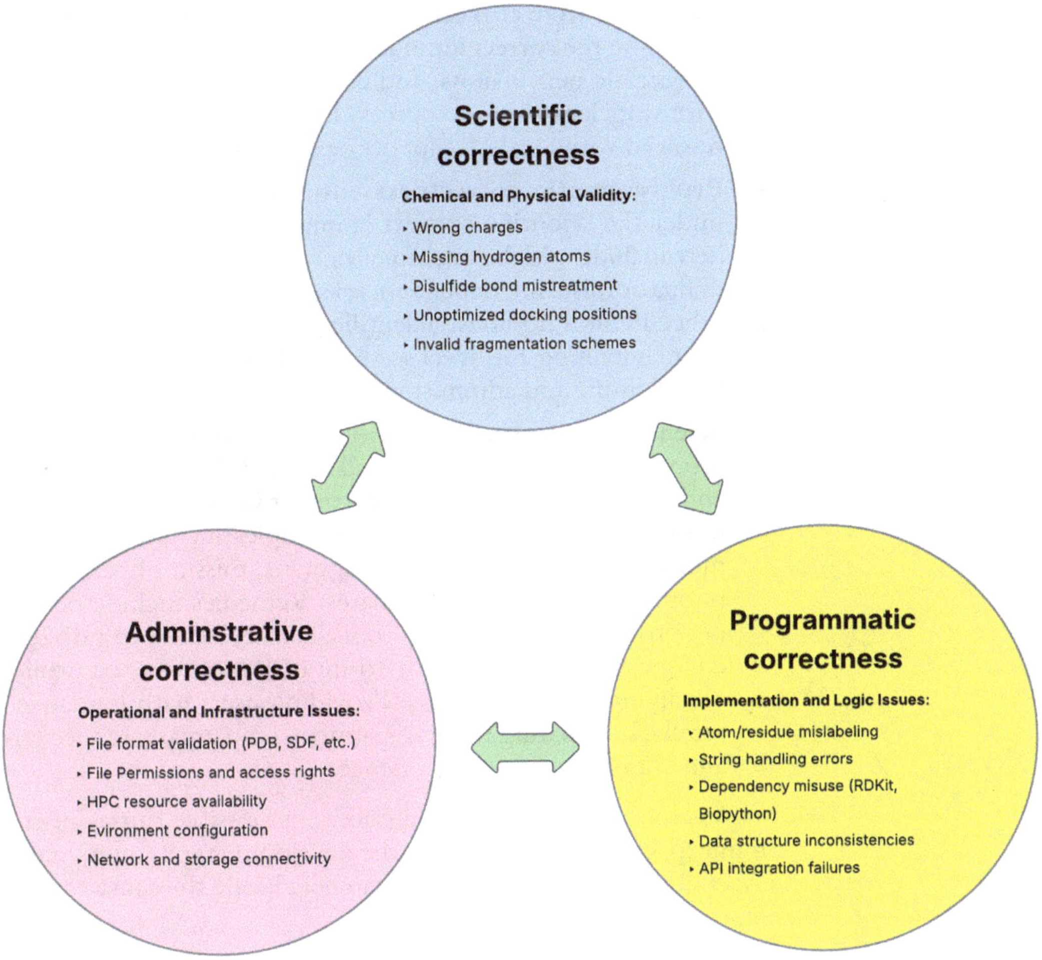

Fig. 2 Abstracted FMO processing flow and the respective tools

Second, the systems of interest are often very large—comprehensive biological assemblies where proteins, peptides, and chemical ligands contain assorted elements with diverse coordinates and structural features. To generate accurate results, the system details must be provided down to the atomic level in the input file. Without automation, this manual preparation becomes laborious and error-prone [4]. This is why automation is critical if FMO is to be integrated into high-throughput screening pipelines.

A well-prepared input file rewards the user with a graceful exit of GAMESS executions and reliable simulation outcomes. However, the process is not always perfect. When failures or unexpected results occur, it is useful to classify problems into three categories of correctness from a scientific programmer's standpoint: **administrative, programmatic**, and **scientific** correctness (see Fig. 2).

- **Administrative correctness** involves operational steps such as providing the correct input file format (PDB, SDF, etc.), setting correct file permissions, and ensuring a working environment with sufficient HPC resources. Issues at this level can often be resolved with local IT support or clear user documentation.
- **Programmatic correctness** ensures that the business logic under the scientific context is implemented correctly. Errors here include mislabeling atoms or residues, incorrect string handling, or misusing dependencies such as RDKit, Biopython, or other Python libraries. Scientific programmers are responsible for maintaining this level so that end users need only focus on the scientific and administrative aspects.
- **Scientific correctness** refers to whether the experiment itself is properly designed and reflected in the GAMESS input. A scientifically incorrect system will trigger GAMESS errors or yield nonsensical results. Common examples include mismatched multiplicity and charge in a fragment, missing hydrogens, or unoptimized atomic coordinates. Remedies include performing protonation (e.g., with tools like **PROPKA** or **pdb2pqr**), energy minimization prior to input generation, or explicitly specifying disulfide bonds. Providing users with guidance or, ideally, automating these adjustments, can greatly reduce errors and improve the user experience.

These three categories can cover any possible causes of errors or unexpected behavior during the development or employment of FMO. By process elimination, it can help locate the cause easily and improve efficiency.

3 There is No Shortcut for Output Parsing and Analysis

GAMESS is written in Fortran, a language that dates back to the 1970s. Its authors have long adhered to the principle of providing a **single source of output** from which scientists can extract all information. This means results are embedded as human-readable plain text rather than structured files.

Several tools can extract data from GAMESS output files:

- Open-source Python libraries like *cclib*, *MDAnalysis*, and *chemtools* can extract topologies and energies from GAMESS output [5].
- Specialized software such as GaussSum and MacMolPlt also support GAMESS outputs [6].

For full flexibility or special scientific requirements, custom scripts are often the best approach. If you go this route, be aware of a few caveats:

- Output files can be gigantic, especially for systems with slow SCF convergence.
- Results typically appear near the end of the file; reading from the bottom upward can greatly improve performance in large files. Batch-processing chunks in a bottom-up direction can further speed parsing.
- Fortran-specific quirks persist. For example, tabular data is printed as plain text. If a number exceeds the field width, Fortran replaces it with asterisks—indicating extremely large (positive or negative) values, often corresponding to unstable or transitional states of a fragment.
- GAMESS enforces legacy constraints such as limiting file names to alphanumerics, underscores, hyphens, and dots, and restricting keyword lengths to eight characters with 80-character line limits. Violating these can cause crashes

4 "Keep the Bricks Even After the Mansion is Built"

Another legacy behavior inherited from decades-old computing is GAMESS's reliance on writing many intermediates to disk. Because RAM was scarce in the 1980s and 1990s, large calculations had to use temporary disk storage—"scratch" space—to relay intermediate results.

A successful GAMESS execution requires:

- Syntactically correct input with chemically sensible content.
- Well-prepared systems (manual, semi-automatic, or fully automated).
- Adequate computing resources.
- Sensible analytic pipelines and visualization.

Errors in any of these aspects can cause failed runs or unusable results. During early-stage development of FMO workflows, retaining intermediates and byproducts from each step is invaluable for debugging [7]. This allows you to systematically verify administrative, scientific, and programmatic correctness in a back-and-forth manner.

5 User Interfaces Should Adapt to Users' Evolving Understanding

Becoming proficient in FMO can take years of experience. An FMO expert combines knowledge of quantum mechanics, chemistry, and physics, capable of translating between molecular-level phenomena and quantum-level causes and effects.

Beginners, however, may find it challenging to design fragmentation schemes, choose solvation models, select basis sets, and decide on energy-calculation methods. Without proper guidance, they can easily be overwhelmed [8].

To support adoption, software interfaces should expose sophistication progressively—offering sensible defaults, clear guidance, and training materials. For instance, when users ask, "What are the interaction energies between residues and ligands in my system?" the software should guide them through questions about fragmentation, solvation, and basis sets without overwhelming them. Each new feature release should be paired with training workshops or tutorials so that flexibility grows alongside user understanding.

6 Scaling Should Be the Next Technical Bottleneck to Break Through

In my experience, scaling GAMESS to gain performance on modern hardware has been a dead end. Two key factors limit performance gains:

- **Manual memory assignment**: GAMESS still relies on user-specified memory keywords. "MWords" specifies the replicated memory per core, and "MEMDDI" specifies distributed memory across nodes. Finding the optimal combination of nodes, cores, and memory per core is delicate—underestimation slows computation dramatically; overestimation often crashes the program. Even core developers have noted the lack of obvious scaling gains from simply allocating more resources.
- **Algorithmic rigidity**: GAMESS parallelism is built on algorithms designed before modern distributed-memory HPC systems. As a result, it does not always scale efficiently.

If parallelism and scalability become bottlenecks, it is worth exploring newer software such as *ORCA* [9], *PSI4*, or *Q-Chem*, which are designed with more modern computational architectures in mind. NVIDIA has also experimented with containerizing GAMESS to leverage advanced GPUs. According to NVIDIA benchmarks (2017), GPU-enabled GAMESS achieved up to 8× performance speedup on Pascal/Volta systems [10].

7 Conclusion

A successful GAMESS-FMO workflow hinges on robust input preparation, careful resource management, efficient output parsing, and user-friendly interfaces. Automation, clear error categorization, and a focus on progressive user education can make FMO a powerful tool for drug discovery. While GAMESS remains a cornerstone of quantum chemistry, scaling limitations mean researchers should also keep an eye on newer software packages and GPU-accelerated approaches as the field evolves.

References

1. Barca GMJ, Bertoni C, Carrington L et al (2020) Recent developments in the general atomic and molecular electronic structure system. J Chem Phys 152:154102. https://doi.org/10.1063/5.0005188
2. Fedorov DG, Kitaura K (2014) Efficient and accurate fragmentation methods. Acc Chem Res 47:2768–2775
3. Gordon MS, Schmidt MW (2005) Advances in electronic structure theory: GAMESS a decade later. In: Dykstra CE, Frenking G, Kim KS, Scuseria GE (eds) Theory and applications of computational chemistry: the first forty years. Elsevier, Amsterdam, pp 1167–1189
4. O'Boyle NM, Tenderholt AL, Langner KM (2008) cclib: a library for package-independent computational chemistry algorithms. J Comput Chem 29:839–845
5. MDAnalysis Project (2023) MDAnalysis: a Python package for the analysis of molecular dynamics simulations. https://www.mdanalysis.org
6. GaussSum Project (2023) GaussSum 3.0: Open-source GUI for quantum chemistry output files. https://sourceforge.net/projects/gausssum/
7. Iowa State University (2017) GAMESS user manual (latest release). https://www.msg.chem.iastate.edu/gamess/
8. Suenaga M (2008) Development of GUI for GAMESS/FMO calculation. J Comput Chem Jpn 7:33–54
9. Neese F (2018) Software update: the ORCA program system. WIREs Comput Mol Sci 8:e1327
10. NVIDIA Developer Blog (2023) Accelerating GAMESS with GPUs. https://developer.nvidia.com/gamess

Check for updates

Chapter 13

Quantum Methods in Drug Discovery: An Introduction to the Theoceptor (Theoretical Receptor) Approach

Nikeel Cull, Yixin He, Sylvia Kaempf, Andrew G. Leach, Mauno Lius and James Osbiston

Abstract

In this chapter, we introduce the theoceptor method. The method seeks to achieve a fully quantum mechanical computational description of protein–ligand binding. To do this, semi-empirical quantum mechanics levels of theory are applied during geometry optimization, with higher levels of theory applied as single point energy evaluations. The system is represented by subsets of the atoms of the protein receptor selected according to proximity to the ligand; backbone and C_β atoms are kept fixed during optimization with other atoms being free to move. The method has been applied to enzymes, GPCRs, ion channels, and inhibitors of corrosion at metal surfaces. Examples described reveal that the method complements structural biology well by enhancing and correcting interpretations, can be applied to X-ray crystallography- or cryo-electron microscopy-derived structures, can be applied equally to covalent and noncovalent binders, can provide excellent predictions for the changes in activity between matched molecular pairs, and can assign mode of action, such as agonist or antagonist.

Keywords GFN2-xTB, DFT, Quantum mechanics, Drug discovery

1 Introduction

The forces that determine protein–ligand binding geometries and energies are quantum mechanical in nature. In this chapter, we describe our theoceptor approach that seeks to benefit from correctly describing these forces. As reflected in the other chapters in this volume, we are not alone in working towards this aim.

The term "theoceptor" is short for "theoretical receptor," in analogy to the theozymes (theoretical enzymes) proposed by Houk [1–5]. The theoceptor methodology aims to create a robust model for the prediction of geometry and relative affinity of ligands binding to protein (or other) receptors. To achieve this, ligands first undergo a conformational search and low-energy conformations are subjected to molecular docking to identify candidate three-dimensional (3D)

Alexander Heifetz (ed.), *Quantum Mechanics Applications in Drug Design*, Methods in Molecular Biology, vol. 3061,
https://doi.org/10.1007/978-1-0716-5420-0_13,

orientations within the active site [6, 7]. The theoceptor, representing the region of the receptor in close proximity to where the ligand binds, is generated from available structures (typically protein structures from either X-ray diffraction or cryo-electron microscopy studies) by identifying all atoms that are proximate to either the native ligand or to a set of docked poses. Receptor atoms that are viewed as playing a mainly scaffolding role are assigned to be fixed in space, and others are allowed to move during subsequent geometry optimizations with quantum mechanical (QM) calculations (usually semiempirical or DFT levels of theory). Following geometry optimization, energies can be evaluated at higher levels of theory. The energies of the free ligand, the receptor, and the ligand–receptor complex can be calculated and hence a binding energy obtained [8, 9]. We have applied the method to a diverse range of protein targets, including enzymes, GPCRs, and ion channels. We are also exploring alternative applications of the method, and this currently includes the interaction of organic compounds with metal surfaces to understand corrosion inhibition. The theoceptor methodology provides valuable insights into ligand–receptor interactions (including assigning tautomer and ionization states of protein–ligand complexes) and corrects errors in deposited structures and assigns modes of action. Many of the studies described in this chapter have not yet been described in the literature, and thus, it represents an overview of our very latest findings.

1.1 Theoceptor Workflow

An overview of the workflow is provided in Fig. 1 and described in the next few sections.

A fundamental challenge for the theoceptor approach lies in bridging the gap between QM calculations, which typically generate electronic energies in the gas or continuum solvated phase, and experimental observations governed by free energies in aqueous solution. The relationship can be expressed as shown in Eq. 1

$$\Delta G_{\text{bind(aq)}} = \Delta E + \Delta H_{\text{corr(gas)}} - T\ \Delta S_{\text{(gas)}} + \Delta\Delta G_{\text{solv}}, \qquad (1)$$

where $\Delta G_{\text{bind(aq)}}$ is the binding free energy in aqueous solution. ΔE represents the change in electronic energy in the gas phase, as described in Eq. 2. $\Delta H_{\text{corr(gas)}}$ accounts for thermal corrections to enthalpy in the gas phase. $T\Delta S_{\text{(gas)}}$ denotes the entropic contribution to binding at temperature T. Finally, $\Delta\Delta G_{\text{solv}}$ describes the change in solvation free energy, which is necessary to connect the gas- or continuum-phase to the experimental aqueous environment. The electronic interaction energy (ΔE) corresponds to the difference in electronic energy between the ligand–receptor complex (E_{complex}) and its separate components (E_{ligand} and E_{receptor}):

$$\Delta E = E_{\text{complex}} - E_{\text{ligand}} - E_{\text{receptor}}. \qquad (2)$$

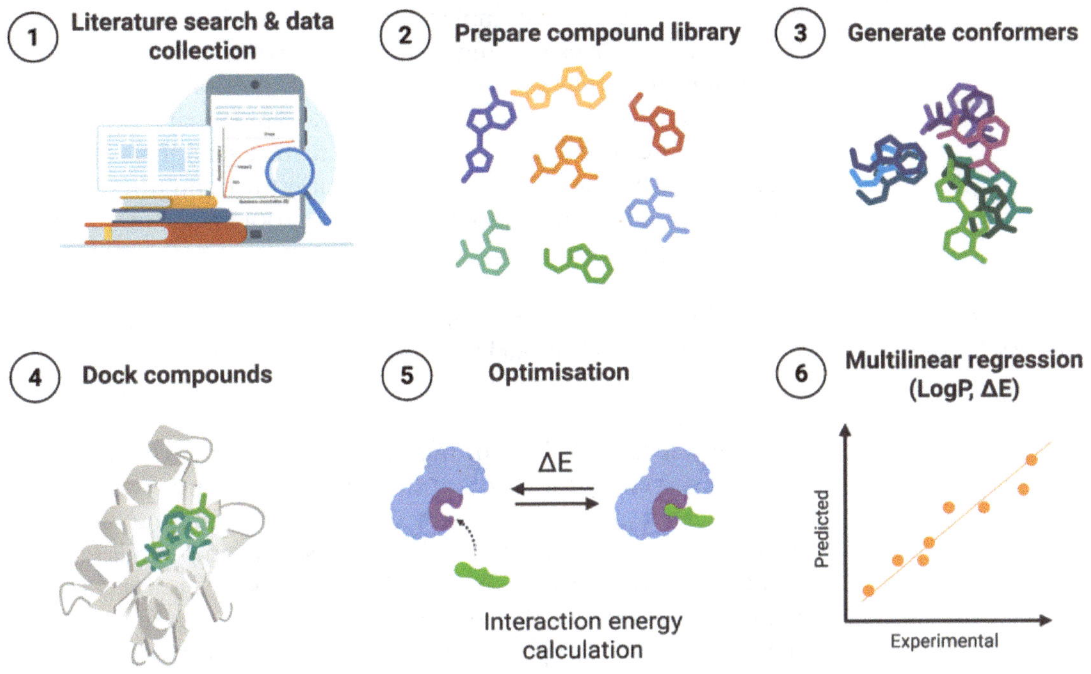

Fig. 1 The general workflow of the theoceptor approach

Computing all terms in Eq. 1 accurately remains computationally demanding and is often impractical for large-scale virtual screening applications. This is particularly true for the calculation of solvation free energies ($\Delta\Delta G_{solv}$), which requires sophisticated implicit or explicit solvent models. To address this challenge, the theoceptor approach proposes that a ligand's lipophilicity, as quantified by its octanol–water partition coefficient (logP), along with fitted coefficients α, β, and γ, can serve as an effective surrogate for the enthalpic, entropic, and solvation contributions. This led to the proposal shown in Eq. 3.

$$\text{Affinity} = \alpha\Delta E + \beta \log P + \gamma. \tag{3}$$

The proposed model expressed in Eq. 3, includes system-specific parameters α, β, and γ, which are determined through training on experimental binding affinity data (pIC_{50}, K_i, K_d). This methodology offers several key advantages: (1) computational efficiency, as only gas-phase electronic energy calculations are required from quantum mechanics; (2) logP values are readily available in chemical databases or can be reliably computed using established methods; and (3) physically sound, as the model maintains a clear alignment to fundamental principles governing protein–ligand binding.

Therefore, to build and test the model, we need to (1) calculate gas-phase electronic interaction energies (ΔE) for a sufficiently large set of ligand–receptor complexes, then (2) obtain or compute logP values for the corresponding ligands. Following that, we can (3) fit a multilinear regression model for affinity to obtain the values for system-specific parameters (α, β, γ). We can validate the model's predictive performance on independent test sets. This approach represents a balance between computation time and physical accuracy, potentially providing a more efficient tool for the prediction of ligand-receptor binding.

1.1.1 Steps 1 and 2: Compound Selection

Given the foregoing, it is clear that a minimum of three compounds is required to train a model to obtain the three coefficients. The more compounds that can be included in the training set, the better. It also implies that even where only two compounds are available, their relative values of ΔE might highlight compounds with enhanced binding driven by forces other than hydrophobicity. During compound selection, careful consideration should be given to whether a model that applies to a chemically diverse set of compounds (a global model) or a model specific to one chemical series (a local model) is desired or achievable. As described below, a test set including matched molecular pairs can bridge between these two approaches [10]. In general, as wide a range of affinity values as possible should be included when training the model.

1.1.2 Step 3: Conformer Generation

To maximize the chances of obtaining the bioactive conformation, a conformational sampling process is undertaken [11]. In the theoceptor approach, a short (~1 ns), hot (500–2000 K) molecular dynamics simulation uses xTB while the SHAKE algorithm constrains all bond lengths [12–15]. The resulting structures are optimized at the GFN0-xTB level. Isostat within Molclus organizes these into clusters, extracting one structure per cluster for subsequent geometry optimization at the GFN2-xTB level and a second round of clustering [13, 16]. Low-energy conformations (typically those expected to be present to at least ~1 % in a simple Boltzmann distribution or those within 5 kcal/mol of the lowest energy conformation) are passed forward to docking.

1.1.3 Step 4: Docking

A docking protocol generates starting geometries for subsequent QM calculations. This requires adequate preparation of the receptor by adding missing residues, removing nonstructural water molecules, and assigning protonation states using established tools (e.g., PyMOL, UCSF Chimera, H++, PROPKA) [17–20]. Docking is usually performed with AutoDock Vina using default parameters unless otherwise specified [6, 7]. For most systems, the search space was defined by a 20 Å × 20 Å × 20 Å grid box centered on the coordinates of the co-crystallized ligand; box dimensions can be adjusted as needed to encompass the potential active site. The exhaustiveness parameter was set to 32 to enhance sampling of binding modes. All docked poses can be propagated through to theoceptor calculations, or similar poses can be clustered to reduce the number of calculations.

1.1.4 Step 5: Theoceptor Calculation

The theoceptor is constructed from protein residues based on their spatial proximity to the ligand. For any particular protein–ligand complex, an in-house script iterates through every atom in the ligand and identifies any receptor atoms within a 5 Å radius. In protein receptors, sidechain atoms trigger inclusion of the complete sidechain (including C_α and C_β that will be fixed during optimization, while other sidechain atoms are free to move), while main chain atoms trigger construction of an $AcNHCH_2CONHMe$ unit (that will also be fixed). Any remaining protein atoms not within the radius of any ligand atom are excluded from the final theoceptor model. The resulting theoceptor structure is optimized using the GFN2-xTB method to calculate the electronic energy of the unbound receptor [13].

Meeko is employed to enable hydrogen atom placement after removal by Vina, and poses are incorporated into the theoceptor model to generate a ligand–receptor complex structure ready for geometry optimization [21]. This complex is optimized (respecting the constraints noted above) using GFN2-xTB to obtain the electronic energy of the complex. Subsequently, other levels of theory can be applied to this optimized geometry. The same level of theory is applied to the free ligand and the empty receptor; optionally, the solvation settings can be different to those used for the complex.

1.1.5 Step 6: Statistical Model Building

logP is widely used in drug discovery [22]. In Lipinski's rule of five, logP is one of the essential parameters [23, 24]. logP measures a compound's partitioning between n-octanol and water, which reflects the molecule's hydrophobicity. This property has been shown to be an important indicator of solubility, oral bioavailability, receptor binding, and toxicity. For computer-aided drug design, several methods for calculating logP based on molecular constitution have been established. These calculations largely rely on an additive method in which the contributions to logP by fragments of a molecule (atoms and functional groups) are summed. In this chapter, logP values were usually computed by the Virtual Computational Chemistry Laboratory (VCCLAB) ALOGPS v2.1 web server using ligand SMILES/SDF structures as inputs [25].

In the final step, logP values are combined with the values of ΔE computed in Step 5 (using either semi-empirical or density functional levels of theory). A multilinear regression model in the form of Eq. 3 is built that requires both terms (ΔE and logP) to be statistically significant; we note that this requires theoceptor models to pass a test that is frequently not checked with other approaches.

2 Theoceptor Applications 1: Enzymes

Enzymes are, in a sense, the most familiar application for the theoceptor method, given that it stems from the Houk group's "theozymes" [1, 3, 4].

2.1 Initial Forays Using DFT for Geometry Optimization

Our group's prototypical study using a theoceptor focused on a binding site that included a heme cofactor in inducible nitrous oxide synthase (iNOS) [9]. The study obtained binding energies at the QM level following manual selection of active site residues. This was the first instance where a statistical treatment relating calculated properties to experimental values was applied. Binding energies of a handful of ligands were correlated with pIC_{50} and LLE (lipophilic ligand efficiency; defined as pIC_{50} − clogP) using Hartree–Fock theory and a range of DFT functionals. These QM-level values were compared to those obtained via docking force fields. An overall improvement in predictions of both pIC_{50} and LLE was seen when the iNOS system was treated quantum-mechanically. The geometries obtained for the low-spin doublet state are in better agreement with those observed experimentally, supporting this as the relevant spin state, which is not easily obtained from X-ray crystallography or force-field methods alone.

The fullest published study that describes the use of a theoceptor is of the lactate dehydrogenase A enzyme [8]. In this example, fully quantum mechanical geometry optimizations (M06/6-31+G**) were performed that provided a remarkable correspondence between experimental affinity and those computed. More surprisingly, this study highlighted that when the binding energy for a compound containing a tetrahydropyran ring is computed using the structure of the ligand deposited in the protein databank, a poor prediction of the binding energy is obtained [26]. This structure (with protein databank code 4r69) included a high-energy twist-boat conformation of the ring that, when corrected to the more stable chair conformation, led to a computed ΔE consistent with the experimentally observed IC_{50}. When the energy that corresponds to a particular ligand placement or conformation disagrees with that expected for their pIC_{50}, a theoceptor naturally identifies incorrect ligand geometries in structural biology.

2.2 Covalent Inhibition in Enzymes

Given that the theoceptor approach treats molecules with quantum mechanics, theoceptors can model both covalent and noncovalent enzyme inhibition. Computationally, covalent inhibition is a poorly treated aspect of enzymatic modeling because bond-breaking/-making processes are not well-represented in most force fields. This relegates accurate modeling of covalent inhibition to QM/MM or other multiscale methods that are computationally costly, especially from a high-throughput perspective [27]. There are methods to predefine a covalent link in docking programs such as AutoDock Vina, but such implementations are not straightforward and do not treat the interaction accurately through omission of the underlying quantum mechanical effects.

To address this, the theoceptor method has been adapted to model boronic acid inhibitors of the autotaxin (ATX) enzyme, a type of pyrophosphatase, where a number of structures feature covalent binding of boron to the protein (PDB: 2XRG, 3WAX, 3WAY, 7Z0N) [28–30]. Expecting the formation of a reversible covalent link between the T209 alkoxide and the boron atom, accurate binding geometries to an RMSD range of <1 Å were obtained through a relatively trivial addition of an external potential along the B-O distance through the GFN2-xTB method. This implementation has also been semi-automated and can screen for potential covalent sites at no increase in cost. As shown in Fig. 2, a docking pose that is in the expected region (in magenta carbons) can be optimized through the theoceptor workflow to give a structure (in cyan carbons) that is in excellent agreement with the deposited crystal structure (in green carbons). Experimental evidence shows that certain boron-containing compounds do not form a covalent bond with T209 [28]. The use of quantum mechanics permits both binding modes to be evaluated and the preferred one to be identified (the one with the best ΔE). This approach has also been successfully applied to covalent inhibitors of FGFR4 and its mutants, applying the same external potential to drive the reaction between C552 and an α,β-unsaturated carbonyl group. As with ATX, this significantly improved the accuracy of predicted binding poses with the covalent link enabling the remainder of the ligand structures to better match crystal observations.

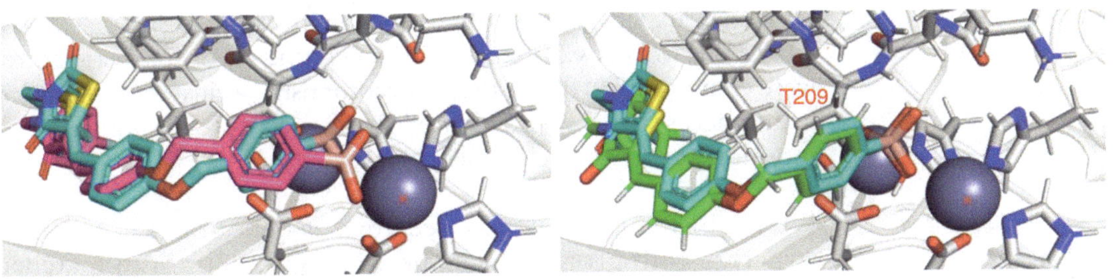

Fig. 2 Comparison between docked (magenta), theoceptor-optimized (green) and crystal structure (cyan) bound ligand structures for HA155, a covalent boronic acid ATX inhibitor. A portion of the theoceptor residues is shown for clarity, which was generated from the 2XRG PDB crystal structure

3 Theoceptor Applications 2: Cardiac Ion Channels

3.1 Ion Channels and Matched Molecular Pairs

Cardiac ion channels play a pivotal role in maintaining the electrophysiological activities of the heart by modulating the generation and propagation of action potentials (AP) [31, 32]. In drug development, these channels serve both as critical pharmacological targets and key safety indicators [33, 34]. The voltage-gated sodium channel $Na_V1.5$ and potassium channel hERG/$K_V11.1$ were selected for initial investigation due to their central roles in shaping the cardiac AP.

The drug binding sites of $Na_V1.5$ and hERG channels are generally located within the central cavity formed by the four pore-forming domains, beneath the selectivity filter and above the intracellular gate [35, 36]. This cavity encompasses multiple interaction hotspots rather than a single defined binding pocket [37, 38]. Therefore, in constructing a theoceptor model, we include all residues that possibly contribute to ligand binding. Importantly, we ensured incorporation of experimentally validated key residues for drug interactions (F1760 and Y1767 in $Na_V1.5$ [39, 40]; Y652 and F656 in hERG [41, 42]).

Matched molecular pair analysis (MMPA), first introduced by Kenny and Sadowski, is a widely used technique in drug design [10, 43, 44]. The method involves identifying sets of pairs of compounds that differ by the same, well-defined structural transformation(s) and associates this change in structure with a change in activity using experimental data. In this work, MMPA was applied to compounds from the ChEMBL database to investigate how a small structural change can affect the pIC_{50} against the ion channels. We will discuss two matched pairs for each channel and their structure–activity relationships with respect to ion channel inhibition. These pairs are also used as an additional validation test for the theoceptor models; the greater the number of correctly predicted matched pairs, the more relevant to drug design a predictive model is likely to be, as it can reproduce experimentally observed activity changes.

In this work, two methods for generating theoceptor structures were employed: (1) define the theoceptor using a cluster of all docked ligand poses (hERG) and (2) define the theoceptor using chemical intuition and literature studies ($Na_V1.5$).

3.2 Studies of the hERG Potassium Channel

For the hERG channel, a theoceptor was built from the cryo-EM structure with protein databank code 8zyo [36]. Ten matched pairs were selected from the ChEMBL database. We will discuss two of these matched pairs, astemizole and CHEMBL4205490, and quinidine and quinine. Their 2D chemical structures and 3D xTB optimized theoceptor complex structures are shown in Fig. 3. The theoceptor model correctly predicts which compound is a stronger hERG binder when the calculated binding energies (ΔE)

are considered, even for these pairs that cannot be rationalized by lipophilicity (logP).

Where astemizole contains a *para*-fluorobenzyl substitution on the benzimidazole, CHEMBL4205490 has an octyl group, as shown in Fig. 3. CHEMBL4205490 has a slightly higher logP than astemizole, indicating it is slightly more hydrophobic. Typically, higher logP corresponds with increased hERG binding affinity (higher pIC_{50}) [45, 46]. In this case, astemizole is the more potent hERG blocker despite a lower logP. Turning to the calculated xTB binding energies, ΔE, the difference between this matched pair is 7.04 kcal/mol. The optimized theoceptor complex ligand pose reveals that the overall binding mode is similar. Both poses place the piperidine, which is positively charged at physiological pH, in the center of the pore, the same region where potassium ions flow. The anisole group is also positioned similarly in both compounds. However, the benzimidazole moiety differs in placement between the two compounds. Due to the fourfold symmetry of the hERG channel, either orientation would be consistent with the cryo-EM structure for astemizole [36], where both placements for benzimidazole are proposed. Further analysis of the poses reveals that the fluorobenzene group in astemizole allows for favorable π–π stacking with Y652, whereas in CHEMBL4205490, this is not possible with the long alkyl chain. This example illustrates how a change in chemical structure can affect hERGbinding affinity through specific drug–channel interactions, which can be captured by the theoceptor model.

In the case of quinidine and quinine, the ligands are stereoisomers and share the same logP value, see Fig. 3B. This means that in the theoceptor model, the observed change in potency against the channel can only be explained by the ΔE term. The calculated binding energy difference between quinidine and quinine is only 2.96 kcal/mol, indicating that the model captures a subtle effect, such as increased hydrogen bond donation from the quinidine NH group. Examination of the lowest energy xTB binding poses shows that the main difference is in the orientation of the hydroxyl group, likely modulating interactions within the channel.

3.3 Studies of the $Na_V1.5$ Sodium Channel

MMPA generated 11 matched pairs from an input set of $Na_V1.5$ inhibitors from the ChEMBL database. Although the chemical structures among these pairs are diverse, they can be broadly classified into two categories based on their conformational space: (1) large and flexible molecules and (2) small and rigid molecules. The former contains at least one flexible moiety, such as alicyclic six-membered rings, while the latter possesses rigid molecular cores and their flexibility is confined to rotatable bonds in the substituents. The $Na_V1.5$ theoceptor model was constructed using a cryo-EM structure (with protein databank code 6lqa) [35]. Residues facing the central cavity are incorporated into the model, including the inner surfaces

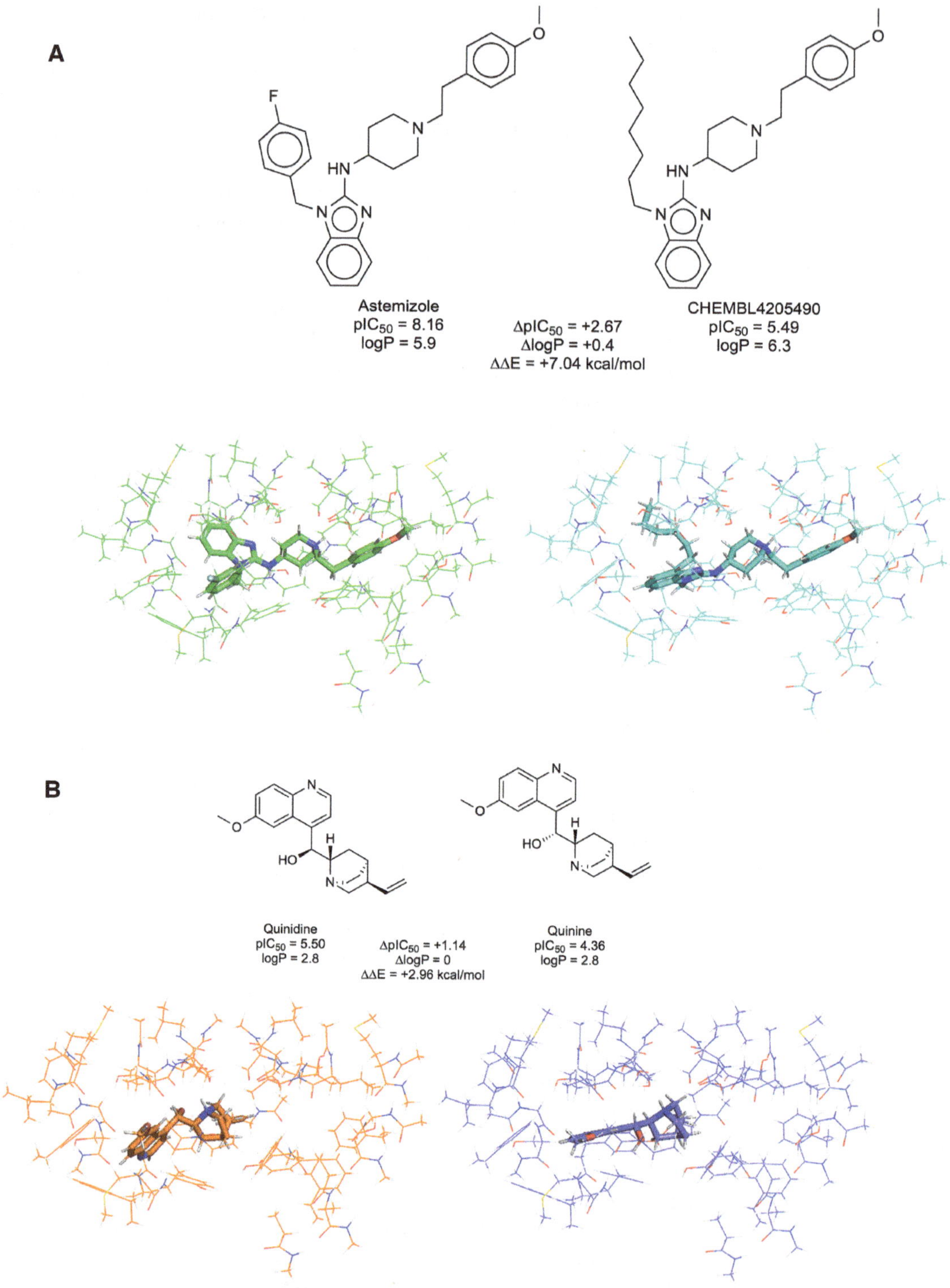

Fig. 3 (**A**) Chemical structures of the matched pair astemizole and CHEMBL4205490 in the top panel. The bottom panel shows 3D conformations of the lowest xTB optimized theoceptor complexes with the hERG channel, with astemizole in green and CHEMBL4205490 in blue. Ligands are shown in sticks and the theoceptor in lines. (**B**) Chemical structures of the matched pair quinidine and quinine in the top panel. The bottom panel shows 3D conformations of the lowest xTB optimized theoceptor complexes with quinidine in orange and quinine in purple. Ligands are shown in sticks and the theoceptor in lines

of four fenestrations (the windows formed between two adjacent pore-forming domains).

As illustrated in Fig. 4, two matched pairs from each category were selected as examples to discuss their predicted binding modes. The binding energies (ΔE) calculated from the theoceptor model correctly distinguished the stronger binder (with higher pIC_{50}), suggesting that the poses with the lowest ΔE likely represent their stable binding conformations.

For the flexible pair (CHEMBL2012247 and CHEMBL2012259: Fig. 4, left), the structural difference arises from an additional fluorine substituent in CHEMBL2012259. Their lowest-energy binding poses are nearly identical. In both compounds, the piperidine ring is positioned near the intracellular gate of the channel, and the adjacent thiazolyl moiety forms a π–π interaction with Y1767—one of the validated key residues. This interaction behavior likely contributes to their moderate-to-strong inhibitory activity, as both compounds directly occupy the ion conduction pathway. Their phenyl groups extend into the DII-III fenestration, engaging in hydrophobic and van der Waals interactions with surrounding residues. As the steric difference between a phenyl and an ortho-fluorophenyl group is small, and the DII-III fenestration is relatively large and hydrophobic

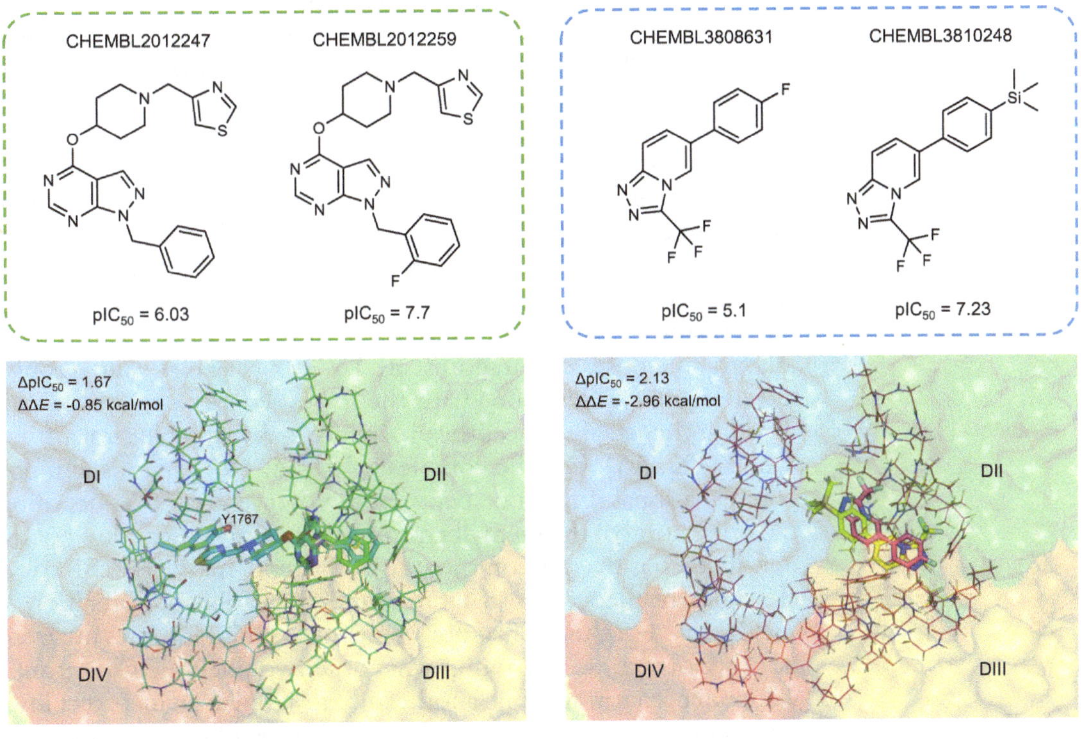

Fig. 4 Chemical structures and predicted binding modes of two matched molecular pairs from the $Na_V1.5$ inhibitor set. The 3D conformations shown are GFN2-xTB optimized geometries with the lowest binding energy (ΔE). Ligands are shown in sticks and the theoceptor in lines

in nature, the pocket can accommodate both substituents, resulting in the lowest-energy binding poses for the two molecules with ΔE values differing by only 0.85 kcal/mol.

For the rigid pair (CHEMBL3808631 and CHEMBL3810248), the structural difference lies in the substituents on their phenyl rings. CHEMBL3810248 features a bulkier and more hydrophobic trimethylsilyl (–$SiMe_3$) group compared with the fluorine substituent in CHEMBL3808631. Owing to their compact molecular scaffolds, both compounds can access multiple binding regions within the pocket, including the central pore and fenestrations. For both molecules, the lowest-energy binding poses are located near the DII and DII-III fenestration. The bulky trimethylsilyl group of CHEMBL3810248 enables additional interactions with residues in DI, while CHEMBL3808631 interacts more prominently with the DII residues.

4 Theoceptor Applications 3: GPCRs

The complex biology of the G-Protein Coupled Receptors (GPCRs) makes them a challenge for the theoceptor methodology. Our initial interest in this area concerned the adrenergic receptors, for which there are approved and widely used drugs that are both agonists and antagonists [47–49]. The study of these receptors spanned the period from the iNOS and LDHA models described previously, where full DFT geometry optimization was employed in the calculations (see above), to more recent work where xTB is widely used.

There are a number of available X-ray crystal structures of the β_1 isoform of the adrenergic receptor with both agonists and antagonists bound [50]. The strategy was to employ an exemplar protein structure from the Protein Databank for each mode of action. Thus, structure with PDB code 7bu7 was used to represent antagonist binding and 7bu6 agonist binding [50]. When comparing the crystallographic poses of antagonist BI-167107 (in 7bu7) and agonist norepinephrine (in 7bu6) with the lowest energy poses obtained through the usual theoceptor procedure, a remarkable similarity is found (Fig. 5A). For the complex of BI-167107, there is a minor conformational difference between the two structures concerning the orientation of the methylbenzene substituent. When the electron density map is overlaid on these two (magenta mesh, Fig. 5B), it is clear that the theoceptor structure is an equally valid interpretation of the density. For the agonist nor-epinephrine, initial studies involved docking of a conformation in which the two phenolic OH groups were oriented towards one another and resulted in an incorrect lowest energy orientation (not shown). This level of detail concerning the orientation of hydrogen bonds is valuable for those seeking to design drugs or other molecules that bind to proteins. A remarkable

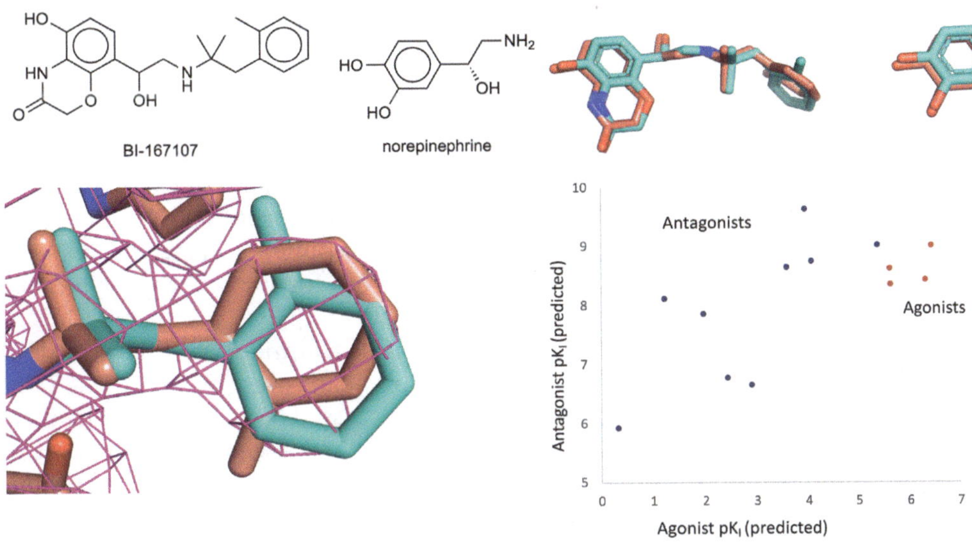

Fig. 5 (**A**) The chemical structure of antagonist BI-167107 and agonist norepinephrine and their lowest energy pose obtained from theoceptor calculations (peach-colored carbons) compared to the crystallographic pose (cyan-colored carbons). Protein atoms have been deleted for clarity. (**B**) The lowest energy pose from the theoceptor (pink) and crystallographic pose (cyan) overlaid with the mesh for the $_c$ electron density ($2F_0$-F_c). (**C**) The predicted pK_I based on a theoceptor using an antagonist-bound protein conformation (7bu7) plotted against the predicted pK_I using an agonist-bound conformation (7bu6). The identity of the compounds as antagonists (blue) or agonists (red) is shown

possibility with theoceptor calculations is to compute the binding energy of agonists to antagonist conformations of the protein and vice versa. This results in an interesting separation of the two modes of action (Fig. 5C). The ability to identify modes of action according to differences in binding to different protein conformations is being further explored in current research.

5 Theoceptor Applications 4: Corrosion Inhibitors

Throughout this chapter, the theoceptor method has been discussed as a means of describing protein–ligand binding. However, it is also applicable to the binding of potential corrosion inhibitors to metal surfaces, specifically by modeling the interaction with large (100 s of atoms) iron clusters. Organic corrosion inhibitors are regarded as a highly successful approach to corrosion control, while also sharing many characteristics with drug-like compounds. Indeed, paracetamol (or acetaminophen) is a viable candidate for reducing the onset of both headaches and rust [51, 52], and so, for the purpose of the theoceptor methodology, corrosion inhibitors can be treated in the same manner as described previously for ligands (and will be referred to as such).

What remains is to treat large iron clusters as if they were protein target structures. Evidently, the application of the theoceptor method diverges from previous descriptions at this stage, as in lieu of crystallographic iron structures and suitable docking software, ligands are placed into computationally generated iron complexes (using CrystalMaker software [53]) in a semi-automated process. The GFN0-xTB method has been selected for the theoceptor creation and optimizations in this application [54].

In evaluating the application of the theoceptor approach to describe corrosion inhibitor-metal binding, MMPA has been employed in a similar manner as discussed for cardiac ion channels. One example identified in MMPA, which ostensibly improves corrosion inhibition performance, is *N*-phenylation, as shown for two azo-derivatives, A and B, in Fig. 6 [55].

The validation of the theoceptor method here is twofold. Firstly, a visual inspection of the lowest energy optimized outputs shows that reasonable binding modes are being produced. Secondly, the binding affinities calculated from the theoceptor method correlate with the

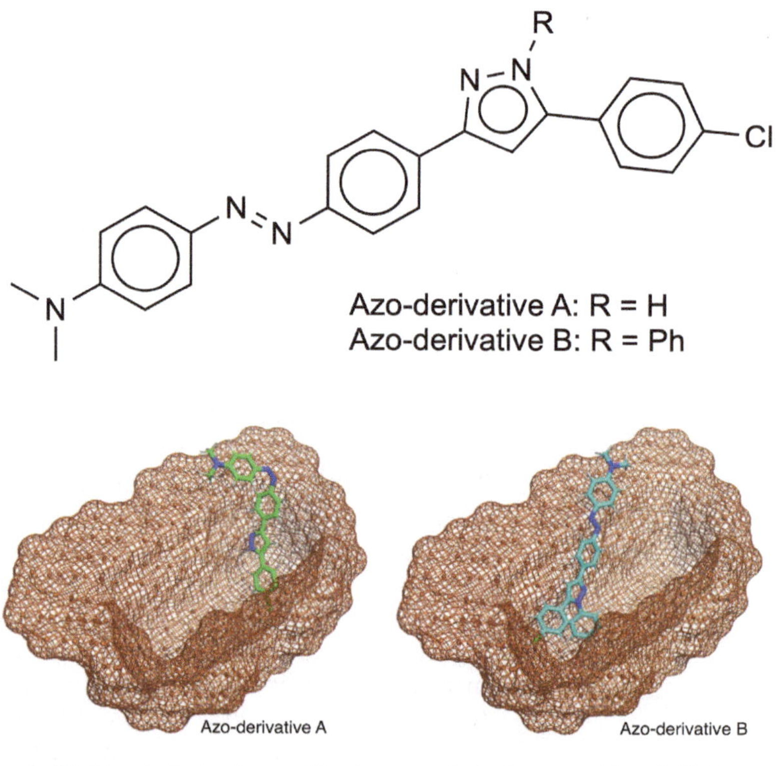

Fig. 6 (**A**) Chemical structures of the two azo-derivatives considered. (**B**) Lowest energy theoceptor structures of the two azo-derivatives interacting with a small crevice in an iron surface

MMPA; in this example, Azo-derivative B has a higher inhibition score and a stronger binding affinity by the theoceptor method. While the intricacies of this variation of the theoceptor method are still being refined, these early results indicate that the approach is promising. This could go some way to describing corrosion inhibitor–metal interactions in a field where molecular-scale modeling is the exception instead of the norm.

6 Concluding Comments

The theoceptor method, in which a sequence of conformational expansion, docking and optimization with xTB and optional DFT single points has proved an efficient and effective method to explore potential energy surfaces for receptor–ligand binding. In studies of the enzyme LDHA, we were able to show that apparently poor predictions of binding energy can flag poor crystal structure assignments. Studies of autotaxin have revealed that covalent binding can be treated equivalently to noncovalent. In studies of ion channels, we have shown that the method can be applied to cryo-electron microscopy-derived structures even when the ligands involved span a wide chemical diversity. Studying matched molecular pairs enables a clear interpretation to be obtained that can guide molecular design. When investigating agonists and antagonists binding to the adrenergic receptors, we have found that the mode of action of drugs can be assigned by computing binding energy with multiple protein conformations. Finally, we are exploring the transfer of the method to studying how corrosion inhibitors interact at the surface of metals in order to exert their effect. The development of the method is ongoing.

References

1. Kiss G, Celebi-Olcum N, Moretti R, Baker D, Houk KN (2013) Computational enzyme design. Angew Chem Int Ed Engl 52(22):5700–5725. https://doi.org/10.1002/anie.201204077
2. Kiss G, Röthlisberger D, Baker D, Houk KN (2010) Evaluation and ranking of enzyme designs. Protein Sci 19(9):1760–1773. https://doi.org/10.1002/pro.462
3. Tantillo DJ, Chen J, Houk KN (1998) Theozymes and compuzymes: theoretical models for biological catalysis. Curr Opin Chem Biol 2(6):743–750. https://doi.org/10.1016/s1367-5931(98)80112-9
4. Zhang X, DeChancie J, Gunaydin H, Chowdry AB, Clemente FR, Smith AJ, Handel TM, Houk KN (2008) Quantum mechanical design of enzyme active sites. J Org Chem 73(3):889–899. https://doi.org/10.1021/jo701974n
5. Dechancie J, Clemente FR, Smith AJ, Gunaydin H, Zhao YL, Zhang X, Houk KN (2007) How similar are enzyme active site geometries derived from quantum mechanical theozymes to crystal structures of enzyme-inhibitor complexes? Implications for enzyme design. Protein Sci 16(9):1851–1866. https://doi.org/10.1110/ps.072963707
6. Eberhardt J, Santos-Martins D, Tillack AF, Forli S (2021) AutoDock Vina 1.2.0: new docking methods, expanded force field, and python bindings. J Chem Inf Model 61(8):3891–3898. https://doi.org/10.1021/acs.jcim.1c00203
7. Trott O, Olson AJ (2010) AutoDock Vina: improving the speed and accuracy of docking

with a new scoring function, efficient optimization, and multithreading. J Comput Chem 31(2):455–461. https://doi.org/10.1002/jcc.21334
8. Lukac I, Abdelhakim H, Ward RA, St-Gallay SA, Madden JC, Leach AG (2019) Predicting protein-ligand binding affinity and correcting crystal structures with quantum mechanical calculations: lactate dehydrogenase A. Chem Sci 10(7):2218–2227. https://doi.org/10.1039/c8sc04564j
9. Leach AG, Olsson L-L, Warner DJ (2013) A monomeric form of iNOS can rationalise observed SAR for inhibitors of dimerisation: quantum mechanics and docking compared. MedChemComm 4(1):180–186. https://doi.org/10.1039/c2md20159c
10. Kenny PW, Sadowski J (2005) Structure modification in chemical databases. In: Oprea TI (Ed.) Chemoinformatics in drug discovery. Wiley-VCH, Weinheim, 271–285 https://doi.org/https://www.wiley-vch.de/en/areas-interest/natural-sciences/chemoinformatics-in-drug-discovery-978-3-527-30753-1
11. Sliwoski G, Kothiwale S, Meiler J, Lowe Jr EW (2014) Computational methods in drug discovery. Pharmacol Rev 66(1):334–395. https://doi.org/10.1124/pr.112.007336
12. Pracht P, Bohle F, Grimme S (2020) Automated exploration of the low-energy chemical space with fast quantum chemical methods. Phys Chem Chem Phys 22(14):7169–7192. https://doi.org/10.1039/c9cp06869d
13. Bannwarth C, Ehlert S, Grimme S (2019) GFN2-xTB-an accurate and broadly parametrized self-consistent tight-binding quantum chemical method with multipole electrostatics and density-dependent dispersion contributions. J Chem Theory Comput 15(3):1652–1671. https://doi.org/10.1021/acs.jctc.8b01176
14. Ryckaert J-P, Ciccotti G, Berendsen HJC (1977) Numerical integration of the cartesian equations of motion of a system with constraints: molecular dynamics of n-alkanes. J Comput Phys 23(3):327–341. https://doi.org/10.1016/0021-9991(77)90098-5
15. Pracht P, Grimme S, Bannwarth C, Bohle F, Ehlert S, Feldmann G, Gorges J, Muller M, Neudecker T, Plett C, Spicher S, Steinbach P, Wesolowski PA, Zeller F (2024) CREST-A program for the exploration of low-energy molecular chemical space. J Chem Phys 160(11), 114110. https://doi.org/10.1063/5.0197592
16. Lu T (2023) Molclus Program, Version 1.12. http://www.keinsci.com/research/molclus.html
17. Pettersen EF, Goddard TD, Huang CC, Couch GS, Greenblatt DM, Meng EC, Ferrin TE (2004) UCSF Chimera—a visualization system for exploratory research and analysis. J Comput Chem 25(13):1605–1612. https://doi.org/10.1002/jcc.20084
18. Anandakrishnan R, Aguilar B, Onufriev AV (2012) H++ 3.0: automating pK prediction and the preparation of biomolecular structures for atomistic molecular modeling and simulations. Nucleic Acids Res 40:W537–W541. https://doi.org/10.1093/nar/gks375
19. Olsson MH, Sondergaard CR, Rostkowski M, Jensen JH (2011) PROPKA3: consistent treatment of internal and surface residues in empirical pKa predictions. J Chem Theory Comput 7(2):525–537. https://doi.org/10.1021/ct100578z
20. DeLano W The PyMOL Molecular Graphics System, Version 2.5.0, Schrödinger, LLC
21. Santos-Martins D, He Y, Eberhardt J, Sharma P, Bruciaferri N, Holcomb M, Llanos MA, Hansel-Harris A, Barkdull AP, Tillack AF, Bianco G, Paulsen M-L, Mato J, Taneja I, Forli S (2025) Meeko: molecule parametrization and software interoperability for docking and beyond. J Chem Inf Model 65(24):13045–13050. https://doi.org/10.1021/acs.jcim.5c02271
22. Waring MJ (2010) Lipophilicity in drug discovery. Expert Opin Drug Discov 5(3):235–248. https://doi.org/10.1517/17460441003605098
23. Lipinski CA (2004) Lead- and drug-like compounds: the rule-of-five revolution. Drug Discov Today Technol 1(4):337–341. https://doi.org/10.1016/j.ddtec.2004.11.007
24. Lipinski CA, Lombardo F, Dominy BW, Feeney PJ (1997) Experimental and computational approaches to estimate solubility and permeability in drug discovery and development settings. Adv Drug Deliv Rev 23(1-3):3–25
25. Tetko IV, Gasteiger J, Todeschini R, Mauri A, Livingstone D, Ertl P, Palyulin VA, Radchenko EV, Zefirov NS, Makarenko AS, Tanchuk VY, Prokopenko VV (2005) Virtual computational chemistry laboratory—design and description. J Comput Aided Mol Des 19(6):453–463. https://doi.org/10.1007/s10822-005-8694-y
26. Labadie S, Dragovich PS, Chen J, Fauber BP, Boggs J, Corson LB, Ding CZ, Eigenbrot C, Ge H, Ho Q, Lai KW, Ma S, Malek S, Peterson D, Purkey HE, Robarge K, Salphati L, Sideris S, Ultsch M, VanderPorten E, Wei B, Xu Q, Yen I, Yue Q, Zhang H, Zhang X, Zhou A (2015) Optimization of 5-(2,6-dichlorophenyl)-3-hydroxy-2-mercaptocyclohex-2-enones as potent inhibitors of human lactate dehydrogenase. Bioorg Med Chem Lett 25(1):75–82. https://doi.org/10.1016/j.bmcl.2014.11.008

27. Li S, Du H, Zhang H, Zhang X, Ye Y, Shen C, Hou T, Pan P (2025) Accelerating covalent drug discovery: recent advances in covalent docking tools. WIREs Comput Mol Sci 15(6), e70062. https://doi.org/10.1002/wcms.70062
28. Clark JM, Salgado-Polo F, Macdonald SJF, Barrett TN, Perrakis A, Jamieson C (2022) Structure-based design of a novel class of autotaxin inhibitors based on endogenous allosteric modulators. J Med Chem 65(8):6338–6351. https://doi.org/10.1021/acs.jmedchem.2c00368
29. Kawaguchi M, Okabe T, Okudaira S, Nishimasu H, Ishitani R, Kojima H, Nureki O, Aoki J, Nagano T (2013) Screening and X-ray crystal structure-based optimization of autotaxin (ENPP2) inhibitors, using a newly developed fluorescence probe. ACS Chem Biol 8(8):1713–1721. https://doi.org/10.1021/cb400150c
30. Hausmann J, Kamtekar S, Christodoulou E, Day JE, Wu T, Fulkerson Z, Albers HM, van Meeteren LA, Houben AJ, van Zeijl L, Jansen S, Andries M, Hall T, Pegg LE, Benson TE, Kasiem M, Harlos K, Kooi CW, Smyth SS, Ovaa H, Bollen M, Morris AJ, Moolenaar WH, Perrakis A (2011) Structural basis of substrate discrimination and integrin binding by autotaxin. Nat Struct Mol Biol 18(2):198–204. https://doi.org/10.1038/nsmb.1980
31. Nerbonne JM, Kass RS (2005) Molecular physiology of cardiac repolarization. Physiol Rev 85(4):1205–1253. https://doi.org/10.1152/physrev.00002.2005
32. Grant AO (2009) Cardiac ion channels. Circ Arrhythm Electrophysiol 2(2):185–194. https://doi.org/10.1161/CIRCEP.108.789081
33. Priest BT, McDermott JS (2015) Cardiac ion channels. Channels (Austin) 9(6):352–359. https://doi.org/10.1080/19336950.2015.1076597
34. Kaczorowski GJ, McManus OB, Priest BT, Garcia ML (2008) Ion channels as drug targets: the next GPCRs. J Gen Physiol 131(5):399–405. https://doi.org/10.1085/jgp.200709946
35. Li Z, Jin X, Wu T, Huang G, Wu K, Lei J, Pan X, Yan N (2021) Structural basis for pore blockade of the human cardiac sodium channel Nav1.5 by the antiarrhythmic drug quinidine**. Angew Chemie 133(20):11575–11581. https://doi.org/10.1002/ange.202102196
36. Miyashita Y, Moriya T, Kato T, Kawasaki M, Yasuda S, Adachi N, Suzuki K, Ogasawara S, Saito T, Senda T, Murata T (2024) Improved higher resolution cryo-EM structures reveal the binding modes of hERG channel inhibitors. Structure 32(11):1926–1935 e1923. https://doi.org/10.1016/j.str.2024.08.021
37. Tao E, Corry B (2025) Drugs exhibit diverse binding modes and access routes in the Nav1.5 cardiac sodium channel pore. J Gen Physiol 157(2), e202413658. https://doi.org/10.1085/jgp.202413658
38. Vandenberg JI, Perozo E, Allen TW (2017) Towards a structural view of drug binding to hERG K(+) channels. Trends Pharmacol Sci 38(10):899–907. https://doi.org/10.1016/j.tips.2017.06.004
39. Ragsdale DS, McPhee JC, Scheuer T, Catterall WA (1996) Common molecular determinants of local anesthetic, antiarrhythmic, and anticonvulsant block of voltage-gated Na+ channels. Proc Natl Acad Sci U S A 93(17):9270–9275. https://doi.org/10.1073/pnas.93.17.9270
40. Ragsdale DS, McPhee JC, Scheuer T, Catterall WA (1994) Molecular determinants of state-dependent block of Na+ channels by local anesthetics. Science 265(5179):1724–1728. https://doi.org/10.1126/science.8085162
41. Chen J, Seebohm G, Sanguinetti MC (2002) Position of aromatic residues in the S6 domain, not inactivation, dictates cisapride sensitivity of HERG and eag potassium channels. Proc Natl Acad Sci U S A 99(19):12461–12466. https://doi.org/10.1073/pnas.192367299
42. Mitcheson JS, Chen J, Lin M, Culberson C, Sanguinetti MC (2000) A structural basis for drug-induced long QT syndrome. Proc Natl Acad Sci U S A 97(22):12329–12333. https://doi.org/10.1073/pnas.210244497
43. Griffen E, Leach AG, Robb GR, Warner DJ (2011) Matched molecular pairs as a medicinal chemistry tool. J Med Chem 54(22):7739–7750. https://doi.org/10.1021/jm200452d
44. Leach AG, Jones HD, Cosgrove DA, Kenny PW, Ruston L, MacFaul P, Wood JM, Colclough N, Law B (2006) Matched molecular pairs as a guide in the optimization of pharmaceutical properties; a study of aqueous solubility, plasma protein binding and oral exposure. J Med Chem 49(23):6672–6682. https://doi.org/10.1021/jm0605233
45. Kawai Y, Tsukamoto S, Ito J, Akimoto K, Takahashi M (2011) A risk assessment of human ether-a-go-go-related gene potassium channel inhibition by using lipophilicity and basicity for drug discovery. Chem Pharm Bull (Tokyo) 59(9):1110–1116. https://doi.org/10.1248/cpb.59.1110
46. Waring MJ, Johnstone C (2007) A quantitative assessment of hERG liability as a function of lipophilicity. Bioorg Med Chem Lett 17(6):1759–1764
47. Grogan A, Lucero EY, Jiang H, Rockman HA (2023) Pathophysiology and pharmacology

of G protein-coupled receptors in the heart. Cardiovasc Res 119(5):1117–1129. https://doi.org/10.1093/cvr/cvac171
48. Fung S (2025) Consider the cardiovascular and pulmonary effects of treatments in patients with both asthma and cardiovascular disease. Drugs Therapy Persp 41(6):254–260. https://doi.org/10.1007/s40267-025-01166-0
49. Motiejunaite J, Amar L, Vidal-Petiot E (2021) Adrenergic receptors and cardiovascular effects of catecholamines. Ann Endocrinol (Paris) 82(3-4):193–197. https://doi.org/10.1016/j.ando.2020.03.012
50. Xu X, Kaindl J, Clark MJ, Hubner H, Hirata K, Sunahara RK, Gmeiner P, Kobilka BK, Liu X (2021) Binding pathway determines norepinephrine selectivity for the human beta(1)AR over beta(2)AR. Cell Res 31(5):569–579. https://doi.org/10.1038/s41422-020-00424-2
51. Merimi C, Hammouti B, Zaidi K, Elmsellem H (2023) Comparative study of inhibitory efficacy of drug (Acetaminophen) in 1 M HCl medium applied to carbon and mild steels. Mater Today: Proc 72:3890–3895. https://doi.org/10.1016/j.matpr.2022.10.214
52. Al-Gorair AS, Abdallah M (2021) Expired paracetamol as corrosion inhibitor for low carbon steel in sulfuric acid. Electrochemical, kinetics and thermodynamics investigation. Int J Electrochem Sci 16(7), 210771. https://doi.org/10.20964/2021.07.73
53. CrystalMaker, CrystalMaker Software Ltd, Oxford, England. www.crystalmaker.com.
54. Pracht P, Caldeweyher E, Ehlert S, Grimme S (2019) A robust non-self-consistent tight-binding quantum chemistry method for large molecules. ChemRxiv. https://doi.org/10.26434/chemrxiv.8326202.v1
55. Zahraa TK, Mehdi Salih S (2023) Study of some azo derivatives as corrosion inhibitors for mild steel in 1 M H_2SO_4. Surf Eng Appl Electrochem 58(6):708–719. https://doi.org/10.3103/s1068375522060151

Chapter 14

Quantum-Driven High-Throughput Docking: Advances in Molecular Scoring and Drug Discovery

Nicolás Arrupe and Claudio N. Cavasotto

Abstract

Computational methods are consolidated tools in early drug lead discovery. Among them, high-throughput molecular docking (HTD) is widely used to identify bioactive compounds within large chemical libraries by assessing likely binding modes and assigning docking scores. These methods also inform how hits may be modified to improve protein–ligand interactions, thus guiding lead optimization while reducing time and cost compared with experimental screening. For more than a decade, advances in methodology and computing power have renewed interest in applying quantum mechanics (QM) to macromolecular systems. A QM description of molecular association can provide improved accuracy over classical molecular mechanics (MM)-methods, since phenomena such as covalent bond formation, electronic polarization, charge transfer, bond rearrangement, and halogen bonding require explicit treatment of electrons. Here, we review recent developments in QM-based molecular docking and HTD, focusing on cases where QM is used explicitly in docking workflows.

Keywords Molecular docking, Quantum mechanics, Semiempirical methods, Drug discovery, High-throughput docking

1 Introduction

The discovery of new drugs remains one of the greatest challenges for the pharmaceutical industry. Despite decades of progress, the process of bringing a novel lead from concept to market is still lengthy and costly, with high failure rates throughout the development pipeline [1–3]. To address these challenges, computational approaches have become indispensable, offering the means to explore chemical space more efficiently and to guide experimental efforts with greater precision. Indeed, the global investment in computer-aided drug design reflects not only its economic importance but also its growing role in shaping modern discovery strategies.

Within this landscape, molecular docking has emerged as one of the most widely applied in silico strategies, especially for early-stage hit identification and prioritization [4]. By simulating how small

Alexander Heifetz (ed.), *Quantum Mechanics Applications in Drug Design*, Methods in Molecular Biology, vol. 3061, https://doi.org/10.1007/978-1-0716-5420-0_14,

molecules interact within protein binding sites, docking enables the rapid identification of candidates with potential biological activity. This approach involves systematic sampling of ligand conformations within the active site and scoring the resulting poses according to their steric and energetic complementarity [5]. The ability of docking scores to discriminate potential small-molecule binders from non-binders is a determining factor in the success of rational drug design, since it directly affects the ability to prioritize molecules that can advance to experimental testing.

Experimental high-throughput screening (HTS) has long served as a cornerstone of early discovery, operating as the first large-scale filter for identifying promising hits [6]. HTS relies on four well-established pillars: access to relevant and diverse libraries, the availability of robust and automatable assays, advanced robotics for execution, and rigorous data analysis pipelines [7]. Despite advances in this technique since the 1990s, key limitations remain, including incomplete chemical diversity and the cost and reproducibility of assays [8]. Complementarily, high-throughput docking (HTD) has become a sort of virtual analog of HTS. HTD screens large chemical libraries in silico, explores possible binding modes, and assigns to each ligand a docking score (typically, a computationally cheaper approximation of binding free energy, ΔG_{bind}) [9]. These scores are then used to rank molecules for downstream experimental validation. Much like HTS, HTD is built on a funnel logic; it is efficient at generating large numbers of candidates, but its reliability depends critically on the accuracy of the scoring function, the treatment of solvation and protein flexibility, the inclusion of water molecules, and the quality of receptor structures [10]. Both approaches, therefore, share common challenges: the quality and diversity of their libraries, the accuracy of the "assay" (whether experimental or computational), the level of automation, and the inevitable trade-offs between throughput and accuracy [11].

2 Docking Stages and Scoring Functions

Molecular docking workflows involve two main phases: (1) conformational sampling, which explores the configurational space to identify energetically favorable binding poses, and (2) affinity estimation, which ranks molecules according to their likelihood of binding through a scoring function [12]. While some software packages use the same function to evaluate both docking energy and scoring energy, this does not necessarily have to be the case. Therefore, maintaining this distinction remains conceptually important.

Modern scoring approaches can be classified into categories: force field-based, empirical, and knowledge-based [13, 14].

Docking quality is generally assessed across four complementary features [15]: (1) docking power, or the ability to recover the experimentally observed binding pose; (2) scoring power, referring to the correlation between predicted and experimental binding affinities; (3) ranking power, which evaluates whether congeneric ligands can be correctly ordered by affinity once reliable poses are provided; and (4) screening power, often measured as the early enrichment of true ligands within virtual screening subsets. Notably, benchmarks such as CASF have consistently shown that pose recovery tends to be more reliable than scoring, ranking, or screening, the latter being far more sensitive to choices of energy models and solvation treatment. As a result, many practical workflows adopt a hybrid strategy: classical docking is used to generate plausible poses, which are then rescored with more demanding approaches, such as QM-based, to improve ranking and screening outcomes [16]. While progress has been steady, it is fair to admit that all methods have their inherent limitations, and finding the right balance between speed and accuracy remains as much an art as a science.

3 QM-Based Scoring Functions

As discussed above, docking programs are generally effective at recovering near-native poses [17–20]. However, correlating predicted scores with experimental affinities, ranking congeneric ligands, and achieving early enrichment in virtual screening remain more challenging. By explicitly considering electrons, QM naturally accounts for polarization, charge transfer, covalent bonding, and halogen bonding, features often oversimplified in classical scoring.

Yet, the integration of quantum mechanical methods into molecular docking remains a formidable challenge in contemporary computational chemistry [21]. Traditional QM approaches, including density functional theory (DFT) and post Hartree–Fock methods, provide extraordinarily precise descriptions of electronic interactions, but at a computational cost that scales exponentially with system size. This unfavorable scaling constitutes a fundamental barrier to their direct application in docking workflows.

Over the past decade, advances in computational algorithms and hardware performance have facilitated the development of QM-based functions capable of approximating binding affinity and being used as scoring functions in molecular docking. As a result, semiempirical quantum mechanical (SQM) methods, in particular linear scaling ones, have emerged as particularly promising, offering an attractive compromise between the accuracy of quantum-level descriptions and the efficiency required for practical applications, thereby overcoming many of the computational barriers associated with ab initio QM approaches [22–24].

From the perspective of evaluation metrics, the primary goal of SQM methods is to enhance scoring/ranking power (i.e., increase correlation with affinities and improve ordering with poses) without attempting to replace large-scale pose sampling, where QM cost would be prohibitive [25].

Around 2020, the field of QM-based docking reached a moment of consolidation. Cavasotto et al. [26, 27] reported one of the most extensive validations to date in a high-throughput docking scenario, using an end-to-end workflow to ten protein systems and more than 50,000 complexes (>1390 ligands and 48,700 decoys). The pipeline combined classical pose generation with successive refinement steps: local minimization to remove steric clashes, systematic active-site cutouts (~8–12 Å with capped termini and retention of relevant waters), and final SQM rescoring on the truncated model:

$$QMDS = \Delta G_0^{QM} + \Delta G_{conf}^{QM}(P) + \Delta G_{conf}^{QM}(L) - T\Delta S \,, \tag{1}$$

$$\Delta G_{conf}^{QM}(X) = G_0^{QM}(X) - G^{QM}(X), \tag{2}$$

where *QMDS* is the quantum mechanical docking score, $G_0{}^{QM}(X)$ is the energy of the isolated X (P or L) in the conformation of the docked PL complex, $G^{QM}(X)$ is the energy of X in the free unbound state, T is the temperature, and S is the conformational entropy. All QM calculations were carried out with an improved version of the semiempirical PM7 Hamiltonian [28] using the linear-scaling MOZYME module of MOPAC [29]. This method includes dispersion corrections, treats hydrogen and halogen bonding at the parametrization stage, and incorporates refinements beyond PM6, making it a suitable choice for docking-scale applications. Solvation free energies were obtained with the COSMO continuum model [30] (default radii and surface tension parameters), with solvent-accessible areas taken directly from the program output. In this work, several QM docking strategies were tested to assess the role of relaxation and deformation terms. Comparisons with full-protein calculations demonstrated that reduced-site cutouts preserve the essential physics with negligible accuracy loss, while substantially reducing computational cost. Ligand conformational entropy was estimated through Monte Carlo sampling, which proved more reliable than simple rotatable-bond counts. Across the ten targets, the most striking advantage of SQM-based rescoring was observed in early enrichment factors, particularly in systems where Vina underperformed (Fig. 1).

This indicates that QM scoring is especially valuable for docking-based virtual screening, as it prioritizes true ligands at the very top of the ranking (precisely the region of greatest practical interest in experimental follow-up) (cf. Figs. 1 and 2).

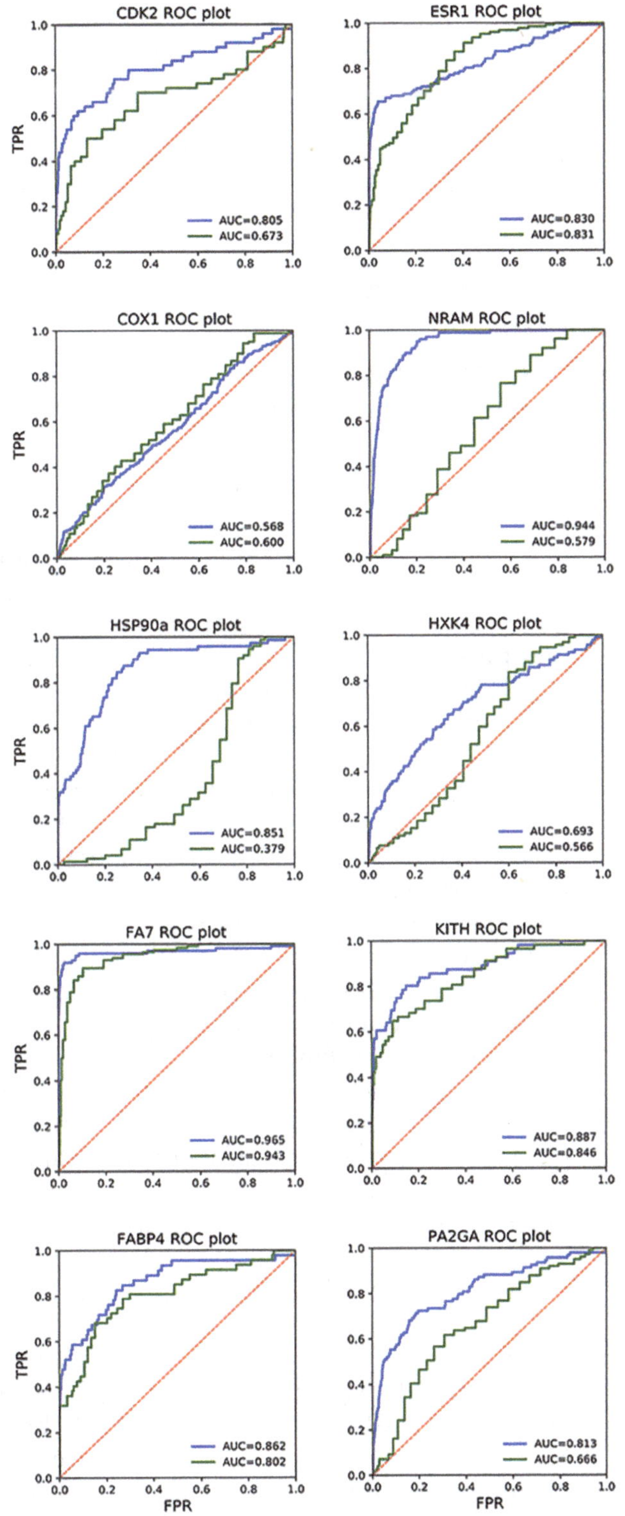

Fig. 1 Receiver operating characteristic (ROC) plots of AutoDock Vina (green line) and QM-scoring QMDS (blue line) for ten different target systems. The dotted red line corresponds to random selection (AUC = 0.5). FPR false positive rate, TPR true positive rate, AUC area under the curve. Original figure was published by Cavasotto and Aucar [27]

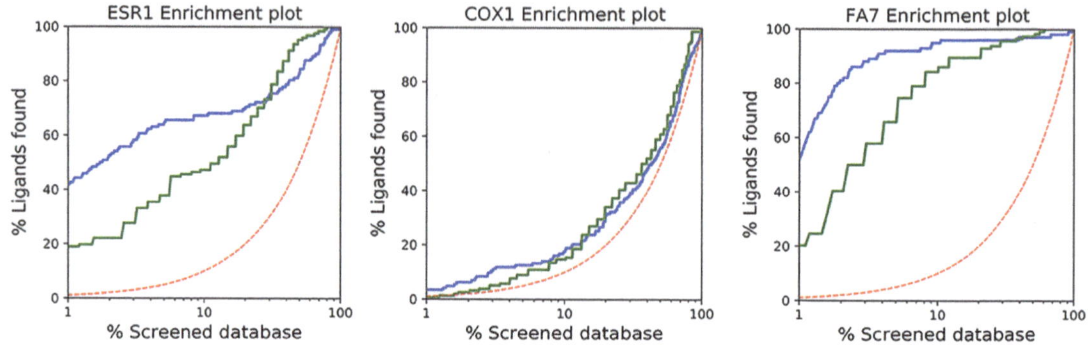

Fig. 2 Enrichment plots for ESR1, COX1, and FA7 using AutoDock Vina (green line) and QM-scoring QMDS (blue line). The red dotted line corresponds to random selection. Original figure was published by Cavasotto and Aucar [27]

From a computational perspective, rescoring reduced models required ~6–8 min per complex on a single core, about one order of magnitude slower than MM scoring, yet already feasible for mid-throughput campaigns. With expected advances in hardware, parallelization, and SQM implementations, the attainable scale of QM methods is projected to expand substantially soon, reinforcing its role as a practical "middle funnel" between fast classical docking and costly MD-based free-energy simulations.

While Cavasotto et al. demonstrated the feasibility of large-scale QM rescoring across more than 50,000 complexes, primarily benchmarked against a single empirical function (AutoDock Vina), subsequent studies explored complementary aspects of method performance. Eyrilmez et al. [31] evaluated a related SQM/COSMO scoring approach on a single protein system encompassing 4541 ligands, extending the comparison beyond Vina to a broad panel of classical scoring functions, including AutoDock4, Vina, Smina, Glide SP/XP, and GOLD (ASP, GoldScore, ChemScore). In parallel, Kříž and Řezáč [32] systematically compared multiple semiempirical Hamiltonians on a smaller benchmark of 15 protein–ligand complexes, identifying the parameterizations that best reproduced experimental binding trends. In every case, the SQM rescoring step markedly improved performance, reducing false positives, recovering crystallographic poses, and enhancing enrichment factors compared to the underlying empirical scores.

Building directly on this foundation, the same group introduced SQM2.20, a decisive step forward [33]. Conceived as a direct evolution of their 2020 framework, SQM2.20 broadened the scope of SQM scoring into a universal, high-performance function. It explicitly integrated all major contributions to binding free energy (Eq. 3) within a consistent semiempirical framework, while upgrading to PM6-D3H4X/COSMO2 [34, 35] with refined corrections for dispersion, hydrogen and halogen bonding, and solvation.

$$\text{SQM2.20} \quad \text{score} = \Delta E_{int} + \Delta\Delta G_{solv} + \Delta G_{conf}\left(L\right) + \Delta \mathrm{G}_{H+} - T\Delta \mathrm{S}, \tag{3}$$

where ΔE_{int} represents the gas-phase interaction energy between the ligand and the protein, $\Delta\Delta G_{solv}$ is the change in solvationfree energy upon complex formation, $\Delta G_{conf}(L)$ denotes the conformational free-energy penalty of the ligand in aqueous solution and ΔG_{H+} accounts for the free energy of proton transfer between the ligand and the buffer.

Unlike the earlier protocol tuned for specific cases, SQM2.20 exhibited consistent accuracy across diverse protein families, ligand chemistries, and binding modes, suggesting a high degree of transferability without system-specific parametrization. Its validation on the PL-REX benchmark (164 complexes, 10 diverse targets) confirmed robust and generalizable accuracy against experimental binding free energies, with an average R^2 of 0.69, well above the 0.20–0.40 range typical of classical scoring functions, approaching DFT fidelity while remaining four orders of magnitude faster. With runtimes of ~20 min per-complex, SQM2.20 established itself as a practical tool for mid-throughput rescoring, refinement, and even lead optimization.

In recent months, Choutka et al. [36] have presented the most comprehensive evaluation of SQM scoring functions to date. Their study systematically compared SQM with empirical MM-, and MD-based methods, while also performing a detailed energy decomposition analysis. The conclusions broadly align with those of Cavasotto [27] and Pecina [33], but add important topics to the discussion.

The analysis confirmed that ΔE_{int} and $\Delta\Delta G_{solv}$ remain the dominant and compensatory terms, showing that the choice of the solvation model can be as decisive as the electronic Hamiltonian itself. Ligand strain emerged as a moderate corrective term rather than a decisive discriminator, echoing the mixed effects reported by Cavasotto when deformation penalties were introduced. In terms of entropy, RRHO-based corrections reduced absolute error, but often degraded rank correlations, underscoring that entropy remains one of the key aspects in need of methodological improvement. Particularly striking was the behavior of halogen-bonding subsets: removing these ligands dramatically improved correlations across all methods, and contrary to expectations, SQM approaches did not outperform force fields in handling σ–hole interactions. Even variants with empirical halogen-bond corrections (e.g., PM6-D3H4X, GFN1) showed no systematic advantage, with AMBER-based protocols performing best on this subset. This highlights that anisotropic interactions remain an open challenge for both classical and quantum scoring functions.

Placed within the broader benchmarking landscape, these results further clarified the positioning of SQM-based approaches. When evaluated against experimental binding free energies, empirical and machine learning single-structure scores typically yield $R^2 \approx$ 0.20–0.36, and MM endpoint methods hover around 0.37–0.38, semiempirical SQM protocols reach ≈0.47, closing much of the gap to MD-based free-energy methods (≈0.50–0.62) [37]. This hierarchy supports the view (anticipated by Cavasotto and later formalized by Pecina in SQM2.20) that SQM rescoring is most effective when protein structures are reliable and binding pockets moderately rigid, while MD remains the preferred option when receptor flexibility is high or crystallographic resolution is limited.

4 Integration with Machine Learning and Hybrid Approaches

Hybrid approaches that combine semiempirical QM with machine learning have recently demonstrated tangible benefits for docking. The central idea is straightforward: SQM provides a physically grounded backbone for long-range effects such as electrostatics and dispersion, while ML supplies local, data-driven corrections where semiempirical methods are weakest, notably in short-range contacts and torsional profiles. This division of labor improves interaction and conformational terms without inflating computational cost and offers a more robust alternative to standalone ML scorers, which often suffer from poor transferability when applied outside their training domain.

Beyond the developments of SQM2.20, one particularly illustrative example is PM6-ML [38], a hybrid scheme where a machine-learning correction is applied directly on top of PM6 to reproduce high-level DFT reference energies. The approach follows a Δ-ML strategy: the ML model is trained on the systematic differences between PM6 and DFT, so that the final energy corresponds to the PM6 prediction plus a data-driven correction. In practice, PM6 provides the transferable long-range physics, while ML targets the local deficiencies of PM6 in describing noncovalent contacts and torsional energetics. This synergy yields accuracy close to DFT, with RMSE below 1 kcal/mol on noncovalent benchmarks, while retaining costs like standard SQM. Crucially, the explicit physical backbone of PM6 (implemented in the linear-scaling MOZYME framework) anchors the ML component, preventing overfitting to small molecules and ensuring scalability to large biomolecular complexes.

Taken together, these efforts underscore a practical lesson: while machine learning alone can provide modest gains, it rarely generalizes reliably across diverse targets. When paired with SQM, however, ML corrections enhance accuracy while preserving transferability, positioning such hybrids as a promising middle ground between empirical scorers and computationally demanding free-energy methods. The effectiveness of hybrid rescoring still depends critically on the quality of the input geometries: high-resolution or carefully relaxed poses yield the greatest benefit, whereas in systems with substantial receptor flexibility, MD-based approaches remain the more reliable choice.

5 Outlook and Conclusions

Quantum mechanical methods for docking have advanced from niche applications to increasingly practical tools, showing consistently improving in scoring power and early enrichment. Semiempirical Hamiltonians combined with continuum solvation and reduced-system models already allow medium-throughput rescoring, and studies on tens of thousands of complexes demonstrate that such approaches can systematically outperform empirical functions when classical scoring struggles. Rather than being prohibitively expensive, QM-based rescoring is becoming a tractable "middle funnel" option that bridges the gap between ultrafast empirical docking and high-accuracy MD-based free energy calculations.

Looking ahead, several areas of active development inspire optimism. The ongoing refinement of semiempirical Hamiltonians is steadily improving their ability to describe key phenomena such as polarization, halogen bonding, and metal coordination. Hybrid strategies that pair QM with lightweight machine-learning corrections are pushing accuracy close to DFT while retaining costs close to SQM, and the synergy of physics-based methods with data-driven refinements offers a uniquely robust framework. The parallelization of linear-scaling algorithms, together with the explosive growth of GPU computing [39], suggests that what is today feasible at medium throughput could soon scale to full high-throughput campaigns.

Challenges remain, but they are increasingly seen as opportunities for innovation. Better protocols for handling receptor flexibility, systematic guidelines for reduced-system construction, and more reliable treatments of ligand entropy will all expand the applicability of QM scoring. At the same time, benchmarking efforts such as CASF and PL-REX are providing a clearer map of where QM already excels and where further progress is needed. The trend across independent studies is converging: QM-based scoring captures critical interaction physics often missing in MM, and when implemented judiciously, it consistently improves ligand ranking and enrichment.

In sum, the trajectory of the field is unmistakably upward. With sustained algorithmic innovation, hardware acceleration, and integration with machine learning, QM scoring is evolving into a versatile, interpretable, and scalable tool for structure-based drug discovery. What was once considered too computationally costly or of pure academic interest is rapidly becoming a practical component of discovery pipelines, promising more exact energetics, better prioritization of leads, and ultimately, a higher success rate in the identification of novel therapeutics.

Acknowledgments

This work was supported by a grant from Agencia Nacional de Promoción Científica y Tecnológica, Argentina (PICT 2021-1129). CNC thanks Molsoft LLC (San Diego, CA) for providing an academic license for the ICM program. Authors would like to thank CCAD (Centro de Computación de Alto Desempeño de la Universidad Nacional de Córdoba) who provided the supercomputing time for this work.

References

1. Fernald KDS, Förster PC, Claassen E et al (2024) The pharmaceutical productivity gap—incremental decline in R&D efficiency despite transient improvements. Drug Discov Today 29:104160
2. Morgan S, Grootendorst P, Lexchin J et al (2011) The cost of drug development: a systematic review. Health Policy 100:4–17
3. Schlander M, Hernandez-Villafuerte K, Cheng C-Y et al (2021) How much does it cost to research and develop a new drug? A systematic review and assessment. Pharmacoeconomics 39:1243–1269
4. Xuan-Yu M, Hong-Xing Z, Mihaly M et al (2011) Molecular docking: a powerful approach for structure-based drug discovery. Curr Comput-Aided Drug Des 7:146–157
5. Bender BJ, Gahbauer S, Luttens A et al (2021) A practical guide to large-scale docking. Nat Protoc 16:4799–4832
6. Janzen WP (2014) Screening technologies for small molecule discovery: the state of the art. Chem Biol 21:1162–1170
7. Broach JR, Thorner J (1996) High-throughput screening for drug discovery. Nature 384:14–16
8. Macarron R, Banks MN, Bojanic D et al (2011) Impact of high-throughput screening in biomedical research. Nat Rev Drug Discov 10:188–195
9. Phatak SS, Stephan CC, Cavasotto CN (2009) High-throughput and in silico screenings in drug discovery. Exp Opin Drug Discov 4:947–959
10. Spyrakis F, Cavasotto CN (2015) Open challenges in structure-based virtual screening: Receptor modeling, target flexibility consideration and active site water molecules description. Arch Biochem Biophys 583:105–119
11. Jorgensen WL (2012) Challenges for academic drug discovery. Angew Chem Int Ed Engl 51:11680–11684
12. Cavasotto CN, Orry AJ (2007) Ligand docking and structure-based virtual screening in drug discovery. Curr Top Med Chem 7:1006–1014
13. Cavasotto CN (2012) Binding free energy calculations and scoring in small-molecule docking. In: Luque FJ, Barril X (Eds.) Eds. Physico-chemical and computational approaches to drug discovery. Royal Society of Chemistry, London, 195–222
14. Kitchen DB, Decornez H, Furr JR et al (2004) Docking and scoring in virtual screening for drug discovery: methods and applications. Nat Rev Drug Discov 3:935–949
15. Li Y, Han L, Liu Z et al (2014) Comparative assessment of scoring functions on an updated benchmark. 2. Evaluation methods and general results. J Chem Inf Model 54:1717–1736
16. Kim M, Cho AE (2016) Incorporating QM and solvation into docking for applications to GPCR targets. Phys Chem Chem Phys 18:28281–28289
17. Slater O, Kontoyianni M (2019) The compromise of virtual screening and its impact on drug discovery. Expert Opin Drug Discov 14:619–637
18. Su M, Yang Q, Du Y et al (2019) Comparative assessment of scoring functions: the CASF-2016 Update. J Chem Inf Model 59:895–913
19. Wang Z, Sun H, Yao X et al (2016) Comprehensive evaluation of ten docking programs on a diverse set of protein-ligand

complexes: the prediction accuracy of sampling power and scoring power. Phys Chem Chem Phys 18:12964–12975
20. Warren GL, Andrews CW, Capelli AM et al (2006) A critical assessment of docking programs and scoring functions. J Med Chem 49:5912–5931
21. Niazi SK (2025) Quantum mechanics in drug discovery: a comprehensive review of methods, applications, and future directions. Int J Mol Sci 26:6325
22. Christensen AS, Kubar T, Cui Q et al (2016) Semiempirical quantum mechanical methods for noncovalent interactions for chemical and biochemical applications. Chem Rev 116:5301–5337
23. Moussa JE (2025) The enduring relevance of semiempirical quantum mechanics. J Phys Chem A 129:8465–8477
24. Pecina A, Haldar S, Fanfrlik J et al (2017) SQM/COSMO scoring function at the DFTB3-D3H4 level: unique identification of native protein-ligand poses. J Chem Inf Model 57:127–132
25. Pecina A, Brynda J, Vrzal L et al (2018) Ranking power of the SQM/COSMO scoring function on carbonic anhydrase II-Inhibitor complexes. ChemPhysChem 19:873–879
26. Aucar MG, Cavasotto CN (2020) Molecular docking using quantum mechanical-based methods. Methods Mol Biol 2114:269–284
27. Cavasotto CN, Aucar MG (2020) High-throughput docking using quantum mechanical scoring. Front Chem 8:246
28. Stewart JJP (2013) Optimization of parameters for semiempirical methods VI: more modifications to the NDDO approximations and re-optimization of parameters. J Mol Model 19:1–32
29. Stewart J JP (1996) Appplication of localized molecular orbitals to the solution of semiempirical self-consistent field equations. Int J Quant Chem 58:133–146
30. Klamt A, Schüürmann G (1993) COSMO: a new approach to dielectric screening in solvents with explicit expressions for the screening energy and its gradient. J Chem Soc Perkin Trans 2:799–805
31. Eyrilmez SM, Kopruluoglu C, Rezac J et al (2019) Impressive enrichment of semiempirical quantum mechanics-based scoring function: HSP90 protein with 4541 inhibitors and decoys. ChemPhysChem 20:2759–2766
32. Kříž and Řezáč J (2020) Benchmarking of semiempirical quantum-mechanical methods on systems relevant to computer-aided drug design. J Chem Inf Model 60:1453–1460
33. Pecina A, Fanfrlik J, Lepsik M et al (2024) SQM2.20: semiempirical quantum-mechanical scoring function yields DFT-quality protein-ligand binding affinity predictions in minutes. Nat Commun 15: 1127
34. Kříž and Řezáč J (2019) Reparametrization of the COSMO solvent model for semiempirical methods PM6 and PM7. J Chem Inf Model 59:229–235
35. Brahmkshatriya P S, Dobes P, Fanfrlik J et al (2013) Quantum mechanical scoring: structural and energetic insights into cyclin-dependent kinase 2 inhibition by Pyrazolo[1,5-a] pyrimidines. Curr Comput-Aided Drug Des 9:118–129
36. Choutka J, Kaminsky J, Wang E et al (2025) End-point affinity estimation of galectin ligands by classical and semiempirical quantum mechanical potentials. J Chem Inf Model 65:762–777
37. Jalaie M, Fanfrlík J, Pecina A et al (2025) Comparative analysis of quantum-mechanical and standard single-structure protein–ligand scoring functions with MD-based free energy calculations. J Chem Inf Model 65:8127–8136
38. Novacek M, Rezac J (2025) PM6-ML: the synergy of semiempirical quantum chemistry and machine learning transformed into a practical computational method. J Chem Theory Comput 21:678–690
39. Dongarra J, Gunnels J, Bayraktar H et al (2024) Hardware trends impacting floating-point computations in scientific applications. arXiv preprint arXiv:241112090.

Check for updates

Chapter 15

Integration of GPCR Modeling with QM for Drug Design: Academic and Industry Collaboration

Alexander Heifetz and Andrea Townsend-Nicholson

Abstract

This chapter summarizes more than a decade of sustained academic–industry collaboration between Professor Andrea Townsend-Nicholson (University College London) and Dr Alexander Heifetz (formerly Evotec, now Sygnature Discovery). Working together with their research teams, they have developed and applied advanced computational strategies to investigate G protein–coupled receptors (GPCRs) and translate molecular insights directly into structure-based drug design. Central to this work is the Fragment Molecular Orbital (FMO) method, a quantum mechanical approach that enables residue-level decomposition of interaction energies, providing quantitative analysis of receptor stability, ligand binding, and activation. Through a series of studies including GPCR–ligand interaction mapping and characterization of conserved inter-helical networks, as reviewed in this chapter, the collaboration established a unified framework for GPCR-focused molecular design. The methodologies developed through this partnership have been successfully deployed across multiple GPCR discovery programs and now form part of the broader computational chemistry toolkit used in both academia and industry. This collaboration also stands as a model of how long-term partnerships between academia and the pharmaceutical industry can accelerate scientific innovation and deliver practical technologies for drug discovery.

Keywords G protein-coupled receptors, Fragment molecular orbital method, Quantum mechanics, Structure-based drug design, Computational chemistry, GPCR–ligand interactions, Inter-helical networks, Drug discovery

1 Introduction and Motivation

G protein-coupled receptors (GPCRs) constitute the largest and most pharmacologically relevant family of membrane proteins, with approximately 800 GPCR genes in the human genome [1–2]. They regulate virtually every domain of human physiology, including endocrine control, neurotransmission, cardiovascular function, metabolism, and immune response. Around 40% of all marketed drugs act via GPCRs, yet these medicines collectively target only about 50–70 GPCRs, leaving over 700 receptors—more than 85% of the

Alexander Heifetz (ed.), *Quantum Mechanics Applications in Drug Design*, Methods in Molecular Biology, vol. 3061, https://doi.org/10.1007/978-1-0716-5420-0_15,

superfamily–currently untargeted [1]. This makes GPCRs not only the most validated target class in pharmacology but also one of the largest untapped opportunities in drug discovery. In recent years, pharmaceutical interest in GPCRs has intensified, especially within previously underexploited "metabolic GPCRs" that regulate appetite, glucose control, obesity, and energy balance. New drug development has shifted toward GPCRs linked to metabolic disease [3]. These emerging clinical successes demonstrate that GPCRs remain high-value targets with massive unexplored therapeutic potential.

Despite this opportunity, rational GPCR drug design still faces major challenges [2,4–6]. Achieving subtype selectivity, controlling signaling bias and tuning binding kinetics (residence time) remain difficult due to receptor conformational flexibility, multiple activation states, and nonclassical interactions that are not captured by classical molecular mechanics (MM). Moreover, many important interactions such as CH–π, cation–π, halogen–π, nonclassical hydrogen bonds, and hydrophobic interactions are difficult to identify visually and are poorly captured by current force fields [7–10]. This limits the ability of conventional structure-based design to rationalize affinity and selectivity in a chemically meaningful way. Even when high-resolution GPCR structures are available, visual inspection and force-field scoring often fail to accurately explain structure–activity relationships (SAR) or guide medicinal chemistry optimization.

Quantum mechanical (QM) approaches provide a more realistic physical description of binding by explicitly accounting for polarization, charge transfer, dispersion, and nonclassical interactions [11–17]. However, their application to biomolecular systems has historically been limited by computational cost. The Fragment Molecular Orbital (FMO) method overcomes this barrier by partitioning the receptor–ligand system into fragments (e.g., residues and ligand moieties), computing pair interaction energies (PIEs), and decomposing them using PIEDA into electrostatics, exchange-repulsion, charge transfer, dispersion, and solvation energy terms [18–21]. FMO provides residue-resolved, quantitative insights into interaction strength (in kcal/mol), the chemical nature of interactions (e.g., electrostatic, hydrophobic), and key energy components such as desolvation and depolarization. See Chap. 4 for more information.

The enormous potential of GPCRs as drug targets [2], combined with the unique capabilities of the FMO method, motivated us to explore their integration for structure-based drug design. In this chapter, we illustrate how we have applied the FMO method in our previously published reports [19,22,23] to perform (1) FMO analysis of GPCR–ligand recognition, establishing a consensus map of binding-pocket energetics and clarifying the respective roles of electrostatic and hydrophobic interactions [19,22], and (2) FMO characterization of inter-helical networks, quantifying conserved transmembrane (TM–TM) interactions that connect ligand binding with receptor activation [23]. All figures used in this chapter are reproduced from our original publications [19,22,23].

2 The FMO Method Applied to GPCR–Ligand Interactions [19,22]

We applied the FMO method to 18 class A GPCR–ligand crystal structures representing diverse receptor families. This analysis revealed key interactions that are often overlooked in classical structural descriptions, including hydrophobic contributions, nonclassical hydrogen bonds, and backbone-mediated interactions. By quantifying protein–ligand energetics at atomic resolution, FMO provides a richer and more accurate description of binding than conventional structure-based methods. Across all systems analyzed, interactions with $|PIE| \geq 3.0$ kcal/mol were considered significant. An example of such analyses is shown in Fig. 1.

2.1 Consensus Set of Key Binding Residues

A consensus set of key binding residues emerged at positions 3.32, 3.33, 6.48, 6.51, 6.52, 7.39, and 7.43, which contributed strongly to ligand binding in more than 70% of receptors (Fig. 2). These residues form the conserved core of the orthosteric pocket and are known to influence ligand affinity and selectivity. Most interactions involving residues 3.33, 6.48, 6.51, 6.52, and 7.43 were predominantly hydrophobic, while 3.32 and 7.39 contributed mainly electrostatic interactions. Although the residue identities at these positions varied across receptors, their energetic roles were conserved, reflecting evolutionary selection for interaction profiles rather than sequence alone. Importantly, FMO revealed that electrostatic and hydrophobic interactions contribute approximately equally to ligand bindingenergetics on average, highlighting the key role of hydrophobic effects, which are often underestimated in MM-based approaches. FMO also detected stabilizing contributions from backbone interactions and structured water molecules, which are typically neglected in classical analyses.

Overall, this work established FMO as a powerful and practical QM method for dissecting GPCR–ligand interactions at chemical resolution. It provides medicinal chemists with actionable energetic insights for ligand optimization and supports rational design by identifying residues that drive affinity, selectivity, and allosteric effects.

3 Quantum Characterization of Conserved GPCR Helical Networks [23]

It has become increasingly clear that the structural stability, functional dynamics, and ligand-binding properties of GPCRs are governed not only by interactions at the ligand-binding site but also by long-range communication within the transmembrane (TM) helical bundle

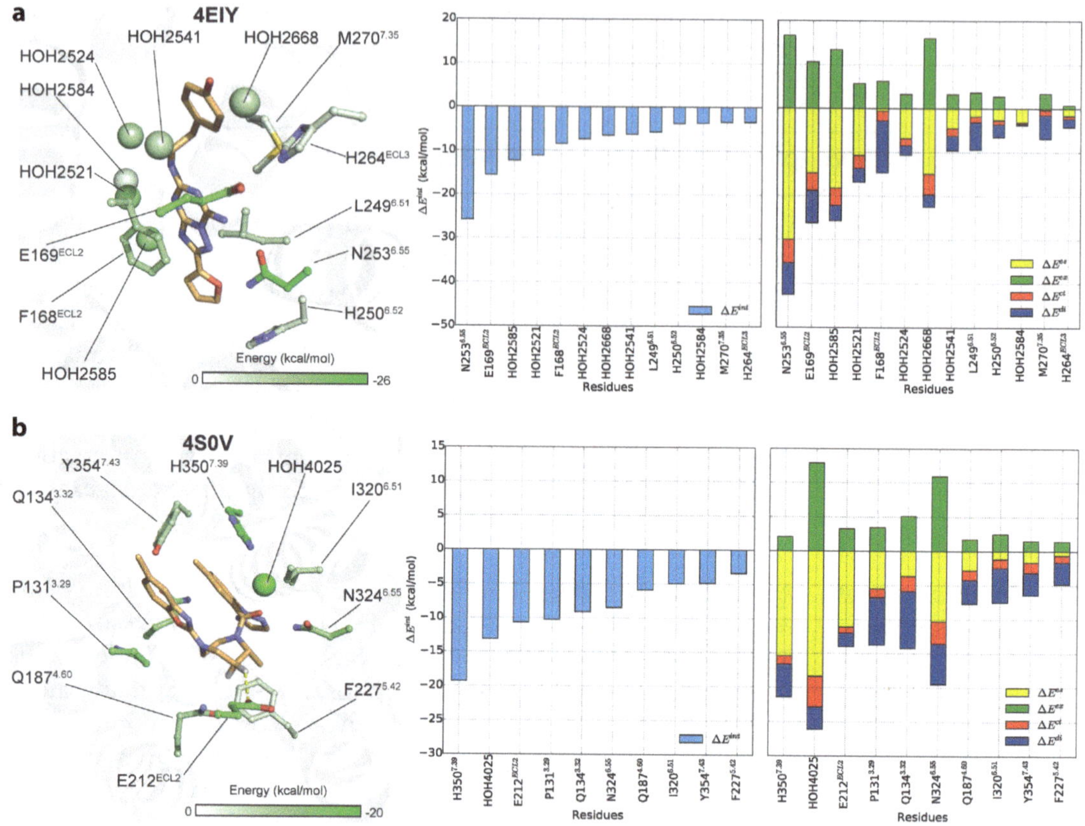

Fig. 1 FMO results [19] for (**a**) Human adenosine A_{2a} receptor in complex with ZM241385 (PDB entry 4EIY) and (**b**) Human orexin OX_2 receptor with Suvorexant (PDB entry 4S0V). The carbon atoms of the ligand are shown in light orange, and for the receptor are colored according to PIE values calculated by FMO. Nitrogen atoms are shown in blue, oxygen in red, sulfur in yellow, and chlorine in light green. The nonclassical hydrogen bond formed between the backbone carbonyl of Glu169 and the ligand is marked as an orange dashed line. The left-hand plot describes the sorted PIE of the most significant residues, and the right-hand plots describe the PIEDA of these key interactions. PIE terms: electrostatics, dispersion, charge-transfer and exchange repulsion are colored coded yellow, blue, red, and green, respectively

[24–27]. Experimental studies have shown that the thermodynamic stability of GPCRs can be modulated by mutating specific TM residues. However, a quantitative understanding of the molecular forces responsible for stabilizing the seven-helix TM architecture and how these forces enable receptor activation had never been described [4, 8, 10, 24, 26, 28, 29].

Using FMO, we analyzed 35 high-resolution crystal structures spanning diverse branches of the class A GPCR family [23]. This work identified 69 topologically equivalent TM residues forming a consensus network of 51 conserved inter-helical interactions (Fig. 3). These findings revealed, for the first time, a comprehensive energetic framework that defines how intramolecular forces maintain GPCR

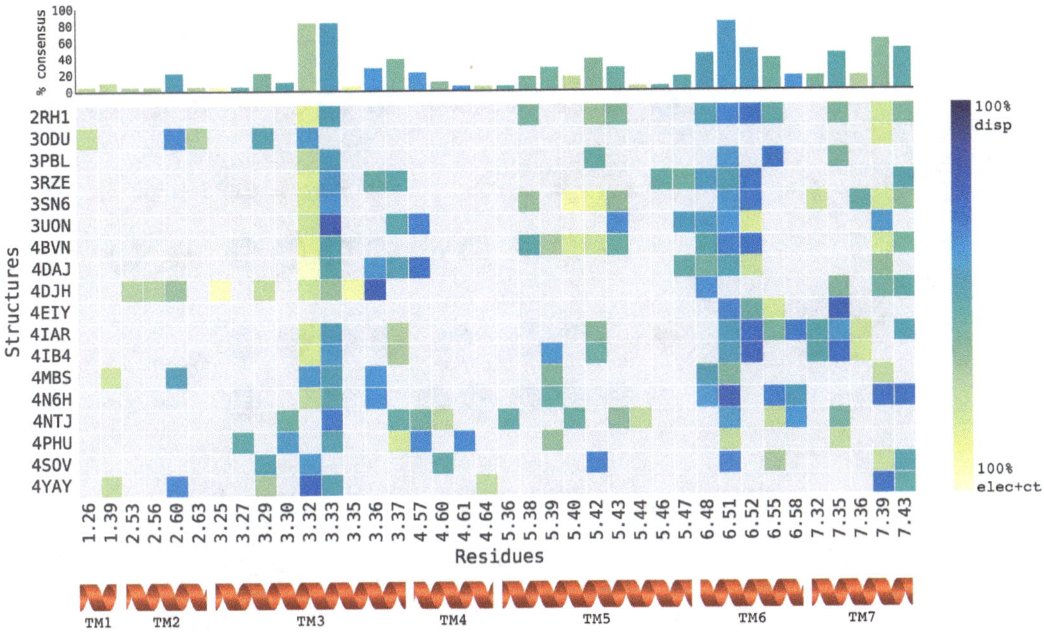

Fig. 2 Comparison [19] of residues involved in ligand-binding. Receptor–ligand information is shown as rows, and the Ballesteros–Weinstein numbers of residues that contact the ligand are shown as columns. In the matrix, the presence of a contact between the ligand and the residue is shown as a colored box, and the absence of a contact is shown as gray box. Boxes are otherwise colored according to their PIEDA type: from dark blue (100% of dispersion contribution) to yellow (100% of sum electrostatics + charge-transfer). An equal contribution mix is light blue, as shown by the right spectrum. The level of interaction consensus is shown above, with the bar color-coded according to average PIEDA type, from dark blue (100% of dispersion contribution) to yellow (100% of sum electrostatics + charge-transfer)

architecture, propagate conformational change, and contribute to ligand binding and receptor activation.

As reported in Sect. 2.1, seven residues (at positions 3.32, 3.33, 6.48, 6.51, 6.52, 7.39, and 7.43) contributed to ligand binding in more than 70% of GPCR structures analyzed. We were able to demonstrate that six of these residues (all except position 3.33) were also integral components of the conserved inter-TM network. For example, residue 6.52 simultaneously forms stabilizing contacts with the ligand and with residue 5.47 (Fig. 3e). This dual role illustrates a key principle of GPCR allostery: residues can act as energetic nodes that connect ligand-binding events to global conformational changes through the inter-helical interaction network.

This arrangement explains why mutations located far from the orthosteric binding pocket frequently alter ligand affinity and receptor signaling. As illustrated in Fig. 3a, ligands bind on top of a structurally conserved TM interaction network, enabling them to influence both local interactions within the binding pocket and remote stabilizing interactions across the receptor. Much like a pilot steering an aircraft from the cockpit, ligand binding at the

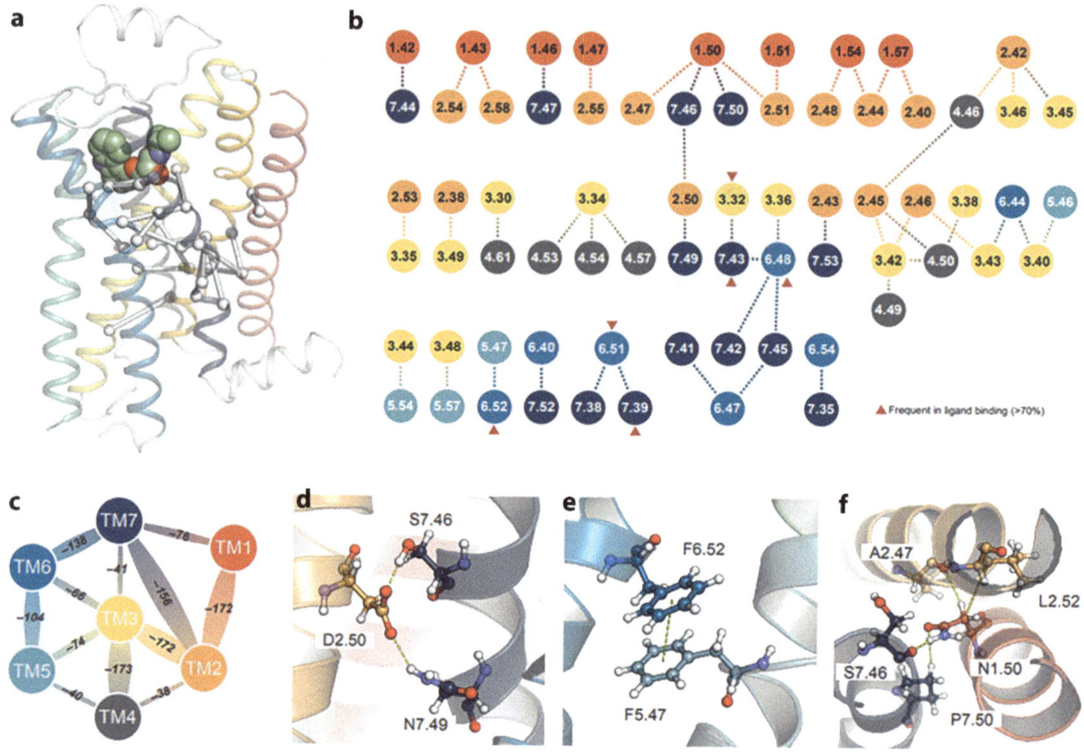

Fig. 3 (**a**) Representative β2 adrenergic receptor (ribbons)-ligand (spheres) complex (PDB code 2RH1). The conserved inter-TM interactions are shown as white tubes. (**b**) The network of 51 conserved inter-TM interactions was formed by 69 residues. The circles represent residues and are color-coded as follows: TM1 red, TM2 brown, TM3 yellow, TM4 grey, TM5 teal, TM6 light blue, and TM7 dark blue. Numbers denote Ballesteros–Weinstein numbering. A dashed line between a pair of circles indicates the presence of a conserved interaction. Residues previously reported as involved in ligand binding in a number of different GPCRs are marked with a red triangle. (**c**) Schematic representation of the TM–TM interaction energies. The line between a pair of circles indicates the total TM–TM pair attraction energy (TAE, in kcal/mol), where the thickness of the line is proportional to the size of the TAE (only interactions <-20 kcal/mol are shown). (**d–f**) Three examples of conserved inter-TM interactions in a representative GPCR (the β2-adrenergic receptor). Nitrogen atoms are shown in blue, oxygen in red, sulfur in yellow, and carbon in green. Major contributions to residue–residue interactions are highlighted with yellow dashed lines

extracellular surface controls the conformation of the entire receptor through this interconnected interaction network.

3.1 Chemical Nature of the Conserved Inter-TM Interactions

Among the 51 conserved inter-TM interactions identified, 15 were predominantly electrostatic, 15 were predominantly hydrophobic, and 21 exhibited a mixed interaction character, with roughly equal electrostatic and hydrophobic contributions. These findings highlighted the critical and previously underappreciated role of hydrophobic forces in stabilizing GPCR helical packing, an aspect often overlooked in traditional structure-based analyses that tend to emphasize polar interactions.

Importantly, this study showed that both the strength and chemical nature of these conserved inter-helical contacts are preserved across evolution, even though the amino acid sequence identity at these 69 positions averages only 46% across the 35 GPCRs analyzed. This means that while the exact residues may vary, the type of interaction is conserved, suggesting that natural selection acts to preserve energetically favorable interaction patterns, rather than specific residue identities.

This observation supports a key concept in GPCR structural biology: evolution conserves interaction energetics, not amino acid sequences. In other words, functionally essential inter-helical contacts are maintained through physicochemical equivalence, for example, by substituting one hydrophobic residue with another—ensuring network integrity despite sequence divergence.

3.2 The Role of Specific TM Helices

Analysis of the conserved interaction network (Fig. 3c) identified TM3 as the central structural hub of class A GPCRs, forming strong energetic contacts with five neighboring helices (TM2, TM4, TM5, TM6, and TM7) and accounting for 39% (20/51) of all conserved inter-TM interactions. The key ligand-contact residue 3.32 also participated in this network, highlighting the dual structural and functional role of TM3.

We found that TM3, TM6, and TM7 interacted more weakly with each other (average TAE −82 kcal/mol) than with the other helices (−114 kcal/mol). This energetically permissive region corresponded to the orthosteric binding site, explaining how GPCRs accommodate chemically diverse ligands within a conserved structural framework.

Furthermore, conserved TM6 interactions were restricted to the extracellular half of the helix (positions 6.40–6.54), while the intracellular half lacked stabilizing interactions, consistent with its conformational mobility during activation and its role in G-protein coupling.

3.3 Comparing Active and Inactive Receptor States

Of the 35 GPCRs analyzed, 6 had crystal structures available in both inactive and active states, enabling direct comparison of their inter-TM interaction networks. Despite substantial conformational changes during activation (Fig. 4a), 46 of the 51 conserved inter-TM interactions were maintained between states (Fig. 4b). FMO analysis also identified 13 state-specific interactions—6 unique to inactive states and 7 unique to active states (Fig. 4c).

State transitions typically involved interaction switching, where a residue changed binding partners rather than losing stabilization entirely; for example, 3.46–6.37 (inactive) was replaced by 3.46–7.53 (active) (Fig. 4d). In other cases, lost interactions were compensated by new ones elsewhere in the network, such as replacement of hydrophobic contacts 3.43–5.54 and 3.43–6.41 (inactive) by a hydrogen bond 2.50–3.39 (active) (Fig. 4f), or loss of a nonclassical hydrogen bond 6.40–7.49 (inactive) compensated by formation of CH–π (3.40–6.44) and π–π (5.47–6.44) interactions (active) (Fig. 4h, i).

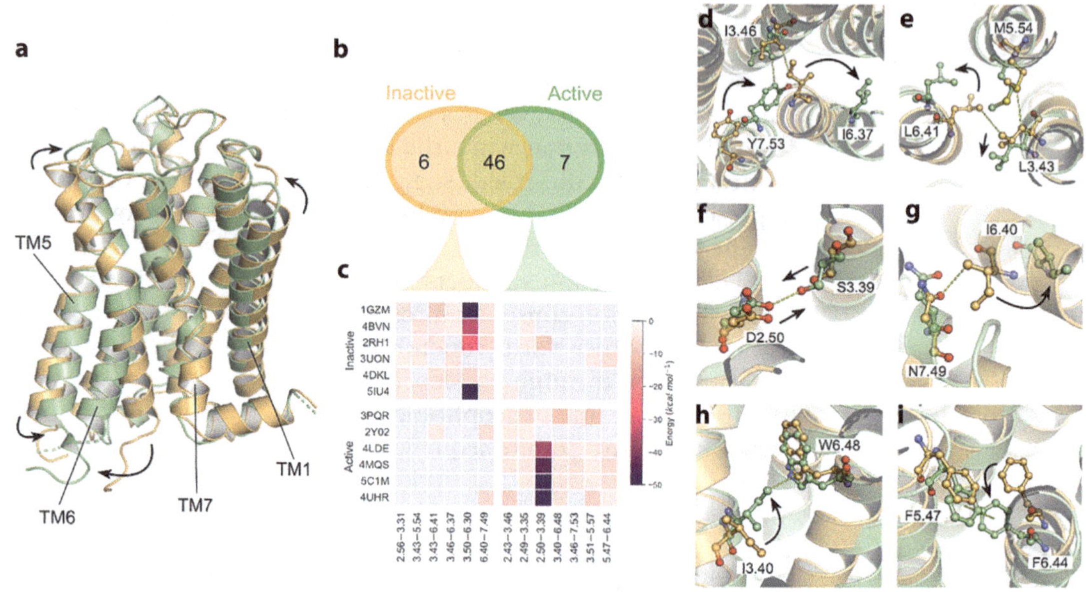

Fig. 4 Comparison [19] of the inter-TM interactions in the inactive and active states for the six proteins that have published crystal structures for both states (PDB codes for the inactive and active structures, respectively, are: Rhodopsin: 1GZM & 3PQR, β1-adrenergic receptor: 4BVN & 2Y02, β2-adrenergic receptor: 2RH1 & 4LDE, M_2 muscarinic receptor: 3UON & 4MQS, μ-opioid receptor: 4DKL & 5C1M, A_{2A} adenosine receptor: 5IU4 & 4UHR). (**a**) Inactive (orange ribbon) and active (green ribbon) structures of the M_2 muscarinic receptor are superimposed (PDB codes 3UON and 4MQS, respectively). (**b**) The overlap in terms of conserved inter-TM interactions between the inactive and active states is shown using a Venn diagram. (**c**) Comparison between state-specific, conserved inter-TM interactions. In the matrix, the size of the PAE between residues is shown as a heat map colored according to the gradient on the right. The absence of an interaction is shown as a grey box. (**d–i**) Examples of conserved changes in the inter-TM interaction network as a result of receptor activation. Nitrogen atoms are shown in blue, oxygen in red, sulfur in yellow, and carbon in green (active state) or in orange (inactive state): (**d–f**) M_2 muscarinic receptor and (**g–i**) β2-adrenergic receptor

Although the number of interactions was largely preserved, the energetics changed significantly. For example, the ionic lock—a strong salt bridge between 3.50–6.30 stabilizing the inactive state—was weakened upon activation and only partially replaced energetically by a new 2.50–3.39 hydrogen bond (Fig. 4f). Overall, the active-state network was ~30 kcal/mol less stable, and this destabilization was typically compensated for by agonist binding.

3.4 Underappre ciated Interactions

While classical interactions such as hydrogen bonds and salt bridges are easily recognized structurally, many additional stabilizing forces are difficult to detect and are poorly described by standard molecular mechanics force fields. FMO analysis revealed that approximately 50% of the 51 conserved inter-TM interactions fall into this "underappreciated" category, meaning they are not reliably captured by non-QM methods.

These include interactions mediated by backbone carbonyls, such as nonclassical hydrogen bonds (Fig. 5a, b, e, and h), CH–π interactions (Fig. 5c, d, and h), and carbonyl–sulfur (C=O···S) contacts (Fig. 3g). Many of these interactions occur at TM junctions, where they act as energetic bridges that stabilize inter-helical packing. For example, we were able to show that residue 3.40 mediates communication between TM5 and TM6 by forming a CH–π interaction with F6.44 and a nonclassical hydrogen bond with the backbone carbonyl of S5.46 (Fig. 5h).

Hydrophobic interactions also made a significant contribution to network stability, with 15 of the 51 conserved inter-TM interactions being predominantly hydrophobic. Since current force fields lack accurate energetic treatment of hydrophobic effects, these too should be considered underappreciated yet essential contributors to GPCR architecture.

In summary, FMO revealed that GPCR structure and function are governed by a conserved inter-helical interaction network that links ligand binding to receptor activation. This network is energetically preserved across evolution, with TM3 acting as a central stabilizing hub and TM6–TM7 providing conformational flexibility. Many key stabilizing forces are missed by classical methods, underscoring the essential role of QM analysis in understanding GPCR energetics.

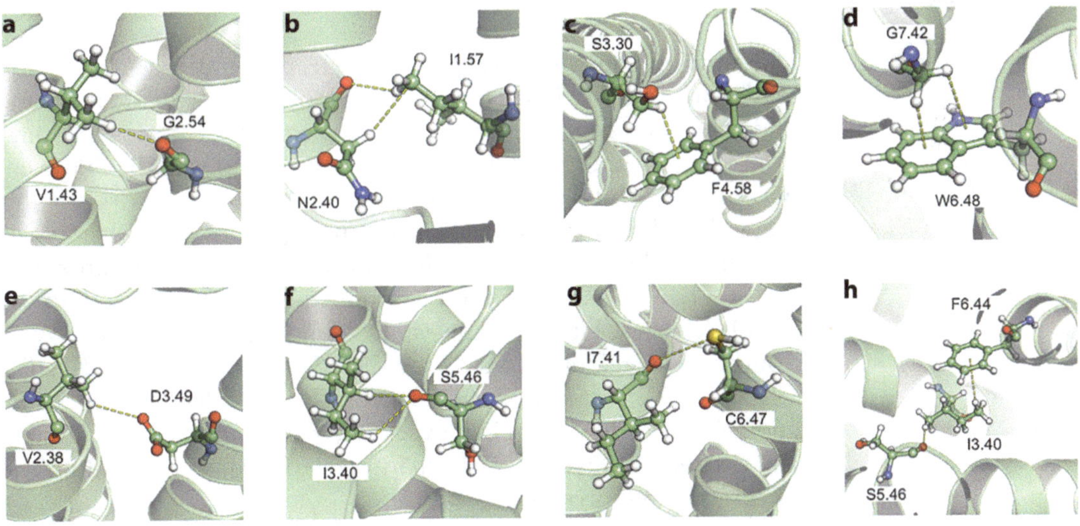

Fig. 5 Examples [19] of "underappreciated" interactions. Nitrogen atoms are shown in blue, oxygen in red, sulfur in yellow, and carbon in green. (**a–g**) Active state of the β2-adrenergic receptor (PDB code 4LDE): (**a**) nonclassical hydrogen bond between the sidechain of V1.43 and the backbone carbonyl of G2.54, (**b**) nonclassical hydrogen bond between the sidechain of I1.57 and the backbone carbonyl of N2.40. These two residues also form an additional hydrophobic interaction, (**c**) CH–π interaction between S3.30 and F4.58 (**d**) CH–π interaction between S3.30 and F4.58, (**e**) sidechain–sidechain nonclassical hydrogen bond between V2.38 and D3.49, (**f**) two nonclassical hydrogen bonds formed between I3.4 and S5.46, (**g**) carbonyl (backbone)-S interaction between I7.47 and C6.47, (**h**) dopamine D_3 receptor (PDB code 3PBL): I3.40 forms two nonclassical interactions with F6.44 (CH–π interaction) and with S5.46 (non-classical hydrogen bond with the backbone carbonyl)

4 Discussion and Conclusions

The studies reviewed in this chapter demonstrate how QM applied at scale through the FMO method has progressed from a theoretical construct to a practical, predictive technology for GPCR drug discovery. By integrating FMO with GPCR modeling, molecular dynamics, docking, and broader structure-based design workflows, this collaborative effort has established a robust and scalable framework for quantifying receptor–ligand and receptor–receptor energetics, and for translating these insights into actionable medicinal chemistry strategies. Beyond its scientific impact, this body of work highlights the value of sustained collaboration between academia and industry, where methodological innovation and real-world application continually reinforce one another. As FMO continues to evolve through increased automation, high-performance computing, and deeper integration with machine learning, it is poised to become a core component of next-generation GPCR discovery platforms, enabling the rational design of selective, efficacious, and mechanistically informed therapeutics.

Acknowledgments

The authors would like to express their deepest gratitude to Dr Dmitri Fedorov (AIST, Japan) for his invaluable support, guidance, and enduring friendship, which have been a constant source of inspiration. His pioneering and foundational contributions to the Fragment Molecular Orbital (FMO) method laid the groundwork for this research and continue to shape its future directions. The authors also gratefully acknowledge the contributions of our collaborators and colleagues who supported this work, including Dr Iñaki Morao, Dr Ewa I. Chudyk, Dr Laura Gleave, Dr Tim James, Dr Michelle W. Y. Southey, Dr Stefania Monteleone, Dr Matteo Aldeghi, Prof Vadim Cherezov, Dr Jintao Long, Dr David Thorner, Prof Madan Babu, and Dr Mike J. Bodkin. Their expertise, contributions, commitment, and collaborative spirit have been integral to the progress and success of this research program.

References

1. Hauser AS, Attwood MM, Rask-Andersen M et al (2017) Trends in GPCR drug discovery: new agents, targets and indications. Nat Rev Drug Discov 16:829–842
2. Congreve M, De Graaf C, Swain NA et al (2020) Impact of GPCR structures on drug discovery. Cell 181:81–91
3. Recio C, Lucy D, Iveson P et al (2018) The role of metabolite-sensing G protein-coupled receptors in inflammation and metabolic disease. Antioxid Redox Signal 29:237–256
4. Venkatakrishnan AJ, Deupi X, Lebon G et al (2016) Diverse activation pathways in class A GPCRs converge near the G-protein-coupling region. Nature 536:484–487
5. Kenakin T (2018) Is the quest for signaling bias worth the effort? Mol Pharmacol 93:266–269

6. Kitambi SS, Toledo EM, Usoskin D et al (2017) Retraction notice to: vulnerability of glioblastoma cells to catastrophic vacuolization and death induced by a small molecule. Cell 170:407
7. Fukuzawa K, Tanaka S (2022) Fragment molecular orbital calculations for biomolecules. Curr Opin Struct Biol 72:127–134
8. Heifetz A, Morao I, Babu MM et al (2020) Characterizing interhelical interactions of G-protein coupled receptors with the fragment molecular orbital method. J Chem Theory Comput 16:2814–2824
9. Heifetz A, Trani G, Aldeghi M et al (2016) Fragment molecular orbital method applied to lead optimization of novel interleukin-2 inducible T-cell kinase (ITK) inhibitors. J Med Chem 59:4352–4363
10. Heifetz A, Chudyk EI, Gleave L et al (2016) The Fragment molecular orbital method reveals new insight into the chemical nature of GPCR–ligand interactions. J Chem Inf Model 56:159–172
11. Nagata T, Fedorov DG, Ishimura K et al (2011) Analytic energy gradient for second-order Møller–Plesset perturbation theory based on the fragment molecular orbital method. J Chem Phys 135:044110
12. Fedorov DG (2021) Electron density from the fragment molecular orbital method combined with density-functional tight-binding. Chem Phys Lett 780:138900
13. Fedorov DG, Nagata T, Kitaura K (2012) Exploring chemistry with the fragment molecular orbital method. Phys Chem Chem Phys 14:7562
14. Fedorov DG, Nakamura T (2022) Free energy decomposition analysis based on the fragment molecular orbital method. J Phys Chem Lett 13:1596–1601
15. Fedorov DG, Kitaura K (2007) Pair interaction energy decomposition analysis. J Comput Chem 28:222–237
16. Fedorov DG, Kitaura K (2016) Subsystem analysis for the fragment molecular orbital method and its application to protein–ligand binding in solution. J Phys Chem A 120:2218–2231
17. Nishimoto Y, Fedorov DG (2016) The fragment molecular orbital method combined with density-functional tight-binding and the polarizable continuum model. Phys Chem Chem Phys 18:22047–22061
18. Fedorov DG (2020) Analyzing interactions with the fragment molecular orbital method. In: Heifetz A (eds) Quantum ry. Springer US, New York, NY, pp 49–73
19. Heifetz A, Chudyk EI, Gleave L et al (2016) The fragment molecular orbital method reveals new insight into the chemical nature of GPCR–ligand interactions. J Chem Inf Model 56:159–172
20. Heifetz A, Morao I, Babu MM et al (2020) Characterizing interhelical interactions of G-protein coupled receptors with the fragment molecular orbital method. J Chem Theory Comput 16:2814–2824
21. Fedorov DG, Nagata T, Kitaura K (2012) Exploring chemistry with the fragment molecular orbital method. Phys Chem Chem Phys 14:7562
22. Heifetz A, James T, Southey M et al (2019) Characterising GPCR–ligand interactions using a fragment molecular orbital-based approach. Curr Opin Struct Biol 55:85–92
23. Heifetz A, Morao I, Babu MM et al (2020) Characterizing interhelical interactions of G-protein coupled receptors with the fragment molecular orbital method. J Chem Theory Comput 16:2814–2824
24. Venkatakrishnan AJ, Deupi X, Lebon G et al (2013) Molecular signatures of G-protein-coupled receptors. Nature 494:185–194
25. Venkatakrishnan A, Flock T, Prado DE et al (2014) Structured and disordered facets of the GPCR fold. Curr Opin Struct Biol 27:129–137
26. Venkatakrishnan AJ, Deupi X, Lebon G et al (2016) Diverse activation pathways in class A GPCRs converge near the G-protein-coupling region. Nature 536:484–487
27. Flock T, Ravarani CNJ, Sun D et al (2015) Universal allosteric mechanism for Gα activation by GPCRs. Nature 524:173–179
28. Hauser AS, Kooistra AJ, Munk C et al (2021) GPCR activation mechanisms across classes and macro/microscales. Nat Struct Mol Biol 28:879–888
29. Venkatakrishnan AJ, Ma AK, Fonseca R et al (2019) Diverse GPCRs exhibit conserved water networks for stabilization and activation. Proc Natl Acad Sci USA 116:3288–3293

Check for updates

Chapter 16

HessFit: Deriving Molecule-Specific Force Fields from Quantum Mechanical Hessians

Emanuele Falbo and Antonio Lavecchia

Abstract

The development of accurate and transferable molecular mechanics force fields (FMs) is crucial for reliable molecular dynamics (MD) simulations. Conventional force fields such as AMBER, CHARMM, OPLS, and GROMOS prioritize transferability at the expense of maximal accuracy, which can limit their performance when modeling novel chemical moieties or metal coordination environments. To address this issue, we present HessFit, an open-source Python toolkit that refines classical force fields using quantum mechanical (QM) information. HessFit derives molecule-specific bond parameters and atomic charges directly from QM Hessians and potential energy surfaces, combining analytical extraction and fitting procedures to obtain accurate bond, angle, and torsion constants. Benchmarks on several small molecules show better reproduction of vibrational frequencies, geometries, and thermodynamic properties compared to GAFF2 and OPLS. Applications to ligand–protein complexes further demonstrate HessFit's ability to model coordination environments that are typically challenging for transferable force fields. By integrating QM-level precision with classical computational efficiency, HessFit advances automated, data-driven force field parameterization, enhancing both accuracy and reproducibility across complex molecular systems.

Keywords Molecular Hessian, Force field parameterization, Python toolkit, Torsional fitting, QM/MM validation, Ligand–protein modeling

1 Introduction

Classical molecular mechanics (MM) force fields continue to be the backbone of molecular dynamics (MD) simulations for large biomolecular systems on biologically relevant timescales. However, their empirical nature and reliance on transferable atom types may limit accuracy for novel chemistries, metal centers, or nonstandard residues. In contrast, quantum mechanical (QM) calculations yield highly accurate descriptions of potential energy surfaces (PES) for small systems, yet their computational cost makes them impractical for routine MD simulations of large systems. HessFit bridges this gap by deriving bonded force field parameters directly from QM outputs (optimized geometries, atomic charges, Hessians) [1]. The workflow

Alexander Heifetz (ed.), *Quantum Mechanics Applications in Drug Design*, Methods in Molecular Biology, vol. 3061, https://doi.org/10.1007/978-1-0716-5420-0_16,

is fully automated, yielding molecule-specific parameters readily usable in standard MM engines. HessFit combines several methodological innovations: removal of nonbonded contributions from the QM Hessian, projection onto redundant internal coordinates (RICs), where diagonal elements correspond to bonded terms, and a hybrid analytic and least-squares fitting scheme for torsional potentials. This protocol reduces artificial mode coupling and improves vibrational frequency predictions relative to both transferable force fields and other Hessian-based approaches (e.g., the Seminario method) [1]. The toolkit is implemented as a Python command-line application, compatible with Gaussian [2], and generates output files readily convertible to AMBER input formats [3].

Molecular mechanics FFs were developed to describe a wide range of systems, including small drug-like molecules, lipids, carbohydrates, proteins, and nucleic acids. The parameters defining their potential energy functions are derived from experimental measurements and quantum mechanical (QM) calculations, typically performed at high levels of theory such as Hartree–Fock or density functional theory (DFT) [4]. These parameters include atomic properties (mass, radius, charge) and bonded terms-bond lengths, angles, and dihedral torsions-optimized to reproduce experimental and theoretical reference data. Typical experimental sources include X-ray and neutron diffraction (bond lengths and angles), dipole moments, heats of vaporization, and spectroscopic constants. Among the most utilized force fields for small molecules in CADD platforms are OPLS3 [5], GAFF [6], and CGenFF [7]. Each of these depends substantially on QM-derived reference data for parameterization. Likewise, biomolecular force fields such as AMBER ff14SB [8] and CHARMM [9] incorporate QM-derived information to refine the modeling of protein conformations and energetics. Efforts such as ForceBalance [10] and the Open Force Field Consortium [11] further support the robustness and adoption of transferable force fields by enabling systematic and reproducible parameter optimization. Recent toolkits such as QUBEKit [12] enable automated derivation of force field parameters directly from quantum mechanical data, combining Hessian fitting and virtual site schemes, while OpenFF BespokeFit focuses on automated torsion reparameterization within the Open Force Field framework [13]. These frameworks exemplify a growing trend toward QM-driven, data-rich force field refinement, positioning HessFit within the same methodological evolution. These efforts aim to automatically parameterize a larger portion of the chemical space using standardized, well-documented procedures. Significant advances have also been made in nucleic acid modeling, exemplified by the Parmbsc1 force field [14], which was tuned using high-level QM reference data for improved DNA simulations. DNA and RNA adopt diverse tertiary conformations that are crucial for their biological roles and relevance in drug design challenges. Accurate molecular modeling of these systems requires force fields capable of capturing electrostatic effects, phosphate ion interactions, and base pairing and stacking energetics. Recent reviews highlight the increasing importance of quantum-mechanically derived force fields in this context [15], emphasizing the role of Hessian-based parameter extraction as a route to bridge QM precision with MM scalability. In this regard, a polarizable force field for nucleic acids, AMOEBA (Atomic Multipole Optimized Energetics for Biomolecular Applications), was derived from extensive QM calculations [16]. Molecular dynamics simulations using AMOEBA on 20 distinct DNA/RNA molecules reproduced experimental NMR structures with average RMSDs of ~2.0 Å, underscoring its ability to capture conformational variability with high accuracy.

2 Theoretical Foundations of the HessFit Method

HessFit targets FFs of commonly used functional forms (GAFF2/OPLS style): harmonic stretches and bends, Fourier or Ryckaert–Bellemans torsions, Coulombic electrostatics with fixed point charges, and Lennard–Jones (LJ) dispersion. The general molecular mechanics potential is conventionally decomposed into bonded and nonbonded terms:

$$U = U_{\text{bonded}} + U_{\text{nonbonded}}. \tag{1}$$

Bonded terms include harmonic bond and angle contributions, and periodic torsions represented as a truncated Fourier series or Ryckaert–Bellemans expansion.

$$\begin{aligned} U = {} & \sum\nolimits_{bonds}^{*} k_r (r - r_0)^2 \\ & + \sum\nolimits_{\text{angles}}^{*} k_\theta (\theta - \theta_0)^2 \\ & + \sum\nolimits_{\text{dihedrals}}^{*} \sum\nolimits_{i=1}^{4} \frac{1}{2}\left(1 + (-1)^{i+1}\cos i\phi\right) \\ & + \sum\nolimits_{i>j}^{*} \left\{ \frac{q_i q_j}{r_{ij}} + 4\epsilon_{ij} \left[\left(\left(\frac{\sigma_{ij}}{r_{ij}} \right)^{12} - \left(\frac{\sigma_{ij}}{r_{ij}} \right)^{6} \right) \right] \right\}. \end{aligned} \tag{2}$$

A QM Hessian, i.e., the matrix of second derivatives of energy with respect to Cartesian coordinates, encodes local curvature of the PES and thus directly relates to harmonic bonded parameters. The central concept in HessFit involves subtracting electrostatic and van der Waals (nonbonded) contributions from the QM Hessian in order to isolate the pure bonded Hessian (H_{nb}). To obtain the bonded force constant parameters, the classical molecular mechanics Hessian matrix, H_{cc}^{MM} is equated to the quantum mechanical Hessian H_{cc}^{QM}, both expressed in Cartesian coordinates:

$$H_x^{\mathrm{MM}} k = H_x^{\mathrm{QM}}. \tag{3}$$

In the classical formalism, the Hessian can be represented as the sum of all bonded and nonbonded contributions:

$$H_x^{\mathrm{MM}} = \sum_m k_m h_x^{\mathrm{MM}} + H_x^{nb}, \tag{4}$$

here, k denotes the vector of bonded force constants; h_x^{MM} are the Hessian components attributable to bonded interactions; and H_x^{nb} embodies the nonbonded contribution. Since the number of adjustable parameters (M) is typically far smaller than the total number of Hessian elements ($3\,\mathrm{N} \times 3\,\mathrm{N}$, for N atoms), the relation is recast as a least-squares optimization problem. Decompositions of intramolecular and intermolecular interactions from QM Hessians have been previously explored in the literature [1, 17]. Traditionally, these analyses were conducted either in Cartesian or internal coordinates. HessFit, by contrast, utilizes redundant internal coordinates (RICs), which afford a one-to-one mapping between bonded terms and coordinate descriptors, thereby equating the number of equating the number of equations to the number of unknown force constants. In RIC form, the relationship can be expressed as:

$$\sum_m^M k_m h_q^m = H_q^{\mathrm{QM}} - H_q^{nb}. \tag{5}$$

Each term h_q^i is obtained from a calculation where bonded terms are set to one and nonbonded terms to zero. This separation minimizes coupling artifacts, such as stretch–torsion or bend–torsion interactions, thereby ensuring that bond and angle parameters remain physically independent. The nonbonded Hessian H_q^{nb} is computed with nonbonded terms active and all other force constants set to zero, enabling a clean separation of nonbonded effects from bonded contributions. The use of RICs offers a significant advantage: unlike traditional internal coordinates, they are consistently defined for any molecular system, and the resulting Hessian does not depend on the initial coordinate definitions. The QM RIC-based Hessian H_q^{QM} can be obtained from the Cartesian Hessian, H_{cc}^{QM}, through the Wilson B-matrix (W) transformation:

$$H_q^{\mathrm{QM}} = W^T H_{cc}^{\mathrm{QM}} W. \tag{6}$$

Many modern quantum chemistry packages now directly provide the Hessian in RIC form, eliminating the need for this transformation step. Parameterization is realized by matching corresponding diagonal entries of the QM and MM Hessians in RIC representation. Solving the resulting linear system yields the force constants for bond stretching, angle bending, and (stiff) torsional modes. However, it should be noted that Hessian-based extraction assumes harmonicity near equilibrium and may encounter limitations for strongly anharmonic or coupled modes (e.g., stretch–torsion coupling). Future developments could include cross-term treatments or multidimensional Hessian analysis to address such cases. This formulation provides a direct and numerically stable route to determine bonded force constants from quantum mechanical data, minimizing coupling artifacts and ensuring the resulting parameters are fully consistent with the underlying QM potential surface.

3 Validation of Vibrational and Structural Accuracy

To evaluate the accuracy and robustness of the HessFit parameterization scheme, we performed a comprehensive benchmark of its performance against established transferable force fields—GAFF2 and OPLS—and compared all results with QM benchmark data obtained at the DFT level. The evaluation focused on both vibrational frequencies and optimized equilibrium geometries, providing a rigorous test of the force field's ability to reproduce fundamental molecular properties. The comparison of vibrational frequencies was quantified using the mean unsigned error (MUE), defined as $\mathrm{MUE} = \Sigma_i = |\nu_i^{\mathrm{QM}} - \nu_i^{\mathrm{MM}}| \,/\, N$ where N is the total number of vibrational modes and ν_i^{QM} and ν_i^{MM} represent the frequencies obtained from quantum mechanical and molecular mechanical calculations, respectively. Atom typing was performed before parameterization: AMBER [18] atom types were used for amino acids, while GAFF2 types were assigned to general organic molecules. Although HessFit uses the same atomic nomenclature as AMBER or GAFF2, the resulting vibrational frequencies depend entirely on the newly derived bond parameters, which uniquely fit the quantum Hessian. A first benchmark was conducted on a set of rigid aromatic and heterocyclic molecules, including pyridine, furan, oxazole, isoxazole, benzene, imidazole, and pyrazole (Fig. 1A). A direct comparison of HessFit and DFT frequencies revealed excellent agreement, with a MUE of 11.5 cm^{-1}. By comparison, GAFF2 and OPLS produced larger errors, 25.4 cm^{-1} and 17.0 cm^{-1}, respectively. These small deviations reflect the fact that rigid systems, characterized by limited internal flexibility, are well described by the harmonic force field representation implemented in HessFit. A second benchmark focused on a set of flexible organic molecules-n-butane, 1-butanol, methylpropyl sulfide, and n-butoxybenzene, as well as neutral amino acids (glycine, alanine, aspartic acid, leucine, valine, and cysteine) (Fig. 1A). The results demonstrate that HessFit maintains excellent agreement with the

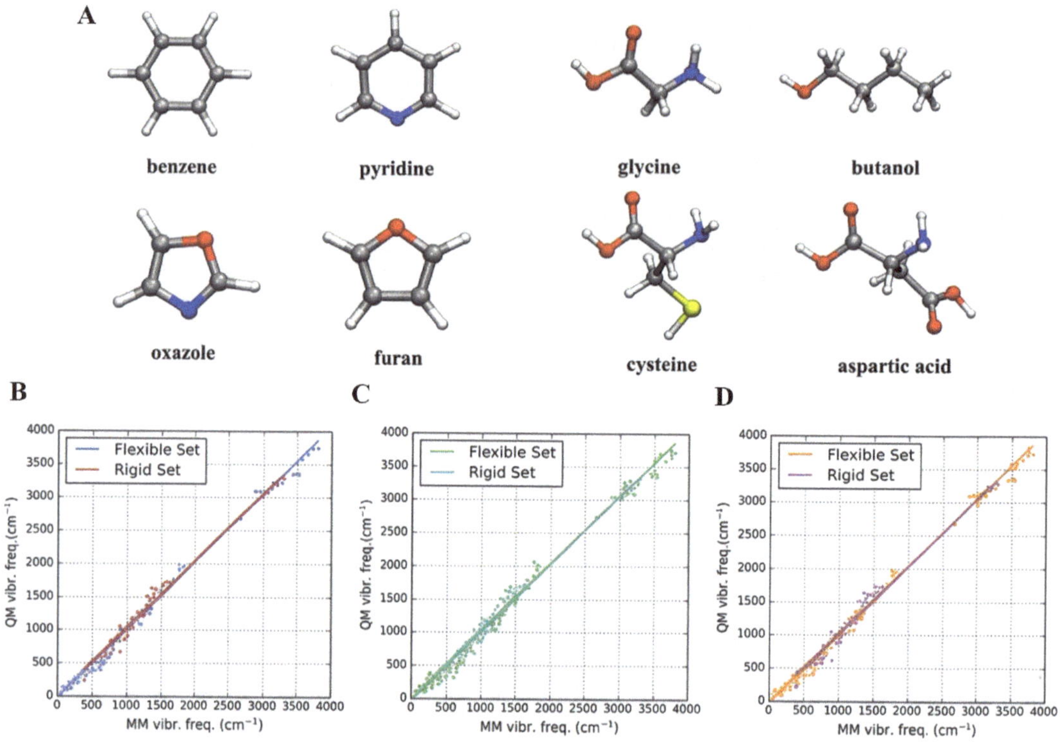

Fig. 1 (**A**) Representative molecular structures from the rigid and flexible datasets used for vibrational benchmarking. (**B–D**) Comparison of vibrational frequencies (QM reference vs. modeled) for representative molecules: (**B**) using HessFit-derived parameters, (**C**) using GAFF2/AMBER default parameters, (**D**) using OPLS parameters. The comparison visualizes how HessFit aligns closely with QM benchmarks across the frequency spectrum

DFT data, producing a MUE of 16 cm^{-1}, compared to 75.0 cm^{-1} for GAFF2 and 30.0 cm^{-1} for OPLS. The slightly higher deviations observed for flexible systems primarily stem from low-frequency torsional modes, which are inherently more sensitive to subtle changes in the potential energy surface. Nonetheless, the remarkably low MUE achieved by HessFit confirms its ability to accurately capture both bound and torsional dynamics, thanks to its direct fitting procedure to the QM Hessian matrix. HessFit's superior performance can be attributed to two main factors: first, the explicit separation of bound and unbound contributions ensures that the fitted bound parameters are not contaminated by long-range interactions, a limitation that often affects empirical force fields. By subtracting the nonbonded component (H_{nb}), the QM Hessian (H^{QM}), HessFit isolates the purely bonded interactions, producing parameters that better reflect intrinsic molecular mechanics. Second, the use of RIC introduces a one-to-one correspondence between diagonal Hessian elements and bound terms, effectively eliminating artificial modal coupling. This approach allows each bond, angle, or torsion to be independently parameterized, improving both numerical stability and transferability. In addition to vibrational frequency validation, equilibrium geometries optimized using HessFit were compared to DFT reference structures. The root mean square deviation (RMSD) was calculated between non-hydrogen atomic positions of the optimized molecular geometries obtained by HessFit, GAFF2, and OPLS. Hydrogen atoms were excluded due to their greater positional sensitivity to potential energy minima. For the set of rigid molecules, the average RMSDs were 0.11 Å for HessFit, 0.23 Å for GAFF2, and 0.16 Å for OPLS, confirming that all methods reproduce DFT geometries with small deviations, although HessFit offers the best overall performance. For flexible molecules, higher RMSDs were obtained: 0.34 Å (HessFit), 0.59 Å (GAFF2), and 0.44 Å (OPLS), demonstrating greater conformational freedom in these systems. These results are consistent with previously reported theoretical studies [19], confirming the accuracy of the method is within the margin of error. This harmonic accuracy confirms the robustness of the HessFit procedure across both rigid and flexible systems, and suggests potential integration into automated drug design workflows where vibrational fidelity correlates with accurate binding energetics.

4 Refinement of Torsional Potentials in Flexible Molecules

Within the harmonic approximation, the potential energy of a molecular system can be locally approximated as a quadratic function of internal coordinates. Thus, diagonal entries of the QM Hessian (H^{QM}) may be used to infer the harmonic torsional constant (k_ϕ) for a dihedral degree of freedom. However, this direct Hessian-based extraction frequently overestimates torsional stiffness, and hence is often restricted to rigid or aromatic systems, where torsional variability is limited, and planarity is required during dynamics. To describe torsional motion more accurately, the potential energy of a dihedral angle ϕ with phase δ is generally developed as a truncated Fourier series:

$$U(\phi) = \sum_{n}^{N_{\max}} \frac{V_n}{2}\left[1 + \cos(n\phi)\right] \tag{7}$$

In practice, only a limited number of terms are included (typically $N_{\max} = 3$–6) to avoid overfitting. The multiplicity, $N_{\max}$, defines the number of minima in the torsional potential, while the coefficients V_n represent the heights of the barriers separating these minima. Recent algebraic approaches, such as the work by Kania et al. [20], provide optimal linear-algebra-based solutions for Fourier torsional coefficient fitting. These methods complement HessFit's LLS strategy by offering analytic convergence and reduced sampling dependence, further validating the algebraic nature of our formulation. HessFit employs two complementary approaches to model torsional profiles near equilibrium: (1) a System-of-Equations (SES) method, analogous to AMBER techniques, where one or two Fourier coefficients are solved analytically from Hessian-derived data; and (2) a Linear Least Squares (LLS) fitting, in which all Fourier coefficients are optimized to best approximate the QM potential energy surface (PES). In practice, in the SES approach, we set $N_{max} = 2$ and assume that two dominant harmonics (n_1 and n_2) suffice to approximate the local torsional profile near equilibrium. These are determined by solving a linear system derived from first derivatives (torque) and second derivatives (curvature) at the equilibrium angle ϕ_0:

$$(-1)^n n_1 V_{n1} \sin(n_1\phi_0) - n_2 V_{n2} \sin(n_2\phi_0) = -\left(\frac{\partial E}{\partial \phi}\right)_{\phi_0}. \quad (8)$$

$$(-1)^n n_1^2 V_{n1} \cos(n_1\phi_0) - n_2^2 V_{n2} \cos(n_2\phi_0) = k_\phi. \quad (9)$$

This equation yields the torsional constant corresponding to the ith harmonic, inferred from the Hessian diagonal in which nonbonded contributions (e.g., electrostatic and van der Waals couplings) have been subtracted or accounted for. For rotations involving sp^3–sp^3 centers, the terms $n_1 = 2$ and $n_2 = 3$ are used, while for sp^2–sp^3 rotations, the same multiplicities are retained but applied to six relevant dihedrals. In contrast, sp^2–sp^2 and other hybridization combinations use $n_1 = 1$ and $n_2 = 2$. The main advantage of this formulation is that it allows the determination of two torsional parameters from purely local information, without the need for extensive sampling. This represents a valuable improvement in the derivation of dihedral parameters, widely considered one of the most challenging aspects of force field development. In some situations, particularly for sp^2–sp^2 rotations, the direct Hessian-based method may yield overly rigid or overly flexible dihedrals. To correct this problem, HessFit applies a scaling adjustment based on a reference rotational barrier of 30 kcal·mol^{-1}·rad^{-2}, approximately corresponding to the rotation of the C=C bond in ethylene [21]. If the calculated barrier exceeds this value, it is reduced proportionally; conversely, if it is too small, the force constant is increased to reach the reference threshold. In most practical applications, particularly biochemical systems, additional accuracy is achieved through LLS refinement. This involves minimizing the difference between the relaxed one-dimensional QM and MM dihedral scans based on the error function:

$$\chi^2 = \left\| \sum_i^M \Delta E_i^{\mathrm{QM}} - \Delta E_i^{\mathrm{MM}} \right\|, \tag{10}$$

where ΔE^{QM} and ΔE^{MM} are the energy differences relative to the lowest-energy conformations at the QM and MM levels, respectively. Specifically, ΔE^{MM} represents the torsional energy contribution calculated for each geometry along the dihedral scan.

QM energies were calculated typically at B3LYP/def2-TZVPP (or in some cases 6–31 G(d)) [22, 23]. Each torsional scan is performed in increments of 15° (or finer), covering the full 2π rotation. During each increment, all other geometrical degrees of freedom are relaxed to minimize energy. The QM scan is fully relaxed for each dihedral rotation, and the corresponding MM energies are calculated for the same geometries to ensure a consistent comparison. This hybrid strategy enables efficient and reasonably accurate torsional parameterization that maintains local geometry. SES offers quick, initial torsional estimates with qualitative agreement to QM PES for many simpler molecules, while LLS refinement aligns the Fourier coefficients more closely to the QM-relaxed torsional profiles. Torsion-fitting is more computationally demanding than Hessian-based extraction (e.g., single-dihedral LLS fits may require a few minutes even on modest CPUs, depending on QM level), yet it is strongly recommended for flexible biochemical motifs where accurate torsional energetics significantly influence the conformational ensemble. Three representative cases were studied to validate the method: butane, n-butanol, and dipeptide glycine. In the butane test, HessFit derived a single torsional parameter for the X–C–C–X dihedral (X being any substituent). The computed average barrier (2.24 kcal $\mathrm{mol^{-1}}$) deviates modestly from the experimental syn barrier (4.54 kcal $\mathrm{mol^{-1}}$) [24] (Fig. 2A), illustrating the limitations of pure Hessian-derived torsional constants for larger rotations. For *n*-butanol, HessFit derived torsional parameters for both the X–C–O–X and C–C–C–O dihedrals. The modest deviation observed in the C–C–C–O torsional profile (Fig. 2B) further substantiates that LLS-based fitting yields robust estimates of torsional barriers consistent with QM reference.

This accuracy stems from the well-structured initial MM profiles, which already capture key features of the torsional landscape, including minima and barrier heights. The same approach was extended to the glycine dipeptide, focusing on the backbone torsional angle ϕ (C–N–Cα–C). In this test, a small number of points were used to evaluate HessFit's ability to recover the correct torsional profile from a limited QM dataset. As shown in Fig. 2C, the fitted LLS potentials reproduce the QM PES of the φ torsion with minimal deviation at the scan edges. To further improve fidelity at the low-energy region, we introduced a Boltzmann-weighted (BW) scheme into the error weighting, prioritizing energetically accessible conformers.

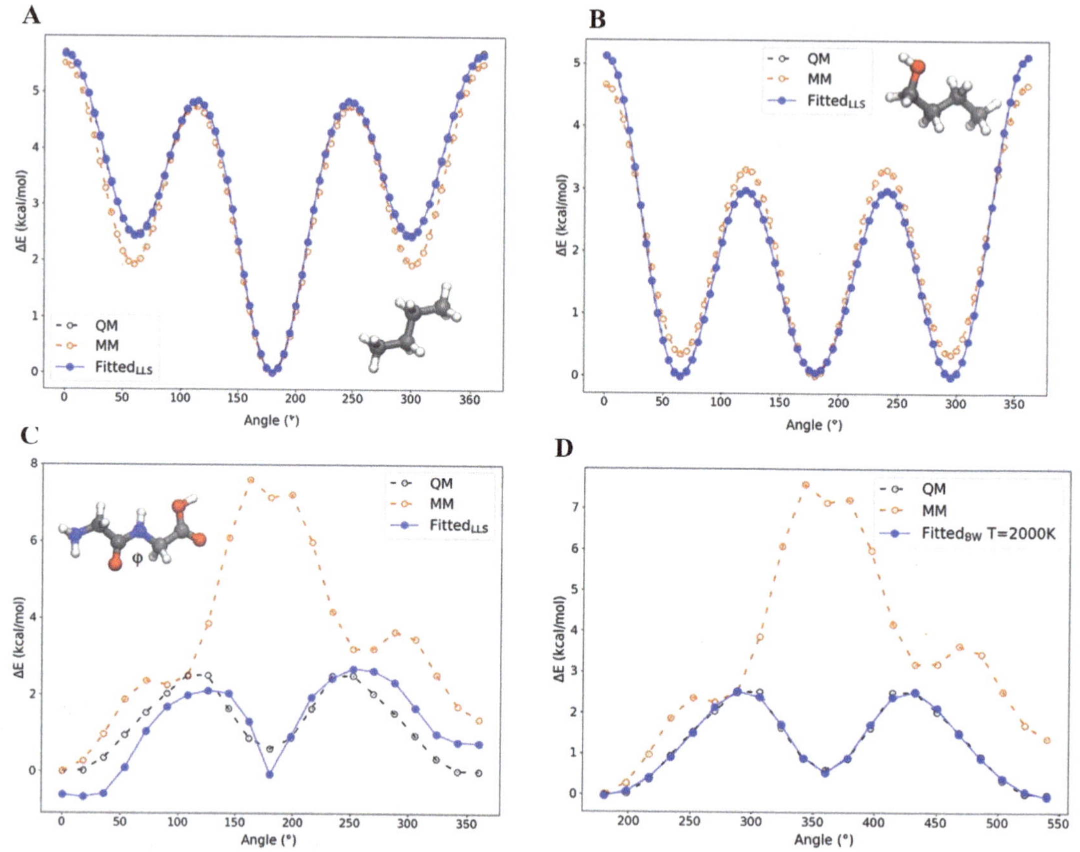

Fig. 2 (**A**) Torsional PES of the C–C–C–C dihedral (n-butane) and (**B**) C–C–C–O dihedral (n-butanol) as computed by QM scans and the fitted HessFit model. (**C**) ϕ dihedral profile for a glycine dipeptide model fitted via LLS; (**D**) same model fitted by the Boltzmann-weighted (BW) method (T = 2000 K)

This nonlinear fitting variant is denoted as the BW method [25]. The minimization procedure was performed on the following error function:

$$\chi^2 = \frac{1}{M}\left\| \sum_{i}^{M} (\Delta E_i^{\mathrm{QM}} - \Delta E_i^{\mathrm{MM}})\, e^{\frac{\Delta E^{\mathrm{QM}}}{k_B T}} \right\|, \tag{11}$$

where ΔE and are still the QM and MM energies relative to their lowest-energy structures, k_B is the Boltzmann constant, and T is the temperature. During the optimization process, the coefficient parameters can be adjusted for T values of 500, 1000, and 2000 K. The choice of Boltzmann-weighting temperature (T = 500–2000 K) is pragmatic rather than physically literal: higher effective temperatures emphasize low-energy, thermally accessible regions of the torsional potential while down-weighting high-energy conformers, improving smoothness and convergence of the fitted profile. For the ϕ angle of the dipeptide, there is a large improvement with respect to LLS model (Fig. 2D), which is due to the advantage of nonlinear fitting BW scheme to capture more complex torsional profiles. A key advantage of HessFit is it obviates the necessity for constructing extensive conformer libraries, a step common in many traditional parameterization workflows [26, 27]. As a result, HessFit is more efficient and scalable for larger, more flexible molecules. This makes the method considerably faster and more practical for larger and more flexible molecules. Moreover, though HessFit supports standard AMBER/GAFF2 atom types, it can also define bespoke atom types, offering additional flexibility in capturing local chemical environments. However, caution is warranted: newly introduced intramolecular parameters should be cross-checked for consistency with intermolecular interactions, to maintain the physical coherence of the force field. Nonetheless, limitations remain: highly anharmonic torsions, multidimensional couplings, or systems with strong electronic delocalization may challenge the Hessian + 1D scan framework. These limitations point to potential future extensions involving multi-torsion coupling or machine-learning-assisted refinement.

5 Case Study: MD Simulations of Human Zn-Proteins

To evaluate the structural and energetic fidelity imparted by HessFit-derived parameters in metalloprotein contexts, we conducted molecular dynamics (MD) simulations of the 2GIV protein, which features a canonical $ZnCys_3His$ coordination center. We benchmarked HessFit-derived parameters against those from ZAFF, a force field specialized for tetra-coordinated Zn^{2+} centers, using comparative MD simulations and structural analyses. The test set included four representative model complexes-$ZnCys_4$, $ZnCys_3His$, and $ZnCys_2His_2$ -previously studied by Ryde and Hu [28]. In these models, cysteine and histidine residues were represented by CH_3S and methylimidazole, respectively. All structures were optimized in a continuum solvent with dispersion corrections, and Lennard–Jones parameters for zinc were explicitly included. The comparison focused on bond lengths, angles, and their corresponding force constants, as these primarily determine the geometry and internal energy of zinc coordination complexes. As shown in Table 1, HessFit generally produces more rigid bonded force constants than ZAFF, particularly for Zn–N in $ZnCys_3His$ ($\approx +75\,\text{kcal}\cdot\text{mol}^{-1}\cdot\text{Å}-^{-2}$) and for the N–Zn–S angular term ($\approx +70\,\text{kcal}\cdot\text{mol}^{-1}\cdot\text{rad}^{-2}$). In contrast, the angular force constants for $ZnCys_4$ and $ZnCys_2His_2$ show comparable magnitudes in both force fields. When benchmarked against QM reference structures, HessFit shows markedly superior geometric agreement: RMSD(HessFit→QM) = 0.11 Å vs. 0.64 Å (ZAFF) for $ZnCys_4$; 0.31 Å vs. 0.78 Å for $ZnCys_3His$; and 0.49 Å vs. 0.93 Å for $ZnCys_2His_2$. These results confirm that HessFit more faithfully reproduces the QM equilibrium geometries of zinc coordination sites.

Table 1

Comparison of bonded force constants (k) and equilibrium parameters (r_{eq}, angles) for Zn coordination models computed via ZAFF vs HessFit

Model	Parameter	ZAFF (*k*)	ZAFF (r_{eq})	HessFit (*k*)	HessFit (r_{eq})
$ZnCys_4$	Zn–S	32.69	2.42	64.68	2.34
	S–Zn–S	35.72	109.47	13.76	109.43
	C–S–Zn	64.39	101.73	19.85	102.72
$ZnCys_3His$	Zn–S	50.99	2.35	70.72	2.36
	Zn–N	35.80	2.14	110.40	2.08
	S–Zn–S	33.79	116.41	43.65	115.45
	C–S–Zn	61.71	101.84	65.68	103.43
	N–Zn–S	29.24	101.21	101.72	102.82
	C–N–Zn	57.36	124.86	89.75	126.01
$ZnCys_2His_2$	Zn–S	67.39	2.30	84.78	2.30
	Zn–N	50.10	2.08	77.53	2.09
	S–Zn–S	22.85	135.46	16.54	133.06
	C–S–Zn	76.59	104.86	20.79	103.50
	N–Zn–S	29.08	103.03	43.35	106.19
	C–N–Zn	47.41	129.66	32.63	130.82
	N–Zn–N	31.65	106.85	16.66	107.81

Values are in kcal·mol^{-1}·Å^2 for bond constants, kcal·mol^{-1}·rad^{-2} for angular constants; r_{eq} in Å and angles in degrees

The HessFit parameters for the $ZnCys_3His$ complex were used to generate the molecular topology and initial coordinates for the MD simulations. The analysis focused on key geometric descriptors of the Zn coordination environment, derived from representative frames extracted from the complete MD trajectories to ensure statistical significance while maintaining computational efficiency [29, 30]. Specifically, we monitored the Zn–N distance involving the imidazole nitrogen of HIS50 and the Zn-S distances corresponding to the thiolate groups of CYS34, CYS37, and CYS54. Additionally, the S–Zn–N and S–Zn–S bond angles were analyzed to characterize the stability of the tetrahedral coordination geometry. As shown in Fig. 3A, both ZAFF and HessFit produced similar mean bond lengths: approximately 2.04 Å for Zn–N interactions and 2.36 Å for Zn–S interactions.

The bond length distributions show narrower standard deviations with HessFit ($\sigma_{Zn-N} = 0.08$ Å vs. 0.32 Å for ZAFF; $\sigma_{Zn-N} = 0.04$ Å vs. 0.20 Å), suggesting tighter, physically consistent sampling around

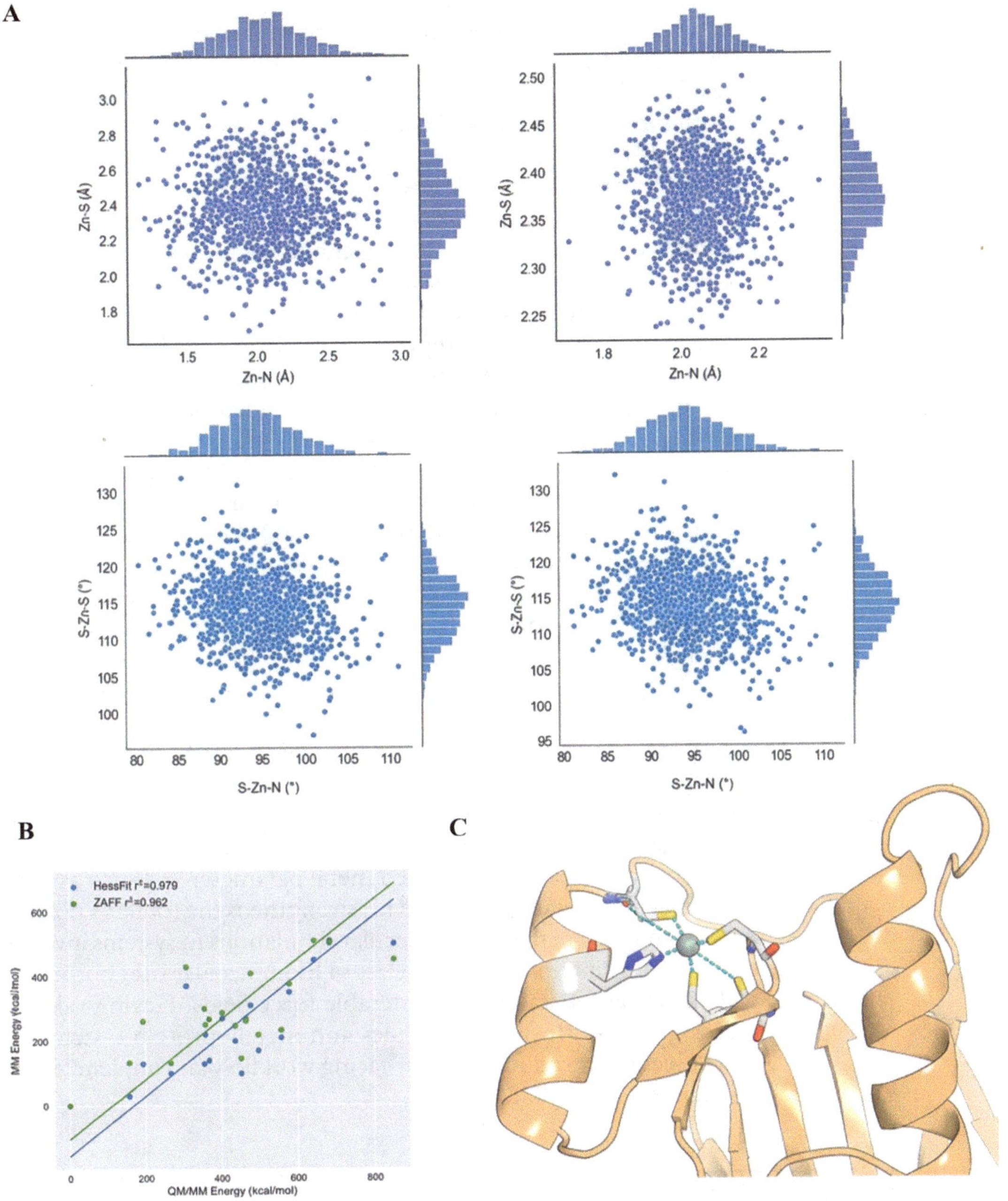

Fig. 3 (**A**) Distributions of Zn–N and Zn–S bond lengths (top row: ZAFF vs HessFit) and S–Zn–N / S–Zn–S angles (bottom row) from MD simulations. (**B**) Correlation between QM/MM and MM energies (normalized to minima) for 20 representative frames using HessFit and ZAFF parameter sets. (**C**) Representative snapshot of 2GIV from the HessFit MD trajectory, depicting the Zn coordination environment

the equilibrium geometry. These results are in close agreement with the reference QM geometry reported in Table 1, confirming that the HessFit parameterization preserves key metal-ligand structural features throughout the MD trajectory. For angular distributions, both force fields produced comparable mean values for the S–Zn–N (≈94.4°) and S–Zn–S (≈114.2°) angles, with slightly narrower distributions observed in the HessFit simulation. Although these values deviate slightly from the QM reference, such differences are expected because the protein backbone imposes conformational coupling effects that are absent in isolated QM models. The observed consistency between HessFit and QM data, however, highlights the method's ability to reproduce realistic coordination geometries even within a complex protein environment. To further probe the energetic consistency of HessFit parameterization, we selected 20 representative frames from the 2GIV trajectory and computed single-point QM/MM energies for each, following the procedure outlined earlier. This comparison provides a direct measure of the coherence between the classical MM and QM energy landscapes and, consequently, the force field's ability to sample the physically relevant conformational space [31–33]. Figure 3B shows that HessFit yields a marginally improved correlation (R^2) between QM/MM and MM energies versus ZAFF. Nevertheless, both force fields display good linearity, indicating that backbone conformational changes dominate the energetic variance in the system. In summary, these results demonstrate that HessFit accurately reproduces both the structural integrity and the energetic profile of the zinc ion coordination site in solution. The narrower bond and angle distributions, together with stronger QM/MM energy correlations, demonstrate that HessFit matches or surpasses specialized metal parameter sets such as ZAFF in modeling metalloproteins. This strengthens the toolkit's potential to extend the reliability of molecular simulations to systems involving metal coordination centers, a class of biomolecular sites notoriously challenging for standard transferable force fields. Having validated HessFit on both small molecules and metalloprotein systems, the next section introduces the user-facing workflow and implementation details of the toolkit.

6 Workflow and Implementation of HessFit

The HessFit toolkit provides an efficient and intuitive workflow for the automated derivation of force field parameters, as shown schematically in Fig. 4. Its theoretical background and computational methodology are described in detail in our recent publication [34]. The workflow begins with a single, user-friendly JSON input file specifying the molecular structure, QM level of theory, basis set, and charge model. After execution via a simple command line

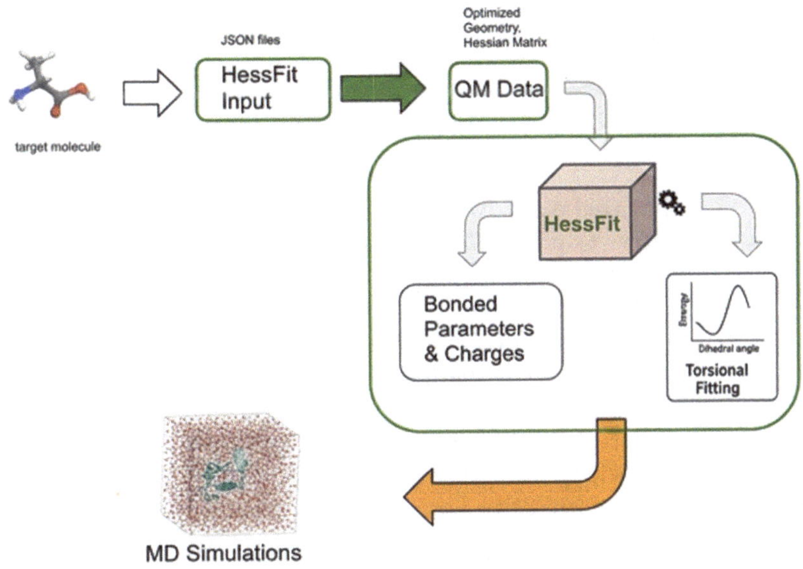

Fig. 4 Schematic representation of HessFit workflow

interface, HessFit launches a fully automated pipeline that handles all subsequent steps. The program first prepares and executes the QM calculations required to obtain the optimized geometry and Hessian matrix. It then processes these outputs to extract bound force constants and atomic charges, using built-in algorithms that separate bound from nonbound contributions. The resulting parameters are automatically formatted for direct use in molecular dynamics (MD) simulations: a.mol2 file containing the refined partial charges and an AMBER-compatible.frcmod file with the corresponding bound terms. Recommended QM level for a balance of accuracy and cost is B3LYP/def2-TZVPP for optimizations and Hessians; smaller basis sets (e.g., 6-31 G(d)) can be used as cost-saving options for initial tests. CM5 charges and Gaussian-embedded AMBER LJ parameters are the default choices for charge/LJ extraction in the HessFit workflow. These choices were used during validation of the model. The modular structure of HessFit allows straightforward integration with existing drug discovery workflows, such as docking and free energy calculations. A schematic illustration could explicitly show how HessFit-derived parameters interface with molecular dynamics or binding free-energy protocols. Further refinements of torsional dihedrals can be obtained by torsion fitting options (LLS or BW) selectable by user flags. This streamlined, input-driven design significantly lowers the technical barrier to parameter derivation, enabling nonexpert users to generate accurate molecule-specific force fields efficiently. The complete toolkit, along with detailed documentation and example workflows, is available free of charge on GitHub at: https://github.com/emanuelefalbo/HessFit.

7 Concluding Remarks

In this work, we presented HessFit, a Python-based toolkit that derives molecule-specific intramolecular and intermolecular force field parameters directly from QM data. The automated procedure aims to generate transferable yet accurate parameters, suitable for condensed-phase MD simulations. Implemented to run seamlessly with the Gaussian program, HessFit leverages redundant internal coordinates to overcome the dependence of diagonalized FFs on user-defined coordinate systems. The use of RIC, maximally independent internal coordinates, provides a consistent and transferable framework for extracting constrained parameters. This approach eliminates the inconsistencies often introduced by arbitrary user-defined coordinates and minimizes human intervention, improving both reproducibility and usability. The harmonic intramolecular contribution of the FF is obtained by fitting the QM Hessian to its classical counterpart, while atomic charges and van der Waals parameters are extracted directly from the Gaussian outputs. This same strategy can be extended to optimized nonstandard ligand, enabling the systematic improvement of existing biomolecular force fields. A key advantage of HessFit lies in its ability to derive multiple dihedral barrier terms directly from the QM Hessian, with optional refinement via relaxed one-dimensional QM potential energy surface (PES) scans. The toolkit also incorporates a built-in implementation of the Seminario method, offering users flexible parameterization options. Comprehensive benchmarking on a range of rigid and flexible organic molecules has demonstrated that HessFit consistently outperforms widely used force fields such as GAFF2 and OPLS, achieving closer agreement with QM benchmark data in terms of vibrational frequencies and geometric accuracy. Importantly, the framework was successfully extended to complex systems, including zinc-based metalloproteins, where HessFit accurately reproduced quantum mechanical ground-state geometries for multiple coordination complexes and demonstrated reliable behavior in full MD simulations of the human protein MYST histone acetyltransferase 1 (2GIV). These findings underscore HessFit's promise in addressing the longstanding challenge of parametrizing metal-containing biomolecules within classical MD frameworks. In addition to metal-centered systems, HessFit could be naturally coupled with machine-learning models that predict bonded coefficients directly from local atomic environments, as demonstrated by recent tools such as *espaloma* [35]. This integration could further accelerate the QM→MM parameterization process. The primary goal of future work is to establish HessFit as the standard, high-throughput computational tool for achieving simulation reliability. We will begin by focusing on high-fidelity automated parameterization. This involves extending the use of HessFit's automated QM-based refinement to a diverse library of pharmacologically relevant ligand-protein systems [36, 37]. Using free energy calculations with advanced sampling techniques, we will validate the improved parameters on refined ligand–protein complexes.

Overall, HessFit provides a unified, automated, and extensible framework for force field development, combining quantum mechanical rigor with classical scalability. By lowering the barrier to accurate parameter derivation, it enables nonexpert users to generate molecule-specific force fields suitable for a wide variety of applications, from small molecule simulations to metalloenzyme modeling and drug–target interaction studies. While current results already outperform traditional transferable force fields, further improvements, such as explicit treatment of bonded coupling terms, optimized 1–4 nonbonded scaling, and integration with polarizable or machine learning-based models, will continue. Future work will also explore the inclusion of polarizable force fields, in line with recent trends highlighted by Pollêto et al. [38], where additive force fields are increasingly complemented by explicit polarization and many-body effects. This would extend HessFit's applicability to electrostatically complex systems, including charged or highly polar drug candidates.

References

1. Seminario JM (1996) Calculation of intramolecular force fields from second-derivative tensors. Int J Quantum Chem 60(7):1271–1277
2. Frisch MJ, Trucks GW, Schlegel HB, Scuseria GE, Robb MA, Cheeseman JR, et al (2016) Gaussian 16, Revision C.01. Gaussian, Inc., Wallingford CT
3. Case DA, Cheatham TE, Darden T, Gohlke H, Luo R, Merz KM et al (2005) The Amber biomolecular simulation programs. J Comput Chem 26:1668–1688
4. Orio M, Pantazis DA, Neese F (2009) Density functional theory. Photosynth Res 102(2):443–453
5. Harder E, Damm W, Maple J, Wu C, Reboul M, Xiang JY et al (2016) OPLS3: a force field providing broad coverage of drug-like small molecules and proteins. J Chem Theory Comput 12(1):281–296
6. Vanommeslaeghe K, Hatcher E, Acharya C, Kundu S, Zhong S, Shim J et al (2010) CHARMM general force field: a force field for drug-like molecules compatible with the CHARMM all-atom additive biological force fields. J Comput Chem 31(4):671–690
7. Vanommerslaeghe K, MacKerell AD (2012) Automation of the CHARMM general force field (CGenFF) I: bond perception and atom typing. J Chem Inf Model 52:3144–3154
8. Maier JA, Martinez C, Kasavajhala K, Wickstrom L, Hauser KE, Simmerling C (2015) ff14SB: improving the accuracy of protein side chain and backbone parameters from ff99SB. J Chem Theory Comput 11(8):3696–3713
9. Huang J, MacKerell AD (2013) CHARMM36 all-atom additive protein force field. Validation based on comparison to NMR data. J Comput Chem 34(2135):2145
10. Wang LP, Martinez TJ, Pande VS (2014) Building force fields: an automatic, systematic, and reproducible approach. J Phys Chem Lett 5(11):1885–1891
11. Mobley DL, Bannan CC, Rizzi A, Bayly CI, Chodera JD, Lim VT et al (2018) Escaping atom types in force fields using direct chemical perception. J Chem Theory Comput 14(11):6076–6092
12. Horton JT, Allen AEA, Dodda LS, Cole DJ (2019) QUBEKit: automating the derivation of

force field parameters from quantum mechanics. J Chem Inf Model 59(4):1366–1381
13. Horton JT, Boothroyd S, Wagner J, Mitchell JA, Gokey T, Dotson DL et al (2022) Open Force Field BespokeFit: automating bespoke torsion parametrization at scale. J Chem Inf Model 62(22):5622–5633
14. Ivani I, Dans PD, Noy A, Pérez A, Faustino I, Hospital A et al (2016) Parmbsc1: a refined force field for DNA simulations. Nat Methods 13(1):55–58
15. Ginex T, Vázquez J, Estarellas C, Luque FJ (2024) Quantum mechanical-based strategies in drug discovery: finding the pace to new challenges in drug design. Curr Opin Struct Biol 87:102870
16. Zhang C, Lu C, Jing Z, Wu C, Piquemal JP, Ponder JW et al (2018) AMOEBA polarizable atomic multipole force field for nucleic acids. J Chem Theory Comput 14(4):2084–2108
17. Barone V, Cacelli I, De Mitri N, Licari D, Monti S, Prampolini G (2013) Joyce and Ulysses: integrated and user-friendly tools for the parameterization of intramolecular force fields from quantum mechanical data. Phys Chem Chem Phys 15(11):3736–3751
18. Cornell WD, Cieplak P, Bayly CI, Gould IR, Merz KM, Ferguson DM et al (1995) A second generation force field for the simulation of proteins, nucleic acids, and organic molecules. J Am Chem Soc 117(19):5179–5197
19. Wang J, Wolf RM, Caldwell JW, Kollman PA, Case DA (2004) Development and testing of a general amber force field. J Comput Chem 25(9):1157–1174
20. Kania A, Sarapata K, Gucwa M, Wójcik-Augustyn A (2021) Optimal solution to the torsional coefficient fitting problem in force field parametrization. J Phys Chem A 125(12):2673–2681
21. Case DA, Cheatham TE, Darden T, Gohlke H, Luo R, Merz KM et al (2005) The Amber biomolecular simulation programs. J Comput Chem 26:1668–1688
22. Papajak E, Zheng J, Xu X, Leverentz RH, Truhlar GD (2011) Perspectives on basis sets beautiful: seasonal plantings of diffuse basis functions. J Chem Theory Comput 7:3027–3034
23. Rassolov VA, Pople JA, Ratner MA, Windus TL (1998) 6-31G* basis set for atoms K through Zn. J Chem Phys 109(4):1223–1229
24. Hoyland JR (1968) Internal rotation in butane. J Chem Phys 49(6):2563–2566
25. Robertson MJ, Tirado-Rives J, Jorgensen WL (2015) Improved peptide and protein torsional energetics with the OPLS-AA Force Field. J Chem Theory Comput 11:3499–3509
26. Marenich AV, Jerome SV, Cramer CJ, Truhlar DG (2012) Charge Model 5: an extension of Hirshfeld population analysis for the accurate description of molecular interactions in gaseous and condensed phases. J Chem Theory Comput 8:527–541
27. Cacelli I, Prampolini G (2007) Parametrization and validation of intramolecular force fields derived from DFT calculations. J Chem Theory Comput 3(5):1803–1817
28. Hu L, Ryde U (2011) Comparison of methods to obtain force-field parameters for metal sites. J Chem Theory Comput 7(8):2452–2463
29. Falbo E, Fusè M, Lazzari F, Mancini G, Barone V (2022) Integration of quantum chemistry, statistical mechanics, and artificial intelligence for computational spectroscopy: the UV–Vis spectrum of TEMPO Radical in different solvents. J Chem Theory Comput 18:6203–6216.
30. Falbo E, Delre P, Lavecchia A (2025) From Apo to ligand-bound: unraveling PPARγ-LBD conformational shifts via advanced molecular dynamics. ACS Omega 10(13):13303–13318
31. Falbo E (2021) Using theoretical chemistry to understand the properties of Polyoxometalates and their potential as energy storage materials [PhD Thesis]. Newcastle University
32. Vreven T, Byun KS, Komàromi I, Dapprich Jr S, Morokuma JAM (2006) K, et al. Combining quantum mechanics methods with molecular mechanics methods in ONIOM. J Chem Theory Comput 2:815–826
33. Falbo E, Penfold T (2020) Redox potentials of polyoxometalates from an implicit solvent model and QM/MM molecular dynamics. J Phys Chem C 124(28):15045–15056
34. Falbo E, Lavecchia A (2024) HessFit: a toolkit to derive automated force fields from quantum mechanical information. J Chem Inf Model 64(14):5634–5645
35. Takaba K, Pulido I, Behara PK, Cavender CE, Friedman AJ, Henry MM et al (2024) Machine-learned molecular mechanics force field for the simulation of protein-ligand systems and beyond. Chem Sci 15(32):12861–12878
36. Laghezza A, Falbo E, Gilardi F, Thomas A, Brunetti L, Leuci R et al (2025) A new potent and selective peroxisome proliferator-activated receptor alpha partial agonist displays anti-steatotic effects In vitro and behaves as a safe hypolipidemic and hypoglycemic agent in a diabetic mouse model. Eur J Med Chem 289:117494
37. Laghezza A, Cerchia C, Genovese M, Montanari R, Capelli D, Wackerlig J et al (2024) A chemical modification of a peroxisome proliferator-activated receptor pan agonist produced a shift to a new dual alpha/gamma partial agonist endowed with mitochondrial pyruvate carrier inhibition and antidiabetic properties. Eur J Med Chem 275:116567
38. Polêto MD, Lemkul JA (2022) Integration of experimental data and use of automated fitting methods in developing protein force fields. Commun Chem 5(1):38

Chapter 17

Fragment Molecular Orbital-Based Molecular Dynamics (FMO-MD): Current Status and Challenges

Yuto Komeiji, Shu Ohno, Shuhei Miyakawa and Kaori Fukuzawa

Abstract

The fragment molecular orbital-based molecular dynamics (FMO-MD) method simulates the time evolution of a molecular system while calculating the system's conformation-dependent electronic state changes using FMO. This method is implemented in programs such as ABINIT-MP and GAMESS; this article focuses mainly on the former. A preliminary test simulation of a micro-RNA (miRNA) and Benzo-*N*,*N*-bis(3-aminopropyl)-2,7-diamino-1,8-naphthyridine (BzDANP) complex highlights many challenges remaining in applying FMO-MD to biological reactions. Finally, possible solutions to these challenges are presented.

Keywords FMO, MD, First principles molecular dynamics, miRNA, BzDANP

1 Introduction

The fragment molecular orbital (FMO)-based molecular dynamics (MD) simulation (FMO-MD, [1]) is a type of first principles MD (FPMD) method that simulates the time evolution of a molecular system while calculating its electronic state using FMO [2]. Since many review articles have been published on FMO-MD [3–6], here we only outline its methodology and applications (Sect. 2). Instead, we focus on the challenges the method faces when applied to biological enzymatic reactions (Sect. 3).

2 Current Status of FMO-MD

FMO-MD has been implemented both in the ABINIT-MP [7, 8] and GAMESS programs [9, 10]. However, we mainly describe studies related to the former and refer to studies related to the latter only when necessary.

Alexander Heifetz (ed.), *Quantum Mechanics Applications in Drug Design*, Methods in Molecular Biology, vol. 3061, https://doi.org/10.1007/978-1-0716-5420-0_17,

Figure 1 illustrates the outline of the FMO-MD method. FMO-MD is an FPMD based on the Born–Oppenheimer approximation; the approximation separates the state of the electrons and the motion of the nuclei, the state of the electrons is solved quantum mechanically by FMO while the motion of the nuclei is solved classically by MD (Fig. 1). Specifically, at each MD step, FMO uses the 3D coordinates of the nuclei of the target molecular system (**r**) to obtain the total energy (*E*) and force acting on each nucleus (**f**). These values are then used to integrate the Newtonian equation of motion. This procedure is repeated for specified time steps. Finally, the obtained trajectory is used to extract some physical and structural properties of the molecular system.

The FMO-MD algorithm consists of three parts: FMO, MD, and the interface between the two. We explain each part separately below, along with a brief review of the method's application calculations.

2.1 FMO Part

The methods of FMO implemented in ABINIT-MP have been reviewed extensively in the literature [7, 8], so they are not repeated here. These FMO methods can be applied to FMO-MD if energy gradient (minus force) is available. The original energy gradient [11] was not fully analytic, but it was later modified to be so [12]. The latter results in a better total energy preservation in FMO-MD than the former [4, 13]. However, if the fully analytic gradient is unavailable, the use of the three-body expansion (FMO3) [14] instead of the two-body expansion (FMO2) is recommended [15]. Most FMO-MD simulations have been performed at Hartree–Fock (HF) or Mϕller Presset second-order perturbation theory (MP2) [16] with the 6-31 G(*) basis set [17, 18]. If FMO3 is too expensive for MP2, a practical alternative is FMO(3)-MP2, which uses FMO3 for HF calculation and FMO2 for MP2 calculations, thereby reducing computation time [18].

2.2 MD Part

The MD part of FMO-MD in ABINIT-MP is similar to the conventional classical MD. The time integration is performed by the standard velocity-Verlet time integration using the forces calculated by FMO. Additionally, the Nosé–Hoover chains algorithm [19] is

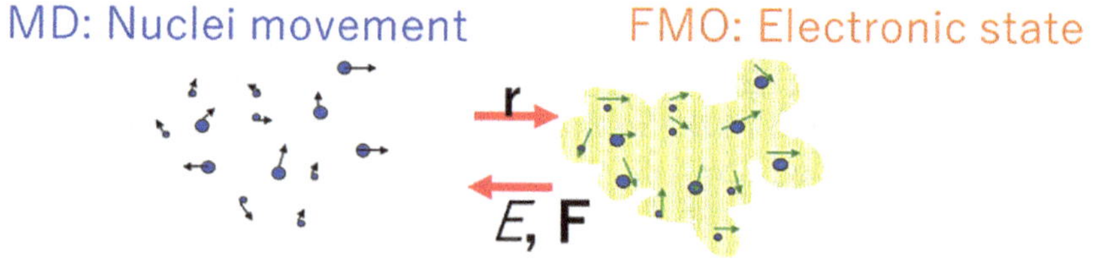

Fig. 1 A schematic of the FMO-MD algorithm. Given 3D coordinates of the molecular system (**r**), FMO solves the electronic structure and calculates the total energy (*E*) and force (**f**), which are then used to integrate the equation of motion of each nucleus by MD. (Reproduced from [6] with permission from the CBI society)

available to achieve a canonical ensemble at a preset temperature. The current version of ABINIT-MP has only free or spherical boundary conditions. In addition to these ordinary MD methods, the blue moon ensemble has been implemented to simulate two-atom or three-atom chemical reactions via constraint dynamics [20, 21].

2.3 Interface Between FMO and MD

The interface between FMO and MD is the most complicated part of the FMO-MD algorithm because the redefinition of fragments is necessary at each MD time step. For example, suppose the following substitution reaction occurred during an FMO-MD simulation, where A, B, and C denote chemical species.

$$A + BC \rightarrow AB + C.$$

A and BC could be separate fragments before the reaction, but AB and C should be separate fragments after the reaction. Thus, an algorithm is necessary to redefine the fragments, especially for simulation of a chemical reaction.

In ABINIT-MP, the redefinition is handled by an algorithm named dynamic fragmentation (DF) [15]. This algorithm determines the distance between atoms and unites them into a fragment if they are close enough based on preset thresholds. The DF algorithm is effective when the target system consists of only small molecules, each of which can be considered a fragment. However, when applied to a system containing a large molecule, such as a protein or polynucleotide, the algorithm results in a fragment that is too large to handle because DF regards it as a single fragment. Such a large molecule should be divided into appropriately sized fragments before FMO-MD. These fragments should be regarded as static fragments (SFs) that do not change during the simulation. To achieve this, we modified DF into "dynamic fragmentation with static fragments (DF/SF)" algorithm [22]. In the DF/SF algorithm, a portion of the molecular system, usually the solute (e.g., protein or DNA), is defined as static, meaning the initial fragmentation data are preserved throughout the simulation. The rest of the system, usually the solvent (water, ions, etc.), is considered "dynamic" and is subject to redefinition by the DF algorithm. There are several modes of this algorithm, but the most useful one only allows H^+-transfer to and/or from static fragments (Fig. 2). The DF/SF algorithm has enabled us to handle heterogeneous molecular systems containing large solutes and solvents.

2.4 Simulation of Chemical Reactions by FMO-MD

As FMO-MD is a type of FPMD, it requires significantly more computational resources than classical molecular mechanics (MM)-MD. Therefore, it is only natural to wish to apply FMO-MD calculations to intrinsically quantum mechanical phenomena, such as chemical and biological reaction simulations, which are difficult to handle with classical MM-MD. Hence, though not numerous, several chemical

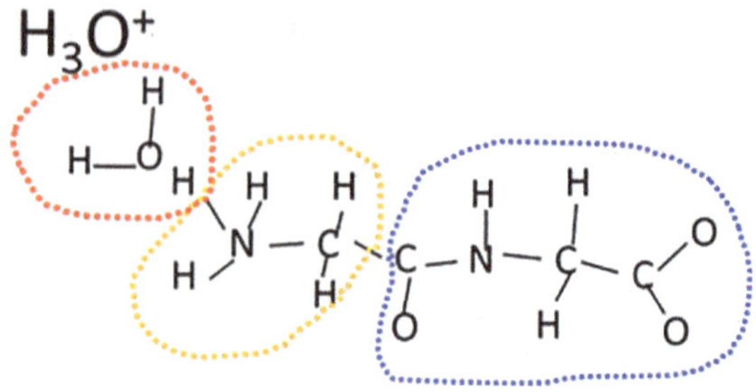

Fig. 2 An example of the results of the DF/SF algorithm. (Gly)2 is used as an example. In this system, H^+ can transfer to water, producing H_3O^+. (Reproduced from [5] with permission from the CBI society)

reaction simulations using FMO-MD have been performed with ABINIT-MP. These simulations have examined the hydrolysis of the methyl diazonium ion [23], the Menschkin reaction [24], and the hydration and amination of formaldehyde [25, 26]. These simulations were performed with small solute molecules in a water solvent consisting of several hundred molecules for a duration of approximately ten picoseconds.

Since then, the FMO-MD functionality of the ABINIT-MP program has only been used to analyze the hydration behavior of small molecules and ions [27–29–] and the vibrational states of ammonia clusters [30], not chemical reactions. In contrast, GAMESS uses density functional tight binding (DFTB) to perform FMO-MD calculations for larger systems, such as hydrated Na or TrpCage protein with durations of 100–200 ps [31, 32]. However, these are not simulations of chemical reactions, either. Thus, as of 2025, chemical reaction analysis using FMO-MD remains largely dormant. Nevertheless, efforts to revive it are currently underway, as discussed below.

3 Challenges

In this section, we address challenges associated with applying FMO-MD to biological reactions. To this end, we present a preliminary FMO-MD simulation from an ongoing, unfinished in silico drug design project. Then, we list the current challenges facing FMO-MD and propose possible solutions.

We present here a test FMO-MD simulation of a micro-RNA (miRNA) and Benzo-*N,N*-bis(3-aminopropyl)-2,7-diamino-1,8-naphthyridine (BzDANP, Fig. 3A) complex. BzDANP is a small synthetic molecule that recognizes the C/U bulge of pre-miRNA-136, thereby inhibiting its processing by Dicer [33, 34]. As part of an in silico project to develop RNA-targeted small molecules, we have begun calculating the interaction between BzDANP and miRNA via MM-MD, FMO, and FMO-MD. Although we intend

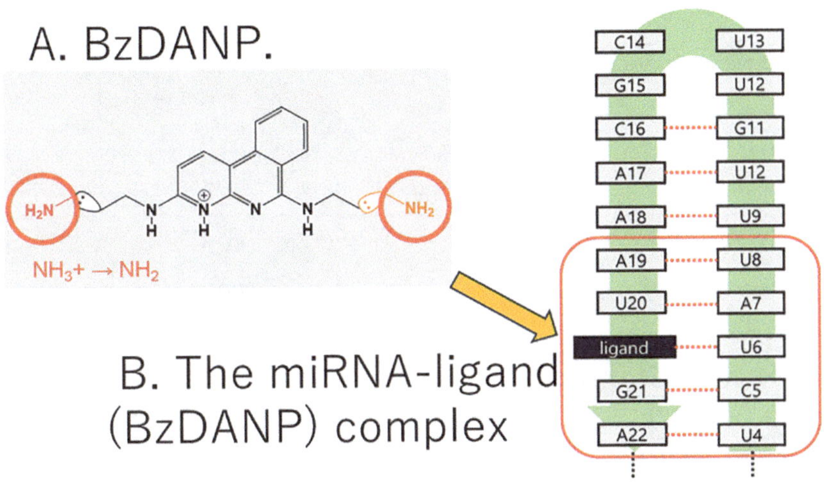

Fig. 3 (**A**) BzDANP. (**B**). The schematic of the simulated miRNA-ligand (BzDANP) complex. The region bounded by a rounded rectangle was simulated by FMO-MD

to advance the project further in the near future, this article only describes preliminary data to illustrate the current status and issues with the FMO-MD method.

The initial molecular configuration was prepared as follows. First, we used the structure of the BzDANP and miRNA complex modeled using MM-MD based on NMR data (Gota Kawai, personal communication). We then excised BzDANP and the interacting nucleotides (Fig. 3B) and immersed them in a spherical water solvent with a radius of 22.2 Å, centered at the centroid of the solutes. This ensured that the minimal distance between the solutes and the water spheres' surface was 6 Å. We capped the phosphate groups of miRNA with H to neutralize the molecules. The amino groups of BzDANP were also neutralized (Fig. 3A). We energy-minimized the constructed molecular model with the AMBER10:EHT force field using the MOE program [35] (Fig. 4).

The initial configuration thus constructed was then subjected to FMO-MD with the ABINIT-MP program on the SQUID supercomputer of The University of Osaka (Intel Xenon Platinum). A total of 512 CPU cores were used while assigning eight cores for each fragment. Each core was allocated 3 GB memory. The BzDANP molecule was divided into three fragments, one consisting of the three benzene rings, while each of the other consists of an amino group (Fig. 3A). Note that the benzene rings were kept intact because division of their π-electron system would have degraded the calculation accuracy largely. The miRNA was fragmented so that a nucleotide comprises a fragment [36]. Each water molecule was considered a fragment. See Table 1 for the sizes of the molecular system and the number of fragments. The DF/SF algorithm was activated, in which

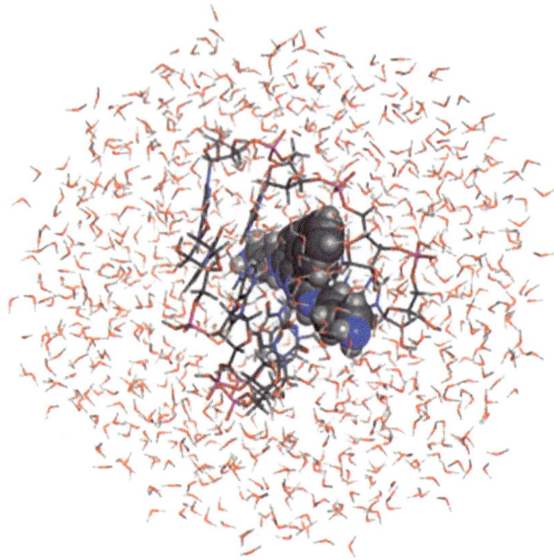

Fig. 4 The initial configuration for FMO-MD. BzDANP is shown in spherical model, miRNA in thick stick model, and water in thin stick model

Table 1
Data related to system size

	Atoms	Molecules	Fragments
miRNA	289	1	18
BzDANP	49	1	3
Water	2877	959	959

BzDANP and miRNA were considered static fragments, while the water molecules were considered dynamic fragments. This algorithm allows for H^+ transfer only between the fragments. The FMO calculation was performed using the HF/6-31 G* basis set. The molecular system was gradually heated to 300 K within 1 ps, and the simulation continued up to 2 ps with a time step of 0.5 fs. The calculation took approximately 13 min per step.

The overlay of the snapshots from the first 1 ps trajectory shows that the molecular structure fluctuated around the initial structure (Fig. 5A). The interfragment interaction energy (IFIE) between U6 and BzDANP showed considerable fluctuation in this short period (Fig. 5B).

We refrain from showing more details of the trajectory as this simulation is a very short, preliminary one. Instead, we discuss present problems of the FMO-MD method revealed through this simulation as well as their possible solutions.

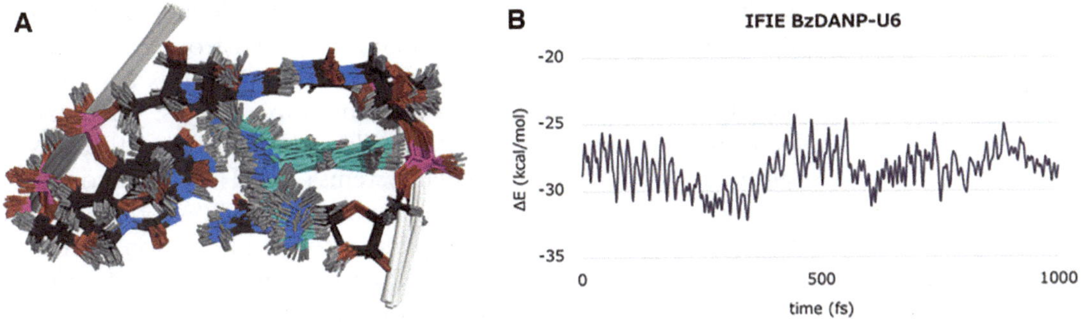

Fig. 5 (**A**) Overlay of snapshots from 0 to 1 ps trajectory. (**B**) IFIE between BzDANP-U6

First, we must accelerate the computation speed. Thirteen minutes per step with 512 CPU cores is too slow for a molecular system of approximately 3,000 atoms. One might argue that the easiest way to increase the speed is to use a highly parallelized supercomputer, such as Fugaku. Unfortunately, this approach may only bring limited success due to the inefficient parallel efficiency. Specifically, the BzDANP fragment is far larger than the others, especially water molecules, with 37 and 3 atoms per fragment, respectively. Thus, calculating BzDANP would become the rate-limiting step, impeding throughput even if the number of cores were increased. There are several solutions to this problem, but one promising solution is balancing the size of the fragments. In the present case, for example, merging several water molecules into one larger fragment would improve both parallel efficiency and accuracy. We plan to implement this capability in the DF/SF algorithm of ABINIT-MP.

Second, the inclusion of counterions is desirable. In the current FMO-MD simulation, charged groups were neutralized by adjusting the number of hydrogen atoms. While this is a useful temporary workaround, many MM-MD and FPMD simulations use counterions to neutralize the simulated system and mimic physiological conditions. While it is possible to include counterions in FMO and FMO-MD calculations, properly handling them is complicated for several reasons, especially charge transfer with surrounding water molecules [37]. However, if the only role of the counterions is neutralization, a classical treatment may suffice. Based on this assumption, several efforts have been made to incorporate classical counterions into FMO (Tatusuya Nakano, Shuhei Miyakawa, et al., in preparation). Classical counterions can also be incorporated into FMO-MD.

Third, an inexpensive calculation method is needed for electron correlation. The FMO-MD test simulation presented here was performed at the HF level, i.e., without electron correlation. However, omitting the correlation results in the omission of dispersion interactions, which play an important role in many organic and biological molecules, such as stacking interactions in polynucleotides [38]. Although neglecting stacking did not have disastrous effects on the current BzDANP/miRNA simulation in as short as 2 ps, longer simulations would be subject to undesirable effects. Fortunately, including dispersion via MP2 has been successful in FMO-MD simulations of smaller molecular systems [18]. There are also several ways to reduce the computational demand of MP2. Another practical method is to add empirical dispersion to the interaction calculated by HF, similar to the Grimme-D2 correction [39]. This correction adds dispersion without significantly increasing computation time and has been proven to improve the accuracy of protein geometry optimization (Miyakawa et al., in preparation). We plan to apply this correction to FMO-MD in the near future.

Thus, many challenges remain for FMO-MD simulations of biological molecules, but we are paving the way.

Acknowledgments

We thank Professor Kazuhiko Nakatani of The University of Osaka and Professor Gota Kawai of Chiba Institute of Technology for providing information and data on BzDANP and miRNA interaction.

References

1. Komeiji Y, Nakano T, Fukuzawa K et al (2003) Fragment molecular orbital method: application to molecular dynamics simulation, "ab initio FMO-MD". Chem Phys Lett 372:342–347
2. Kitaura K, Ikeo E, Asada T et al (1999) Fragment molecular orbital method: an approximate computational method for large molecules. Chem Phys Lett 313:701–706
3. Komeiji Y, Mochizuki Y, Nakano T, et al (2012) Recent advances in fragment molecular orbital-based molecular dynamics (FMO-MD) simulations. In: Wang L (Ed.) Molecular dynamics—theoretical developments and applications in nanotechnology and energy. InTech, Rijeka, 1–22
4. Komeiji Y, Ishikawa T (2021) FMO interfaced with molecular dynamics simulation. In: Mochizuki Y, Tanaka S, Fukuzawa K (Eds.) Recent advances of the fragment molecular orbital method: enhanced performance and applicability. Springer, Singapore, 373–389
5. Komeiji Y (2025) Fundamentals of molecular dynamics simulation. CBI E-book series No 6. 2C edn. CBI Society, Tokyo, (in Japanese)
6. Komeiji Y (2025) Opinion: FMO meets MD—achievements and future directions. CBI J 25:71–78
7. Tanaka S, Mochizuki Y, Komeiji Y et al (2014) Electron-correlated fragment-molecular-orbital calculations for biomolecular and nano systems. Phys Chem Chem Phys 16:10310–10344
8. Mochizuki, Y, Tanaka, S, Fukuzawa, K, 2021. Recent advances of the fragment molecular orbital method: enhanced performance and applicability. Springer, Singapore, pp. 1–605
9. Fedorov DG, Olson RM, Kitaura K et al (2004) A new hierarchical parallelization scheme: generalized distributed data interface (GDDI), and an application to the fragment molecular orbital method (FMO). J Comput Chem 25:872–880
10. Fedorov, DG, 2023. Complete guide to the fragment molecular orbital method in gamess: from one atom to a million, at your service.. World Scientific, Singapore
11. Kitaura K, Sugiki S, Nakano T et al (2001) Fragment molecular orbital method: analytical energy gradients. Chem Phys Lett 336:163–170
12. NagataT Kitaura K, Brorsen KR et al (2011) Fully analytic energy gradient in the fragment molecular orbital method. J Chem Phys 134:12415
13. Brorsen KR, Minezawa N, Xu F et al (2012) Fragment molecular orbital molecular dynamics

with the fully analytic energy gradient. J Chem Theory Comput 8:5008–5012
14. Fedorov GD, Kitaura K (2004) The importance of three-body terms in the fragment molecular orbital method. J Chem Phys 120:6832
15. Komeiji Y, Mochizuki Y, Nakano T (2010) Three-body expansion and generalized dynamic fragmentation improve the Fragment Molecular Orbital-based Molecular Dynamics (FMO-MD). Chem Phys Lett 484:380–386
16. Møller C, Plesset MS (1934) Note on an approximation treatment for many-electron systems. Phys Rev 46:618–622
17. Mochizuki Y, Koikegami S, Nakano T et al (2004) Large scale MP2 calculations with fragment molecular orbital scheme. Chem Phys Lett 396:473–479
18. Mochizuki Y, Nakano T, Komeiji Y et al (2011) Fragment molecular orbital-based molecular dynamics (FMO-MD) method with MP2 gradient. Chem Phys Lett 504:95–99
19. Martyna GJ, Klein ML, Tuckerman M (1992) Nosé–Hoover chains: the canonical ensemble via continuous dynamics. J Chem Phys 97:2635–2643
20. Sprik M, Cicotti G (1998) Free energy from constraint molecular dynamics. J Chem Phys 109:7737–7744
21. Komeiji Y (2007) Implementation of the blue moon ensemble method. CBI J 7:12–23
22. Komeiji Y, Fujiwara T, Okiyama Y et al (2013) Dynamic fragmentation with static fragments (DF/SF) algorithm designed for ab initio fragment molecular orbital-based molecular dynamics (FMO-MD) simulations of polypeptides. CBI J 13:45–57
23. Sato M, Yamataka H, Komeiji Y et al (2008) How does an S_N2 reaction take place in solution? Full Ab initio MD simulations for the hydrolysis of the methyl diazonium ion. J Am Chem Soc 130:2396–2397
24. Komeiji Y, Ishikawa T, Mochizuki Y et al (2009) Fragment Molecular Orbital method-based Molecular Dynamics (FMO-MD) as a simulator for chemical reactions in explicit solvation. J Comput Chem 30:40–50
25. Sato M, Yamataka H, Komeiji Y et al (2010) Does amination of formaldehyde proceeds through zwitterionic intermediate in water? FMO-MD simulations by using constraint dynamics. Chem Eur J 16:6430–6433
26. Sato M, Yamataka H, Komeiji Y et al (2012) FMO-MD simulations on hydration of formaldehyde in water solution with constraint dynamics. Chem Eur J 18:9714–9721
27. Mori H, Hirayama N, Komeiji Y et al (2012) Differences in hydration between cis- and trans-platin: quantum insights by ab initio fragment molecular orbital-based molecular dynamics (FMO-MD). Comput Theor Chem 986:30–34
28. Kato Y, Fujiwara T, Komeiji Y et al (2014) Fragment molecular orbital-based molecular dynamics (FMO-MD) simulations on hydrated Cu(II) ion. CBI J 14:1–13
29. Fujiwara T, Mori H, Komeiji Y et al (2015) Fragment molecular orbital-based molecular dynamics study on hydrated Ln(III) ions. In: Proceedings of computational science workshop 2014, JPS Conf Proc vol 5, p 011001
30. Ninomiya M, Doi H, Matsumoto Y et al (2020) Ab initio fragment molecular orbital-based molecular dynamics (FMO-MD) simulations of (NH3)32 cluster: effects of electron correlation. Bull Chem Soc Jpn 93:553–560
31. Nishimoto Y, Nakata H, Fedorov DG et al (2015) Large-scale quantum-mechanical molecular dynamics simulations using density-functional tight-binding combined with the fragment Molecular Orbital Method. J Phys Chem Lett 6:5034–5039
32. Nishimoto Y, Irle S (2021) The FMO-DFTB method. In: Mochizuki Y, Tanaka S, Fukuzawa K (Eds.) Recent advances of fragment molecular orbital method. Springer, Singapore, 459–485
33. Murata A, Otabe T, Zhang J et al (2016) BzDANP, a small-molecule modulator of pre-miR-29a maturation by Dicer. ACS Chem Biol 11:2790–2796
34. Otabe T, Nagano K, Kawai G et al (2019) Inhibition of pre-miRNA-136 processing by Dicer with small molecule BzDANP suggested the formation of ternary complex of pre-miR-136–BzDANP–Dicer. Bioorg Med Chem 27:2140–2147
35. Molecular Operating Environment, MOE (2019) Chemical Computing Group ULC, Montreal, QC, Canada
36. Fukuzawa K, Watanabe C, Kurisaki I et al (2014) Accuracy of the fragment molecular orbital (FMO) calculations for DNA: total energy, molecular orbital, and inter-fragment interaction energy. Comput Theor Chem 1034:7–16
37. Fujita T, Fukuzawa K, Mochizuki Y et al (2009) Accuracy of fragmentation in ab initio calculations of hydrated sodium cation. Chem Phys Lett 478:295–300
38. Fukuzawa K, Kurisaki I, Watanabe C et al (2015) Explicit solvation modulates intra- and inter-molecular interactions within DNA: electronic aspects revealed by the ab initio fragment molecular orbital (FMO) method. Comput Theor Chem 1054:29–37
39. Grimme S (2006) Semiempirical GGA-type density functional constructed with a long-range dispersion correction. J Comput Chem 27:1787–1799

Chapter 18

How Can QM Methods Be Used to Guide Synthesis Planning in Drug Discovery?

Jemima Haque

Abstract

Quantum mechanical (QM) methods are traditionally used in drug discovery to support molecular design and property prediction; however, they are becoming increasingly important for evaluating synthetic feasibility during the Design–Make–Test–Analyze (DMTA) cycle. As modern tools such as virtual screening and machine-learning workflows rapidly propose many potential candidate molecules, their progression ultimately depends on whether they can be prepared efficiently in the laboratory. QM approaches provide a mechanistically grounded framework for evaluating thermodynamic and kinetic feasibility, predicting selectivity, and identifying competing reaction pathways before experimental effort is invested. This chapter reviews how QM methods can guide synthesis planning by complementing chemist expertise, literature precedent, and computer-assisted synthesis planning (CASP) tools. The application of density functional theory (DFT) to evaluate reaction energetics, transition states, and stereochemical outcomes is discussed, as well as the role of QM-derived reactivity descriptors to enable the quantitative ranking of retrosynthetic disconnections. An automated reaction-path exploration method is also highlighted for analyzing complex pericyclic and stereochemically rich transformations. These approaches demonstrate how QM-driven synthesis planning can support the identification of synthetically accessible drug candidates and advance progress in making the compounds, which is often the rate-limiting step in drug discovery. Some shortcomings of QM methods are also acknowledged, particularly the difficulty of fully modeling complex chemical reactions, which consequently restricts the breadth of their practical applications.

Keywords Quantum mechanics, Density functional theory, Synthetic feasibility, Retrosynthetic analysis, Computer-assisted synthesis planning, Drug discovery, DMTA cycle, Reaction energetics, Reactivity descriptors, Nucleophilicity and electrophilicity, Automated reaction-path exploration, Pericyclic reactions, Stereoselectivity, Natural product synthesis, Medicinal chemistry

1 Introduction: Synthetic Feasibility in the DMTA Cycle

Quantum mechanical (QM) methods are widely used in the drug discovery process, particularly during the design phase of the DMTA (Design–Make–Test–Analyze) cycle, to provide precise predictions of binding interactions between drugs and their biological targets [1]. Equally important is the use of QM approaches to assess whether

Alexander Heifetz (ed.), *Quantum Mechanics Applications in Drug Design*, Methods in Molecular Biology, vol. 3061, https://doi.org/10.1007/978-1-0716-5420-0_18,

proposed drug candidates are synthetically accessible, as this ultimately determines whether they can be evaluated experimentally. High-throughput screening is the dominant method for identifying novel hits [2] and yields thousands of potential molecules. Even after applying drug-likeness filters, such as Lipinski's rules, predicted affinities, ADME-tox properties, and key pharmacophoric features [3], the large number of remaining structures must be prioritized according to their synthetic feasibility. Consequently, the bottleneck for progressing a drug molecule is not only its predicted biological efficacy but also the practicality of its synthesis. This short chapter highlights how QM methods, alongside literature precedent and chemist expertise, can guide synthesis planning, enhancing the efficiency and productivity within the Make stage of the discovery pipeline.

2 QM Metrics for Synthetic Feasibility and Route Evaluation

QM calculations are increasingly being used to validate synthetic routes proposed by machine-learning models and computer-assisted synthesis planning (CASP) tools, by providing quantitative metrics of synthetic feasibility. Reaction feasibility is governed by thermodynamic considerations, which assess the energetic favorability of product formation, and by kinetic factors, which determine whether the activation energy barrier can be overcome. Additional constraints such as chemoselectivity, competing pathways, reagent compatibility, and the accessibility of key molecular conformations further influence route viability. QM methods provide data to evaluate these factors at the atomic level by estimating ground-state energies, locating transition states, mapping conformational ensembles, and computing electronic descriptors such as nucleophilicity and electrophilicity. These predictions enable chemists to triage synthetic pathways and anticipate unwanted side reactions.

3 DFT-Guided Retrosynthesis: Natural Product Case Study

The synthesis of structurally complex molecules is a persistent challenge in drug discovery. Natural products are a prime example, spanning broad and underexplored regions of chemical space while providing privileged scaffolds for modulating challenging biological targets such as protein–protein interactions or addressing difficult disease areas such as antibacterials [4]. In a case study by Newhouse et al., they demonstrated the value of integrating density functional theory (DFT) calculations into retrosynthetic planning through their seminal syntheses of the indole diterpenes paspaline A and emindole PB [5]. These natural products possess highly intricate polycyclic architectures, featuring an indole fused to a diterpenoid core, multiple ring junctions, and several quaternary stereocenters. Both have shown promising antiproliferative and antimetastatic properties, yet their structural complexity has posed synthetic challenges: paspaline A required a 27-step synthesis, while emindole PB had no published synthetic route.

The authors recognized that both targets could, in principle, arise from a late-stage carbocationic rearrangement/cyclisation cascade, analogous to the biosynthetic terpenoid cyclisation (Fig. 1). A key carbocation intermediate was predicted to diverge along three competing pathways: (1) intramolecular Friedel–Crafts attack by the indole to form the paspaline scaffold, (2) a 1,2-methyl shift leading to the emindole framework, or (3) elimination. The feasibility and selectivity of these pathways were therefore central to the retrosynthesis.

To evaluate these possibilities, three substrate variants were proposed that differed only in remote substituents— acyclic, monocyclic tetrahydropyran (THP), and bicyclic ketal—which were spatially distant from the reactive carbocation center (Fig. 2). DFT computations (mPW1PW91/6-31+G(d,p)//B3LYP/6-31 G*) were used to estimate the relative gas phase transition-state energies (ΔΔG‡) for cyclisation versus methyl migration for each precursor. The calculations indicated that the bicyclic ketal substrate, with the most negative ΔΔG‡ value, would favor the cyclization pathway over the methyl shift, despite the substituent being remote from the reactive site. Guided by this prediction, the authors selected this precursor and achieved a nine-step synthesis of paspaline A with the following steps:

1. Construction of the tricyclic terpene core from commercially available starting materials.
2. Installation of the indole and Lewis acid-induced generation of the key carbocation, affording a 44% combined yield of cyclisation and methyl shift products.

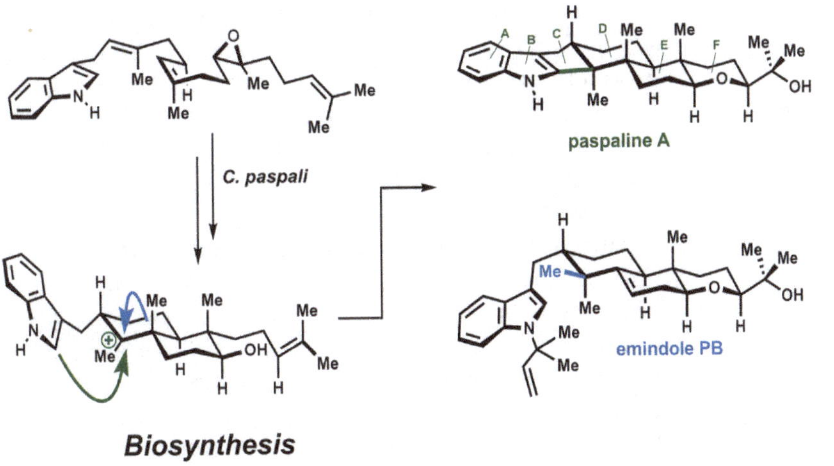

Fig. 1 Terpenoid biosynthesis-inspired step enables divergent access to paspaline A and emindole PB

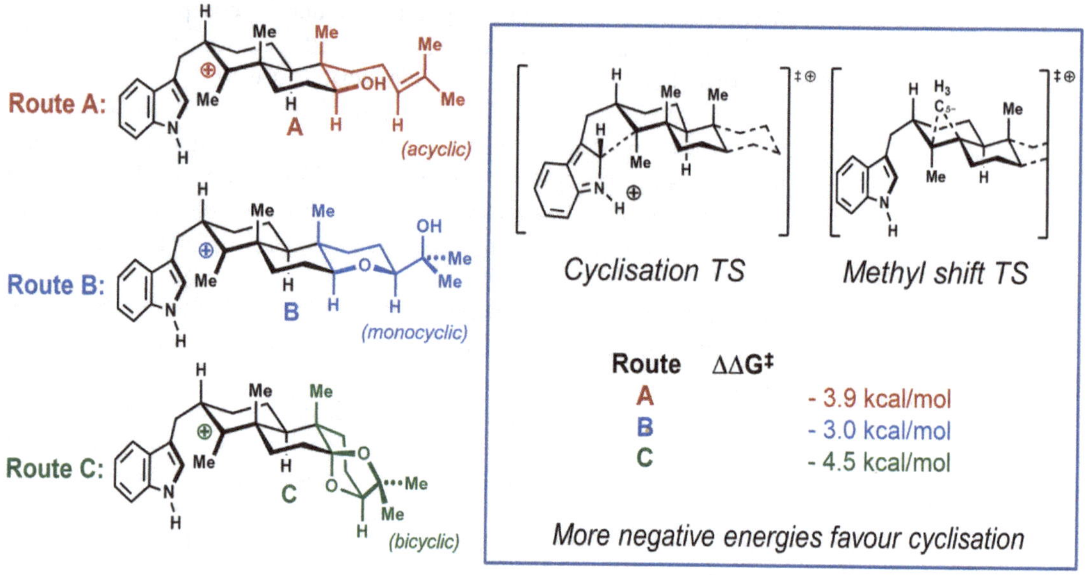

Fig. 2 The relative transition-state free energies (ΔΔG‡) for cyclization versus methyl shift were calculated for three precursor structures. Among these, the bicyclic ketal substrate exhibited the most negative ΔΔG‡ value, indicating the strongest preference for the cyclization pathway

3. Final functional-group interconversions to reach the natural product.

DFT calculations were also applied to predict the stereochemical outcome of the cyclisation, suggesting a preference for the desired trans product by $\geq 1.3\,\text{kcal mol}^{-1}$. Experimentally, however, a 3:1 ratio favoring the methyl-migration product was observed, diverging from the computed preference. The authors proposed that the methyl shift may proceed through a concerted mechanism, aided by the proximal orientation of the migrating group, contrasting with the stepwise mechanism assumed in the computational model. Nonetheless, the monocyclic THP analog, predicted computationally to disfavor cyclization, indeed failed to yield the paspaline-type products experimentally, validating the broader feasibility predictions. Importantly, the same cyclization transformation provided a strategic entry to emindole PB, enabling the first concise 13-step synthesis of this previously inaccessible natural product.

Overall, the authors achieved an elegant synthesis of the two privileged scaffolds by predicting the feasibility of a challenging biomimetic transformation. This study illustrates how DFT calculations can inform retrosynthetic analysis by incorporating energetic feasibility, predicting reactivity and selectivity, and guiding substrate choice and reaction conditions. This can help chemists not only identify which complex molecules are worth making but also how they can be made efficiently using the same starting materials, saving both time and resources. Furthermore, reducing the number of steps in a synthetic route, as achieved in this case study, is a key objective for chemists in industry. Shorter synthetic routes lower operating costs, improve process efficiency, and reduce waste, thereby supporting the broader green chemistry initiatives for sustainable drug development.

4 QM-Derived Reactivity Descriptors for Retrosynthetic Filtering

In addition to providing mechanistic insights that guide the construction of synthetic routes, quantum mechanical calculations can assist in identifying the most reactive sites for strategic disconnections. By estimating intrinsic nucleophilicity and electrophilicity from first principles, QM methods can filter retrosynthetic routes according to chemical selectivity, reducing reliance on heuristic rules. A recent illustration of this concept is ESNUEL (Estimating NUcleophilicity and ELectrophilicity), an automated QM-based workflow developed by Jensen et al. [6], which computes affinity metrics—methyl-cation affinity (MCA) and methyl-anion affinity (MAA)—to quantify site-specific reactivity in neutral organic molecules.

The ESNUEL pipeline proceeds directly from a SMILES string and consists of four main stages:

1. **Reactive-site enumeration**. Candidate nucleophilic and electrophilic centers are identified using cheminformatics rules and SMIRKS patterns. Typical sites include lone-pair donors, anionic centers, carbonyl carbons, Michael acceptors, halogen-bearing electrophiles, and other functionally activated positions.
2. **Conformer generation and screening**. Conformational ensembles are generated using GFNFF-xTB (fast force-field-like tight-binding), with an embedding policy of $min(1 + 3 \times n\_rot, 20)$ conformers. Pre-optimization and energetic screening with GFN1-xTB (ALPB solvation in DMSO) provide rapid, reasonably accurate geometries suitable for downstream QM refinement. These methods offer orders-of-magnitude speed relative to DFT, enabling large-scale automated conformer and transition-state sampling. The method achieves a median wall time of less than 2 min per molecule.
3. **Geometry optimization**. The lowest-energy conformer undergoes a single-point calculation at the r^2SCAN-3c level of theory with SMD(DMSO) implicit solvation to obtain accurate electronic energies for affinity calculations.
4. **Affinity calculations (MCA/MAA)**. Methyl cation affinities (for nucleophilic sites) and methyl anion affinities (for electrophilic sites) are computed as reaction energies for hypothetical additions of CH_3^+ or CH_3^-.

$$\mathbf{Nuc + CH_3^+ \rightarrow MCA\ product,\quad MCA \equiv -\Delta E}$$

$$\mathbf{Elec + CH_3^- \rightarrow MAA\ product,\quad MAA \equiv -\Delta E}$$

Larger MCA/MAA values indicate more favorable site-specific interactions and therefore greater intrinsic reactivity. These scores can be used to rank competing disconnections and flag implausible reactive centers suggested by computer-assisted retrosynthetic routes.

The authors applied ESNUEL to analyze both literature-reported and computationally generated routes to the antiretroviral drug raltegravir (Fig. 3). In the experimentally realized coupling of substrates 5-methyl-1,3,4-oxadiazole-2-carbonyl chloride and 2-(1-amino-1-methylethyl)-N-(4-fluorobenzyl)-5-hydroxy-1-methyl-6-oxo-1,6-dihydropyrimidine-4-carboxamide (raltegravir amine), ESNUEL identified the carbonyl carbon of the acyl chloride as the most electrophilic site (MAA = 94 kcal mol^{-1}) and the primary amine site of the raltegravir amine as the most nucleophilic (MCA = 107 kcal mol^{-1}), correctly predicting the nucleophilic acyl substitution that was carried out experimentally. Across the two coupling partners, ESNUEL evaluated 27 potential nucleophilic and 20 electrophilic sites, yet correctly identified the experimentally used reactive pair as the top-ranked match.

The workflow was used to examine retrosynthetic steps toward raltegravir proposed by the computer-assisted synthesis planning (CASP) platform, Manifold. For the first step, Manifold predicted an ester-amine amidation between methyl 5-methyl-1,3,4-oxadiazole-2-carboxylate and methyl 2-(1-amino-1-methylethyl)-1,6-dihydro-5-hydroxy-1-methyl-6-oxo-4-pyrimidinecarboxylate; however, ESNUEL ranked the reaction between the two ester units as more favorable (MCA = 114 kcal mol^{-1}; MAA = 58 kcal mol^{-1}). This discrepancy flagged the amidation step as chemically unlikely, which would inform the choice of an acid chloride in place of the ester, as used in the experimental synthesis. The second step in the proposed route involves a Chan–Lam coupling of the ester-amidation product with methylboronic acid, requiring deprotonation of the NH group. ESNUEL, combined with pK_a predictions, showed that the neighboring phenol is substantially more acidic and would be deprotonated instead, contradicting the reaction mechanism proposed by the CASP tool. It is important to note that affinity metrics alone cannot capture all bond-breaking energetics involved in substitution reactions and reactions with complex mechanisms; however, they accurately describe polar additions, including Michael-type conjugate additions.

The ESNUEL workflow demonstrates the value of integrating QM-derived reactivity metrics with CASP tools to provide a quantitative filter for chemical selectivity in retrosynthesis.

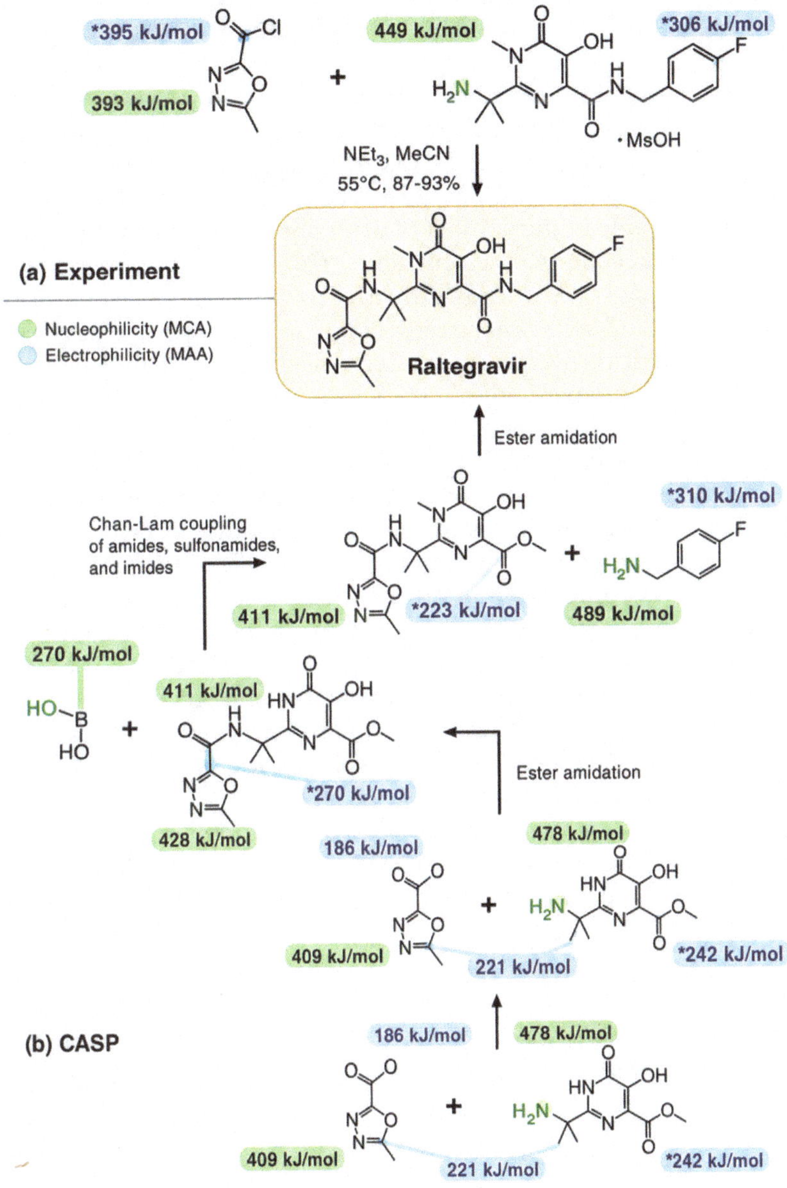

Fig. 3 Retrosynthetic steps toward raltegravir: (**a**) experimental route reported in the literature and (**b**) CASP proposed route. ENSUEL ranking by methyl cation affinities (for nucleophilic sites) and methyl anion affinities (for electrophilic sites) supports the experimental route, while flagging implausible reactive centers suggested by the CASP tool

5 Automated Reaction-Path Exploration for Complex Transformations

QM approaches are also important for analyzing stereoselective pericyclic reactions and catalyst-mediated transformations, where stereochemical complexity and hindered ring architectures can make a target synthetically inaccessible. Such reactions are especially valuable in the context of drug discovery, where efficient, high-yielding reactions are essential for enabling rapid make-test cycles. Pericyclic cascades offer an attractive strategy for constructing multiple bonds and stereocenters in a single step. However, their feasibility is often governed by subtle stereoelectronic effects that are challenging to anticipate using merely rule-based retrosynthetic reasoning.

Mita et al. studied [7] the ability of Artificial Force Induced Reaction (AFIR), implemented within a quantum-chemistry-aided retrosynthetic analysis (QCaRA) workflow, to evaluate pericyclic disconnections and multistep cascades directly from complex target structures. This strategy uses artificial forces to perturb molecular geometries and automatically locate equilibrium structures and transition states (TS), giving the potential reactants starting from a target product, without requiring a template. As a result, the AFIR/QCaRA framework can identify known pericyclic pathways, identify nonobvious cascade sequences involving multiple concerted events and explicitly predict stereochemical outcomes by following intrinsic reaction coordinates.

A notable application of this methodology is to endiandric acid C (Fig. 4), a natural product whose methyl ester was made by Nicolaou via a one-pot thermal cascade of three pericyclic reactions [8].

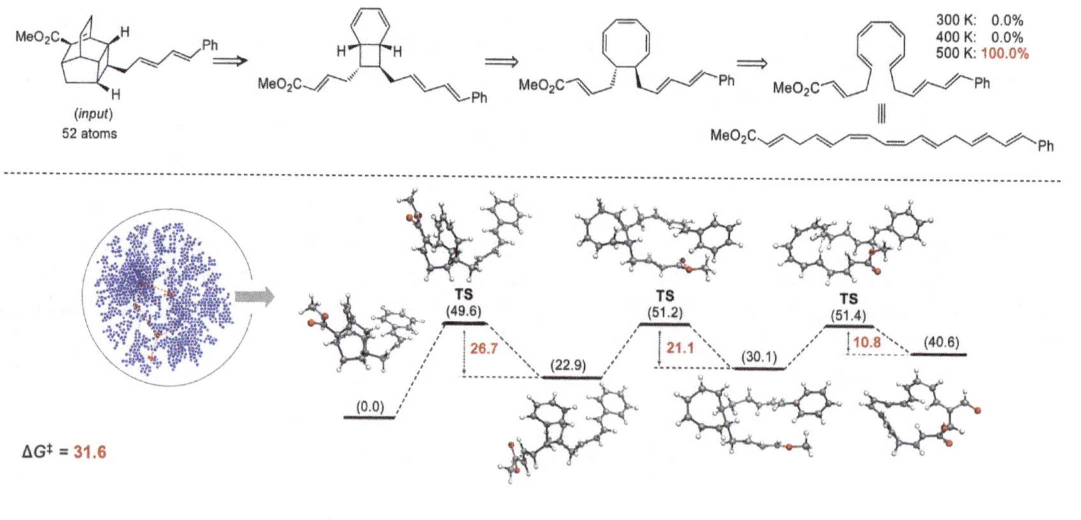

Fig. 4 Retrosynthesis of endiandric acid C: the AFIR/QCaRA workflow correctly reproduced the experimentally established cascade in the correct order. The percentages show the predicted temperature-dependent product distributions

Despite its molecular size of more than 50 atoms, AFIR successfully reconstructed the full retrosynthetic sequence, generating 1000 reaction pathways. The computed network reproduced the experimentally established cascade with correct stereochemical ordering: retro-Diels–Alder ring opening, then a disrotatory 6π electrocyclization, followed by a conrotatory 8π electrocyclic ring opening to the starting polyene. Although the calculated product distribution at 400 K slightly deviated from the experiment due to an overestimated barrier at the initial level of theory, this example demonstrates the role of AFIR calculations in identifying complex, multistep, stereodefined pericyclic sequences and recovering high-energy intermediates with appropriate geometries and relative stabilities.

Importantly, the AFIR/QCaRA workflow was able to predict a reaction that is not compliant with the Woodward–Hoffmann rules. Houk et al. recently rationalized why Prinzbach's vinylogous sesquifulvalene undergoes a thermally forbidden 14π conrotatory electrocyclic reaction to give the trans product, instead of the thermally allowed disrotatory pathway leading to the cis product [9]. Steric and electronic constraints destabilize the allowed transition state, reversing the expected selectivity. QCaRA analysis of both stereoisomeric products (Fig. 5) reproduced these findings: for the cis product, the workflow traced the path back to the reactant and identified the relevant indene intermediate via sequential [1,5]-H shifts, recovering a high activation energy barrier (33 kcal mol^{-1}) consistent with prior DFT calculations. For the trans product, QCaRA located the forbidden pathway with a lower barrier (~23 kcal mol^{-1}) and predicted temperature-dependent product distributions that aligned with experimental observations. The ability to automatically identify such symmetry-forbidden yet experimentally accessible pathways indicates the strength of AFIR in calculations navigating complex potential energy surfaces beyond standard heuristic rules.

This work illustrates how the QM-derived reaction-path workflow informs synthetic feasibility assessment for intricate transformations, enabling chemists to interrogate stereochemical constraints, uncover latent pericyclic cascades, and evaluate the viability of mechanistically unconventional pathways. The method was successfully applied to all categories of pericyclic reactions, including cycloadditions, ene reactions, group-transfer, cheletropic, electrocyclic, and sigmatropic reactions. Despite being limited to reactions with minimal solvent effects, this capability positions QM-driven reaction-path exploration as an increasingly valuable tool for guiding retrosynthetic strategy and expanding the accessible chemical space for the synthesis of complex molecules.

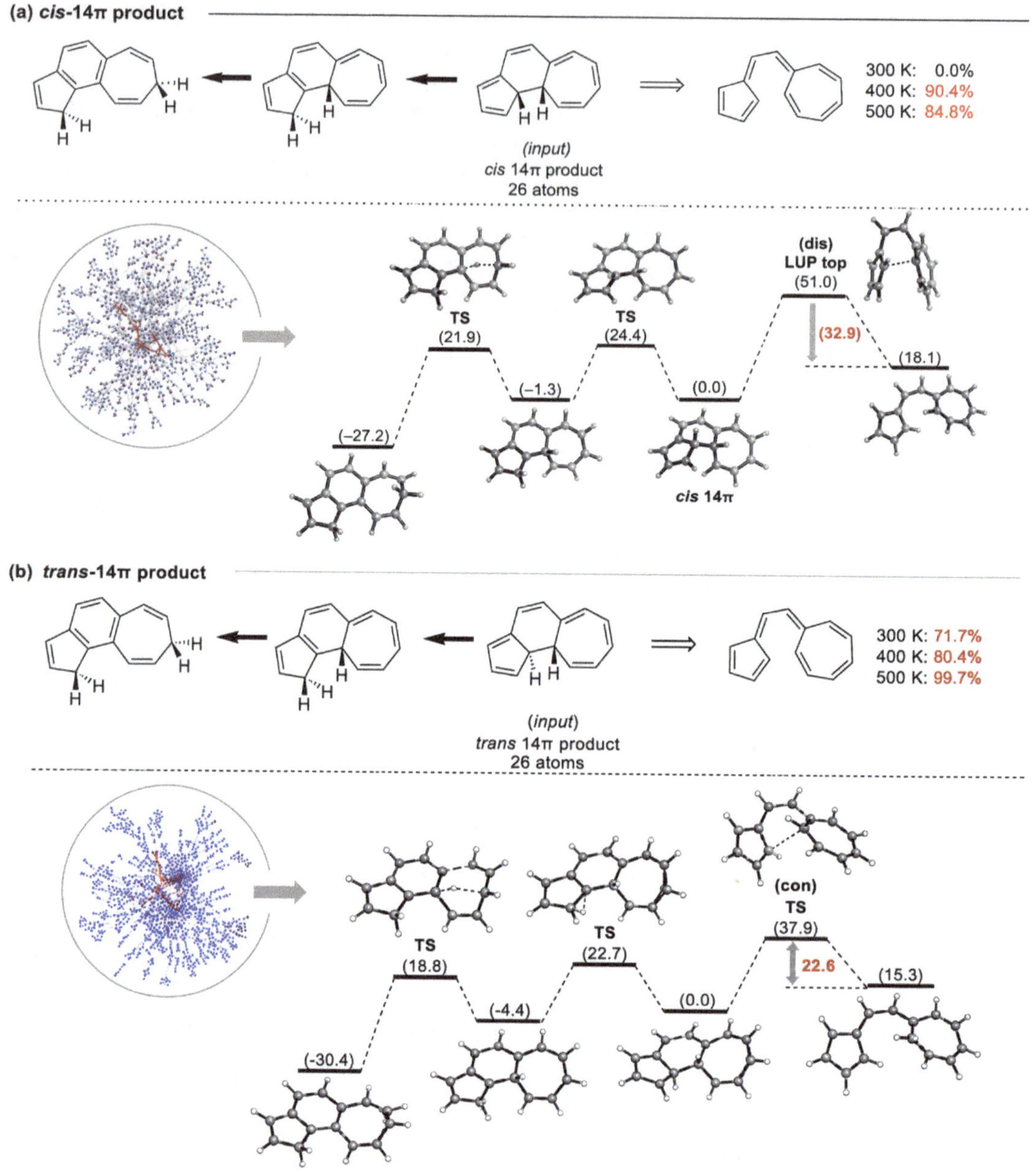

Fig. 5 Retrosynthesis of Prinzbach's vinylogous sesquifulvalene: (**a**) the cis product and (**b**) the trans product. The AFIR/QCaRA workflow correctly identified that the pathway forming the trans product, deviating from the Woodward–Hoffmann rules, has a lower activation energy barrier, and therefore is more feasible than the pathway for the cis product. The percentages show predicted temperature-dependent product distributions

6 Conclusions and Outlook

Quantum mechanical methods are emerging as useful complements to traditional retrosynthesis, machine-learning models, and chemist intuition, providing a data-driven approach for prioritizing which candidate molecules and synthetic routes are most likely to succeed in the laboratory. As the DMTA cycle becomes increasingly constrained by the synthetic Make bottleneck, these approaches can reduce time expenditures and enhance operational efficiency, thereby accelerating the overall cycle progression.

Various types of QM calculations are presented in the literature: DFT-based mechanistic evaluation can resolve subtle stereoelectronic effects that govern key bond-formations, guiding chemists toward precursor designs and reaction conditions that favor the desired pathways, as exemplified by the divergent syntheses of paspaline A and emindole PB. QM-derived reactivity metrics, such as the MCA/MAA affinities implemented in ESNUEL, enable objective filtering of retrosynthetic routes by quantifying intrinsic nucleophilicity and electrophilicity. This provides an efficient means to triage large sets of CASP-generated disconnections and to flag chemically implausible steps before resources are committed. Automated reaction-path exploration methods such as AFIR/QCaRA extend the reach of QM to large and structurally complex systems, uncovering feasible pericyclic cascades, identifying unexpected yet experimentally accessible reaction pathways, and reproducing stereochemical outcomes by computationally probing high-energy intermediates, including reactions that deviate from classical rules such as the Woodward–Hoffmann orbital symmetry constraints. The molecular complexity observed in these examples is particularly encouraging, as it suggests that QM methods may facilitate access to structurally sophisticated drug candidates with enhanced target selectivity and strengthened intellectual property protection.

Overall, these examples demonstrate that QM methods can be used in synthesis planning to inform synthetic strategy, substrate design, and route selection. By integrating energetic feasibility, selectivity predictions, and automated exploration of mechanistic space, chemists can ensure that proposed synthetic routes are both chemically credible and experimentally tractable. Future avenues of exploration should focus on expanding the scope of reaction classes that can be predicted accurately, as QM-guided synthesis planning is currently most useful for polar reactions, rearrangements, pericyclic reactions, and selectivity-driven steps, rather than for all reaction classes. To achieve greater mechanistic accuracy, QM methods must also be extended to include explicit solvent molecules, to account for solvent-mediated interactions and their influence on reaction pathways. Improving the speed of QM calculations would also be beneficial, in order to comply with the ideal lead-optimization timeline of 1–2 weeks. With these further advancements, QM-guided synthesis planning will become a vital component of drug discovery for enabling broader chemical explorations and shortening the DMTA cycle.

References

1. Merz KM (2014) Using quantum mechanical approaches to study biological systems. Acc Chem Res 47:2804–2811
2. Ginex T, Vázquez J, Estarellas C, Luque FJ (2024) Quantum mechanical-based strategies in drug discovery: finding the pace to new challenges in drug design. Curr Opin Struct Biol 87:102870
3. Bonnet P (2012) Is chemical synthetic accessibility computationally predictable for drug- and lead-like molecules? A comparative assessment between medicinal and computational chemists. Eur J Med Chem 54:679–689
4. Morrison KC, Hergenrother PJ (2014) Natural products as starting points for the synthesis of complex and diverse compounds. Nat Prod Rep 31:6–14
5. Kim DE, Zweig JE, Newhouse TR (2019) Synthesis of paspaline A and emindole PB. J Am Chem Soc 141:1479–1483
6. Ree N, Göller AH, Jensen JH (2024) Automated quantum chemistry for estimating nucleophilicity and electrophilicity with applications to retrosynthesis and covalent inhibitors. Digit Discov 3:347–354
7. Mita T, Takano H, Hayashi H, Kanna W, Harabuchi Y, Houk KN, Maeda S (2022) Prediction of high-yielding single-step or cascade pericyclic reactions for the synthesis of complex synthetic targets. J Am Chem Soc 144:22985–23000
8. Nicolaou KC, Petasis NA, Zipkin RE (1982) The endiandric acid cascade. Electrocyclizations in organic synthesis. 4. Biomimetic approach to endiandric acids A–G. Total synthesis and thermal studies. J Am Chem Soc 104: 5560–5562
9. Kukier GA, Turlik A, Xue XS, Houk KN (2021) Violations: how nature circumvents the Woodward–Hoffmann rules and promotes the forbidden conrotatory 4n + 2 electron electrocyclization of Prinzbach's vinylogous sesquifulvalene. J Am Chem Soc 143: 21694–21704

Chapter 19

Fragment Molecular Orbital Drug Design Consortium (FMODD): Bridging Academia and Industry Through Prospective FMO-Driven Drug Design

Alexander Heifetz and Kaori Fukuzawa

Abstract

The Fragment Molecular Orbital Drug Design Consortium (FMODD) was founded in 2014 by volunteers from industry, academia, and government with the goal of promoting the Fragment Molecular Orbital (FMO) method as a practical and mainstream tool for structure-based drug design (SBDD). Although FMO offers superior accuracy and interpretability compared with classical molecular-mechanics approaches, its published use has historically been dominated by methodological development and retrospective analyses. FMODD was established to accelerate the prospective application of FMO in real drug discovery projects through coordinated academia–industry collaboration, access to high-performance computing infrastructure, standardized and automated workflows, and shared community resources such as the FMO Database (FMODB). Equally important, FMODD actively builds and supports a global user community through training, documentation, workshops, and peer support, thereby lowering barriers for new users and enabling the routine adoption of FMO within pharmaceutical SBDD workflows.

Keywords Fragment molecular orbital, Structure-based drug design, Quantum mechanics, Fragment molecular orbital drug design consortium, FMO database, High-performance computing

1 Introduction

The Fragment Molecular Orbital Drug Design Consortium (FMODD [1]; https://fmodd.jp/members-en/) is a Japan-based initiative established in November 2014 with the explicit goal of translating high-level quantum-mechanical (QM) methods into practical drug discovery workflows. FMODD focuses on the Fragment Molecular Orbital (FMO) method, a QM approach that enables residue-level analysis of protein–ligand and protein–protein interactions with a degree of accuracy and physical detail inaccessible to classical molecular mechanics (MM) methods [2–7]. Among available computational approaches, FMO is widely regarded as one of the most promising methods for structure-based drug design (SBDD), which

Alexander Heifetz (ed.), *Quantum Mechanics Applications in Drug Design*, Methods in Molecular Biology, vol. 3061,
https://doi.org/10.1007/978-1-0716-5420-0_19,

remains the dominant and most successful method for advancing compounds beyond the design stage and into pre-clinical validation [8–17]. FMO provides residue-resolved, quantitative insight into interaction strengths (kcal/mol), the chemical nature of interactions (e.g., electrostatic, hydrophobic), and key energetic contributions such as desolvation and depolarization [3, 18, 19]. These physically interpretable descriptors are particularly valuable for efficient SBDD.

One of the key objectives of FMODD is to bridge the long-standing gap between methodological excellence and practical adoption. Although FMO has repeatedly demonstrated superior accuracy and interpretability compared with non-QM scoring functions, its uptake in industrial SBDD has remained limited. FMODD seeks to move FMO beyond proof-of-concept studies toward routine use in real medicinal chemistry projects by combining community building, high-performance computing (HPC), automation, shared data resources, and close academia–industry collaboration.

Despite more than a decade of development, the published FMO literature remains strongly biased toward methodological studies and retrospective analyses (Fig. 1) [20–28]. A bibliometric survey of publications from 2015 to 2025, based on manual inspection of PubMed, Scopus, and FMODB-linked references, indicates that a substantial proportion of studies focus on method development, including advances in fragmentation schemes, solvent and polarization treatments, visualization tools, and computational acceleration [20–27, 29]. Retrospective applications, in which FMO is used to rationalize binding affinities, interpret crystal structures, or explain experimental observations, are also prevalent. In contrast, truly prospective applications, where FMO calculations directly guide

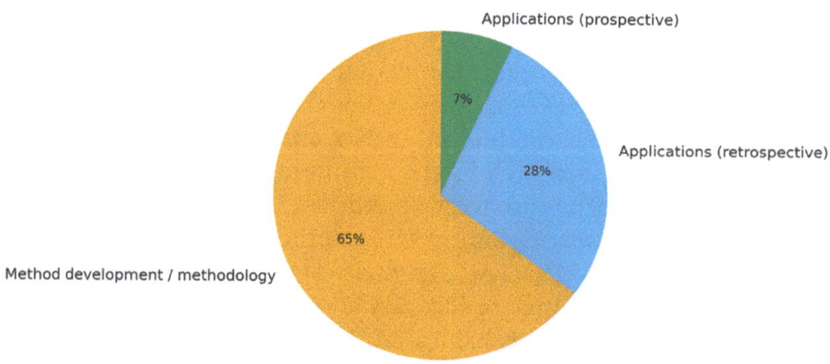

Fig. 1 Estimated distribution of FMO publications by topic (2015–2025). Method development/methodology (~65%), applications—retrospective (~28%), applications—prospective (~7%). Basis: manual PubMed scan (query: "Fragment Molecular Orbital" AND drug–design terms) and synthesis with recent reviews/FMODB references; numbers are preliminary and will be updated with a full bibliometric pass. While these distributions are based on manual classification and should be interpreted as semi-quantitative, they nevertheless capture clear and persistent trends in the published FMO literature

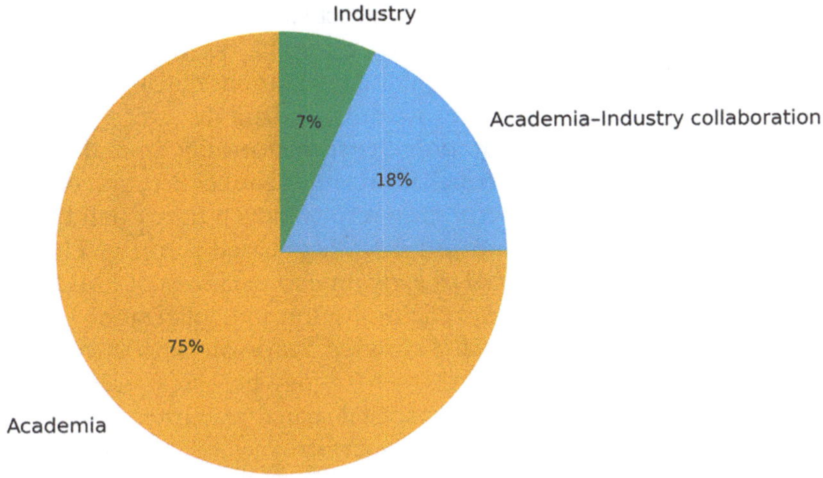

Fig. 2 Estimated distribution by publisher type (2015–2025). Academia (~75%), academia–industry collaboration (~18%), and industry (~7%). Classification by lead/corresponding author affiliation, based on manual inspection of sample records; preliminary. While these distributions are based on manual classification and should be interpreted as semi-quantitative, they nevertheless capture clear and persistent trends in the published FMO literature

compound design and synthesis prior to experimental testing, remain relatively rare in the published literature.

Authorship patterns further reflect this imbalance. Academic groups dominate FMO publications, while academia–industry collaborations are less frequent, and publications led exclusively by industry remain scarce (Fig. 2). Although this distribution may partly reflect publication bias and the confidential nature of industrial research, it nevertheless highlights the slow translation of FMO from academic development into routine use within mainstream pharmaceutical pipelines.

Publicly documented drug–design projects using FMO reinforce this picture. Only a limited number of studies, including FMO-guided pharmacophore modeling for anti-Chagas drug discovery (Yoshino et al.) [30], ITK inhibitor optimization (Heifetz et al. [26]), Darunavir analog design for HIV-1 protease [31], PLK1 inhibitor discovery (Jin et al. [27]) and small molecule drug discovery targeting RNA [32] - demonstrate FMO being used as a genuine design driver rather than as a post hoc explanatory tool.

2 Why FMO Remains Underused

The limited adoption of FMO persists despite extensive evidence that it can outperform classical MM approaches in both accuracy and interpretability [21, 33–37]. FMO has repeatedly demonstrated its ability to explain structure–activity relationships (SARs) that non-QM scoring functions fail to capture [2, 32, 38–47].

Nevertheless, practical reservations remain regarding the broader adoption of FMO in industrial drug discovery. Common concerns include computational cost, historically limited user-friendliness, complexity of installation and maintenance, and challenges in analyzing and interpreting results, which have often been delivered in lengthy and non-intuitive text-based formats. These perceptions are partly rooted in early implementations of FMO that lacked key features such as explicit solvation and polarization treatments. Additional factors include limited hands-on experience within industrial teams; concerns that FMO may be overly accurate or applicable only to high-quality crystal structures rather than docking-derived poses; reliance on high-performance computing (HPC) resources; sensitivity to fragment definitions; and uncertainties associated with protonation and tautomeric states. Furthermore, FMO has often been perceived as an academic methodology, lacking the level of user support, integration, and robustness typically expected of commercial computational chemistry software.

However, these perceptions are increasingly being challenged. At present, several studies are being prepared that explicitly demonstrate the effective integration of FMO analyses with experimental workflows using docking-derived structures, addressing a long-standing concern regarding structural input quality. In parallel, preparations are underway for broader, global deployment of FMO through commercial distribution channels. In particular, the ABINIT-MP program is expected to be made available as a compiled binary by JSOL Corporation by the end of the current fiscal year (March 2026), which will substantially lower barriers related to installation, maintenance, and accessibility. Together, these developments mark an important step towards transitioning FMO from a predominantly academic methodology into a more robust, user-friendly, and industrially deployable technology.

FMODD was established specifically to address these concerns by providing standardized workflows, automation, training, shared infrastructure, and a collaborative support network that lowers barriers to adoption and enables the reliable application of FMO in real-world drug discovery settings.

3 From Consortium to Community: FMODD's Strategy in Practice

FMODD provides a practical and sustainable route for translating the FMO method from an academically established QM approach into a prospective, decision-support tool for SBDD. Rather than focusing solely on algorithmic development, the consortium emphasizes the full ecosystem required for adoption, including infrastructure, standardization, training, and collaboration. Its broad membership spans leading Japanese academic institutions, government laboratories, pharmaceutical companies, biotechnology firms, and contract research organizations (CROs) such as Sygnature Discovery and computational chemistry vendors. This diversity enables direct and continuous feedback between method developers and end users, ensuring that methodological advances are aligned with real-world medicinal chemistry needs.

A cornerstone of FMODD's strategy is access to Japan's High-Performance Computing Infrastructure (HPCI), including flagship systems such as the K computer and supercomputer Fugaku. These resources make it feasible to apply FMO routinely to pharmaceutically relevant systems—GPCRs, kinases, viral enzymes, and large protein–protein interfaces—at scales that were previously impractical. Equally important, FMODD promotes standardized [48] and automated workflows [49, 50], ensuring reproducibility across projects and institutions and lowering the operational barrier to FMO deployment. Together, these elements demonstrate how advanced QM methods can be embedded alongside docking, molecular dynamics, and emerging AI-driven approaches within modern SBDD pipelines.

Beyond computation and data, FMODD is fundamentally a community-driven initiative. Regular symposia, workshops, and training activities provide neutral forums in which scientists from academia, industry, and government laboratories can exchange ideas and experiences. Pharmaceutical researchers present concrete design challenges, academic groups introduce methodological innovations, and infrastructure providers highlight computational capabilities. This sustained dialogue fosters mutual understanding: academics gain insight into the most pressing questions in drug discovery, while industry builds confidence in methods that are being evaluated, validated, and refined by a broader peer community.

Although FMODD originated in Japan, it has evolved into an increasingly international consortium. Active members and collaborators now span multiple regions, supported by designated coordinators to strengthen transnational engagement. For example, Sygnature Discovery in the United Kingdom participates as a European coordinator, helping connect FMODD with pharmaceutical projects across the UK and Europe. Additional collaborations involving researchers from Germany, France, and the United States have emerged through joint publications, invited workshops, and shared case studies. This global engagement reflects both the universal relevance of the FMO method and the recognition that widespread adoption of advanced QM approaches requires dialogue that transcends geographic as well as disciplinary boundaries.

By combining a strong Japanese foundation with expanding international participation, FMODD has become not merely a national consortium but a model for global collaboration in quantum-driven drug design. Importantly, the consortium remains open and inclusive, actively welcoming new academic institutions, industry partners, and government laboratories worldwide. This openness strengthens FMODD's role as a true community of practice—one that lowers barriers to entry, reduces skepticism, promotes shared standards, and ultimately supports the transition of FMO from a specialized technique into a routine and trusted component of modern drug discovery.

4 FMODB and Automation

One of the most significant outputs of FMODD is the FMO Database (FMODB), https://drugdesign.riken.jp/FMODB/index.php), which was released publicly in 2019. FMODB represents the first large-scale repository of quantum chemical data for protein–ligand systems generated using the FMO. Rather than focusing exclusively on counting unique protein–ligand complexes, FMODB is organized around individual FMO calculation datasets. As a result, a single protein–ligand system may contribute multiple entries to the database, for example, when different ligands, protonation or

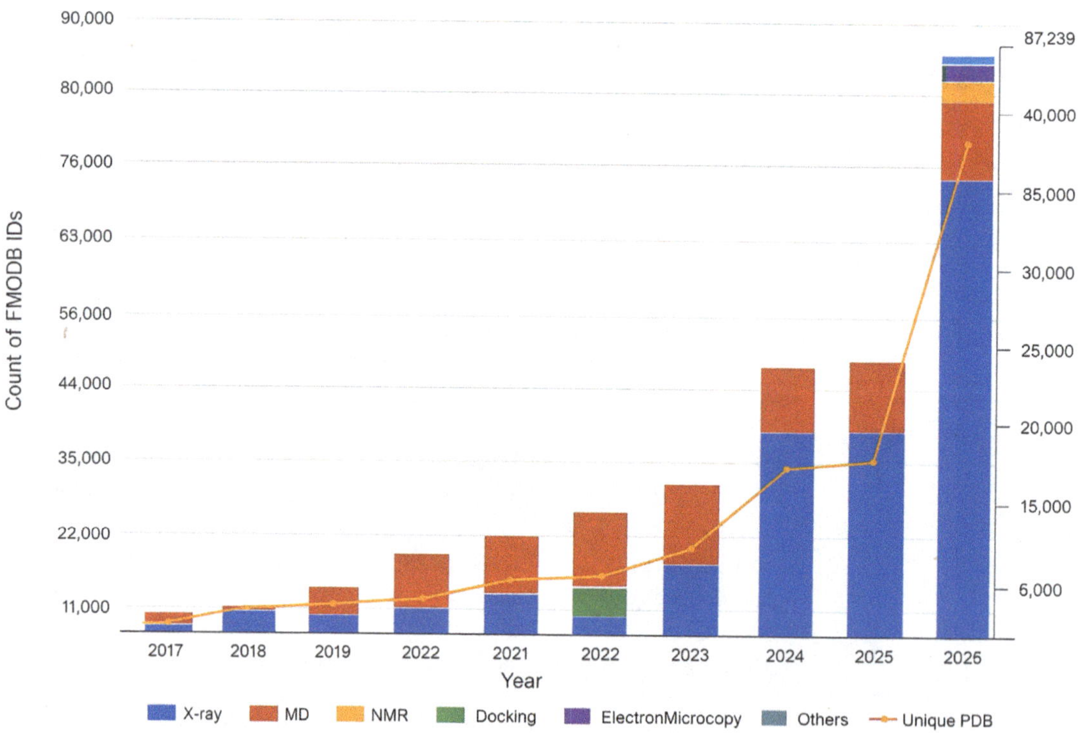

Fig. 3 Growth of FMODB data over time. Stacked bars show annual FMODB entries by structural source, and the line indicates cumulative unique PDB structures. FMODB contains 87,239 entries as of the end of 2025

tautomeric states, fragment definitions, or computational settings are considered, as well as when structures sampled from classical MD calculations are used. As of the end of 2025, FMODB contains 87,239 FMO data entries, reflecting the cumulative output of these calculations (Fig. 3).

The scale and growth of FMODB have been enabled by extensive automation developed [50] within FMODD. Automated pipelines [49] have been implemented to handle structure preparation, protonation and tautomer assignment, FMO input generation, job submission, and post-processing and extraction of results with minimal manual intervention. These workflows significantly lower the barrier to entry for non-expert users, improve reproducibility, and enable FMO calculations to be performed at a scale compatible with systematic and data-driven approaches, including machine learning. Automated FMO protocols and processing workflows have been described previously and are also employed for data collection and analysis in the present work.

Beyond its role as a repository, FMODB serves as a foundation for broader access to FMO technology. In this context, an FMO on-demand service is currently under development. This service will allow users to upload a protein structure file (PDB format) via a web interface and obtain FMO calculation results generated automatically in the background. The aim is to enable not only computational scientists, but also structural biologists and synthetic chemists, to perform FMO analyses on experimentally determined structures or docking-derived models, for example, to assess structural quality or to support early-stage compound evaluation. A prototype of this service is expected by the end of the current fiscal year, with limited public availability potentially initially for FMODD members anticipated by the end of the following year.

References

1. Fukuzawa K, Tanaka S, Yagi Y et al. (2021) FMO drug design consortium. In: Mochizuki Y, Tanaka S, Fukuzawa K (Eds.)Eds. Recent advances of the fragment molecular orbital method. Springer Singapore, Singapore, 127–181
2. Fukuzawa K, Tanaka S (2022) Fragment molecular orbital calculations for biomolecules. Curr Opin Struct Biol 72:127–134
3. Nagata T, Fedorov DG, Sawada T et al. (2012) Analysis of solute–solvent interactions in the fragment molecular orbital method interfaced with effective fragment potentials: theory and application to a solvated Griffithsin–Carbohydrate complex. J Phys Chem A 116:9088–9099
4. Fedorov DG, Inostroza D, Courbiere B et al. (2025) Decomposition analysis for visualization of noncovalent interactions based on the fragment molecular orbital method. J Chem Theory Comput 21:4435–4446
5. Fedorov DG, Nakamura T (2022) Free energy decomposition analysis based on the fragment molecular orbital method. J Phys Chem Lett 13:1596–1601
6. Fedorov DG, Kitaura K (2004) The importance of three-body terms in the fragment molecular orbital method. J Chem Phys 120:6832–6840
7. Tanaka S, Mochizuki Y, Komeiji Y et al. (2014) Electron-correlated fragment-molecular-orbital calculations for biomolecular and nano systems. Phys Chem Chem Phys 16:10310–10344
8. Schneuing A, Harris C, Du Y et al. (2024) Structure-based drug design with equivariant diffusion models. Nat Comput Sci 4:899–909
9. Amzel LM (1998) Structure-based drug design. Curr Opin Biotechnol 9:366–369

10. Ouma RBO, Ngari SM, Kibet JK (2024) A review of the current trends in computational approaches in drug design and metabolism. Discov Public Health 21:108
11. Lin X, Li X, Lin X (2020) A review on applications of computational methods in drug screening and design. Molecules 25:1375
12. Neelam AKN (2025) Advancing drug discovery: the role of computer-aided design and development in modern pharmaceuticals. Discov Pharm Sci 1:8
13. Sadybekov AV, Katritch V (2023) Computational approaches streamlining drug discovery. Nature 616:673–685
14. Mihai DP, Nitulescu GM (2025) Computer-aided drug design and drug discovery. Pharmaceuticals 18:436
15. Wu Z, Chen S, Wang Y et al. (2024) Current perspectives and trend of computer-aided drug design: a review and bibliometric analysis. Int J Surg 110:3848–3878
16. Medina-Franco JL (2021) Grand challenges of computer-aided drug design: the road ahead. Front Drug Discov 1:728551
17. Sabe VT, Ntombela T, Jhamba LA et al. (2021) Current trends in computer aided drug design and a highlight of drugs discovered via computational techniques: a review. Eur J Med Chem 224:113705
18. Nishimoto Y, Fedorov DG (2016) The fragment molecular orbital method combined with density-functional tight-binding and the polarizable continuum model. Phys Chem Chem Phys 18:22047–22061
19. Handa Y, Okuwaki K, Kawashima Y et al. (2024) Prediction of binding pose and affinity of nelfinavir, a SARS-CoV-2 main protease repositioned drug, by combining docking, molecular dynamics, and fragment molecular orbital calculations. J Phys Chem B 128:2249–2265
20. Choi J, Kim H-J, Jin X et al. (2018) Application of the fragment molecular orbital method to discover novel natural products for prion disease. Sci Rep 8: 13063
21. Heifetz A, James T, Southey M et al. (2019) Characterising GPCR–ligand interactions using a fragment molecular orbital-based approach. Curr Opin Struct Biol 55:85–92
22. Paciotti R, Re N, Storchi L (2024) Combining the fragment molecular orbital and GRID approaches for the prediction of ligand–metalloenzyme binding affinity: the case study of hCA II inhibitors. Molecules 29:3600
23. Takaya D (2024) Computer-aided drug design using the fragment molecular orbital method: current status and future applications for SBDD. Chem Pharm Bull (Tokyo) 72:781–786
24. Moriwaki H, Kawashima Y, Watanabe C et al. (2024) FMOe: preprocessing and visualizing package of the fragment molecular orbital method for molecular operating environment and its applications in covalent ligand and metalloprotein analyses. J Chem Inf Model 64:6927–6937
25. Okiyama Y, Nakano T, Watanabe C et al. (2018) Fragment molecular orbital calculations with implicit solvent based on the poisson–boltzmann equation: implementation and DNA study. J Phys Chem B 122:4457–4471
26. Heifetz A, Trani G, Aldeghi M et al. (2016) Fragment molecular orbital method applied to lead optimization of novel interleukin-2 inducible T-cell kinase (ITK) inhibitors. J Med Chem 59:4352–4363
27. Jin H, Kim J, Lee O et al. (2023) Leveraging the fragment molecular orbital method to explore the PLK1 kinase binding site and polo-box domain for potent small-molecule drug design. Int J Mol Sci 24:15639
28. Sattasathuchana T, Xu P, Bertoni C et al. (2024) The effective fragment molecular orbital method: achieving high scalability and accuracy for large systems. J Chem Theory Comput 20:2445–2461
29. Okuwaki K, Watanabe N, Kato K et al. (2024) Geometry optimization using the frozen domain and partial dimer approaches in the fragment molecular orbital method: implementation, benchmark, and applications to protein ligand-binding sites. J Chem Inf Model 64:9449–9458
30. Yoshino R, Yasuo N, Inaoka DK et al. (2015) Pharmacophore modeling for anti-chagas drug design using the fragment molecular orbital method. PLoS One 10:e0125829
31. Chuntakaruk H, Hengphasatporn K, Shigeta Y et al. (2024) FMO-guided design of darunavir analogs as HIV-1 protease inhibitors. Sci Rep 14: 3639
32. Shino A, Otsu M, Imai K et al. (2023) Probing RNA–small molecule interactions using biophysical and computational approaches. ACS Chem Biol 18:2368–2376
33. Kawashita N (2024) Interaction analysis by fragment molecular orbital method for drug discovery research. Chem Pharm Bull (Tokyo) 72:787–793
34. Ginex T, Vázquez J, Estarellas C et al. (2024) Quantum mechanical-based strategies in drug discovery: finding the pace to new challenges in drug design. Curr Opin Struct Biol 87:102870
35. Niazi SK (2025) Quantum mechanics in drug discovery: a comprehensive review of methods, applications, and future directions. Int J Mol Sci 26:6325

36. Arodola OA, Soliman ME (2017) Quantum mechanics implementation in drug-design workflows: does it really help?. Drug Des Devel Ther 11:2551–2564
37. Fedorov DG (2017) The fragment molecular orbital method: theoretical development, implementation in GAMESS, and applications. WIREs Comput Mol Sci 7:e1322
38. Takebe K, Miyakawa S, Sangawa T et al. (2025) Combined structural and FMO-based insights into shaft pilin polymerization mechanism in Streptococcus sanguinis. Int J Biol Macromol 332:148264
39. Ma X, Higashi K, Fukuzawa K et al. (2022) Computational approach to elucidate the formation and stabilization mechanism of amorphous formulation using molecular dynamics simulation and fragment molecular orbital calculation. Int J Pharm 615:121477
40. Handa C, Yamazaki Y, Yonekubo S et al. (2022) Evaluating the correlation of binding affinities between isothermal titration calorimetry and fragment molecular orbital method of estrogen receptor beta with diarylpropionitrile (DPN) or DPN derivatives. J Steroid Biochem Mol Biol 222:106152
41. Takaya D, Watanabe C, Nagase S et al. (2021) FMODB: the world's first database of quantum mechanical calculations for biomacromolecules based on the fragment molecular orbital method. J Chem Inf Model 61:777–794
42. Watanabe C, Okiyama Y, Tanaka S et al. (2021) Molecular recognition of SARS-CoV-2 spike glycoprotein: quantum chemical hot spot and epitope analyses. Chem Sci 12:4722–4739
43. Takaba K, Watanabe C, Tokuhisa A et al. (2022) Protein-ligand binding affinity prediction of cyclin-dependent kinase-2 inhibitors by dynamically averaged fragment molecular orbital-based interaction energy. J Comput Chem 43:1362–1371
44. Fukuzawa K, Kato K, Watanabe C et al. (2021) Special features of COVID-19 in the FMODB: fragment molecular orbital calculations and interaction energy analysis of SARS-CoV-2-related proteins. J Chem Inf Model 61:4594–4612
45. Takebe K, Suzuki M, Kuwada-Kusunose T et al. (2023) Structural and computational analyses of the unique interactions of opicapone in the binding pocket of catechol O-methyltransferase: a crystallographic study and fragment molecular orbital analyses. J Chem Inf Model 63:4468–4476
46. Heifetz A, Chudyk EI, Gleave L et al. (2016) The fragment molecular orbital method reveals new insight into the chemical nature of GPCR–ligand interactions. J Chem Inf Model 56:159–172
47. Yagi Y, Kimura T, Watanabe C et al. (2025) Comprehensive protein-ligand interaction analysis: fragment molecular orbital calculation on the complexes of human protease renin and its inhibitors. Chem-Bio Inform J 25:107–129
48. Fukuzawa K, Kato K, Watanabe C et al. (2021) Special features of COVID-19 in the FMODB: fragment molecular orbital calculations and interaction energy analysis of SARS-CoV-2-related proteins. J Chem Inf Model 61:4594–4612
49. Takaya D, Ohno S, Miyagishi T et al. (2024) Quantum chemical calculation dataset for representative protein folds by the fragment molecular orbital method. Sci Data 11: 1164
50. Watanabe C, Watanabe H, Okiyama Y et al. (2019) Development of an automated fragment molecular orbital (FMO) calculation protocol toward construction of quantum mechanical calculation database for large biomolecules. Chem-Bio Inform J 19:5–18

Chapter 20

Integrating AI-Based Protein Modeling and Fragment Molecular Orbital Analysis for Structure-Based Drug Design

Alessio Atzori, Louise Birch, Colin Sambrook Smith and Alexander Heifetz

Abstract

Recent advances in deep-learning-based protein structure prediction have transformed access to three-dimensional models of biological targets directly from sequence information. Methods such as AlphaFold and emerging co-folding approaches have greatly expanded structural coverage of the proteome, enabling structure-based drug design (SBDD) for targets lacking experimentally determined structures. Integration of AI-derived protein structures with quantum-mechanical methods such as the Fragment Molecular Orbital (FMO) approach provides a powerful and complementary strategy, in which AI-based models deliver rapid structural information, while FMO leverages these structures to enable quantitative, physics-based decomposition of protein–ligand and protein–protein interactions. FMO yields interaction strengths in kcal/mol, characterizes the chemical nature of interactions (electrostatic or hydrophobic), and enables subsystem analysis to evaluate the contributions of different energetic terms, including polarization and desolvation, to binding. Together, these approaches transform predicted structures into energetically interpretable models that directly support structure–activity relationship (SAR) analysis, binding-mode validation, and prospective structure-based drug design.

Keywords Structure-based drug design, Protein structure prediction, AlphaFold, Protein folding, Protein–ligand co-folding, Artificial intelligence, Quantum mechanics, Fragment molecular orbital, Binding energetics, Structure–activity relationships, Virtual screening, Molecular docking, Physics-based methods

1 Introduction

Structure-based drug design (SBDD) remains the dominant computational approach and consistently demonstrates the highest success rates in progressing compounds beyond the design stage [1–10]. However, experimentally determined protein structures are not always available [11, 12]. Determination of a protein's three-dimensional structure has traditionally required substantial experimental effort, relying on techniques such as X-ray crystallography, nuclear magnetic resonance spectroscopy, or cryo-electron microscopy, often

Alexander Heifetz (ed.), *Quantum Mechanics Applications in Drug Design*, Methods in Molecular Biology, vol. 3061,
https://doi.org/10.1007/978-1-0716-5420-0_20,

requiring substantial investments of time and resources. As a result, for many biologically and therapeutically relevant targets, high-quality structural information has historically been unavailable, limiting the broader application of SBDD [11–14].

Over the past decade, and particularly following the introduction of deep-learning-based protein structure prediction methods, this situation has changed dramatically. Today, high-quality three-dimensional protein models can be generated within seconds directly from primary sequence information, fundamentally altering how structural data are accessed and used in drug discovery workflows (Fig. 1).

This transformation did not occur in isolation, but reflects the convergence of advances in machine learning, evolutionary biology, and physics-inspired modeling, enabled by unprecedented growth in computational power and data availability. Together, these developments have elevated protein structure prediction from a niche computational activity to a central component of modern molecular biology and structure-based drug discovery.

The significance of this progress was formally recognized in 2024, when the Nobel Prize in Chemistry was awarded to John Jumper and Demis Hassabis for the development of AlphaFold, alongside David Baker for advances in computational protein design. This recognition underscores the profound impact that AI-driven structure prediction has had on the life sciences and highlights its transformative role in contemporary drug discovery.

While AI-based folding and co-folding models now provide broad structural coverage of the proteome, their value in medicinal chemistry extends beyond structure generation alone. For SBDD, structural models must ultimately support quantitative interpretation of ligand binding, rationalization of structure–activity relationships (SAR), and prospective design decisions. Achieving this requires not

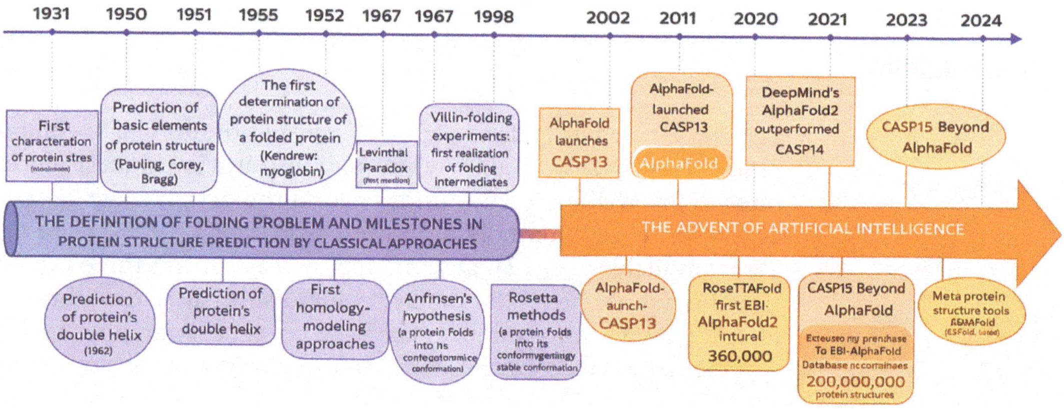

Fig. 1 Timeline illustrating the evolution of structural biology methods, from early experimental approaches to homology- and template-based modeling, culminating in modern deep-learning-driven protein structure prediction. (Adapted from [15])

only accurate geometries but also a reliable description of binding energetics and local physicochemical effects. Consequently, the integration of AI-derived structures with physics-based approaches, particularly quantum-mechanical methods such as the Fragment Molecular Orbital (FMO) method, represents a critical step toward fully realizing the potential of protein structure prediction in drug discovery.

2 The Old Ways of Protein Structure Prediction

Before machine learning transformed structural biology, scientists depended on classical methods that struggled to generalize beyond known protein structures.

Ab initio methods attempted to compute protein structures from first principles, without relying on a protein template. However, as the size of the protein considered increased, the conformational search space expanded exponentially, making these methods computationally prohibitive even for moderately sized systems.

Homology modeling represented the standard approach for computational structure prediction; if the structure of a homologous protein was known, it could be used as a template to generate a model reflecting the evolutionary relationship between the target and template.

The method works reasonably well for proteins with available close homologs but becomes unreliable when proteins have few or no similar templates in existing databases. Below ~30% sequence identity, comparative models are typically unreliable due to alignment and structural divergence errors.

This limitation highlighted a fundamental problem: homology-based methods depended on the availability of experimentally solved structures. With only a small fraction of the ~200 million known proteins being structurally characterized, vast regions of the proteome remained unexplored. In drug discovery, this gap represented a considerable drawback: many promising therapeutic targets lacked structural information, severely limiting structure-based drug design approaches.

3 AlphaFold 2: A Solution from Deep Learning

AlphaFold 2 (AF2), introduced by DeepMind in December 2020, achieved unprecedented accuracy using a methodology fundamentally different from traditional physics-based models. Instead of deriving structures from first-principles physics, the system learned from evolutionary patterns encoded in multiple sequence alignments (MSAs), collections of related protein sequences accumulated across evolution [16].

AF2's approach is based on the observation that amino acids coevolving in related sequences are usually close in 3D space. The model learned to recognize these evolutionary patterns while respecting physical constraints like bond angles and protein geometry. The system was trained on ~170,000 experimentally solved structures and was able to produce accurate predictions even for proteins without close homologs. In CASP14 (a blind assessment competition for structure prediction methods), it managed to achieve near-experimental precision, with an average deviation of about 1.6 Å [16] (see, e.g., Fig. 2).

However, it is worth mentioning that the 3D protein models generated were limited to protein chains alone (monomers), and no ligands, cofactors, or posttranslational modifications could be explicitly modeled.

The introduction of AF2 completely changed how protein structures were predicted, improving both speed and accessibility. Following the release of its open-source code in July 2021, the AlphaFold Protein Structure Database expanded to include nearly all known proteins, increasing structural coverage of the human proteome from 48% to about 76% [17].

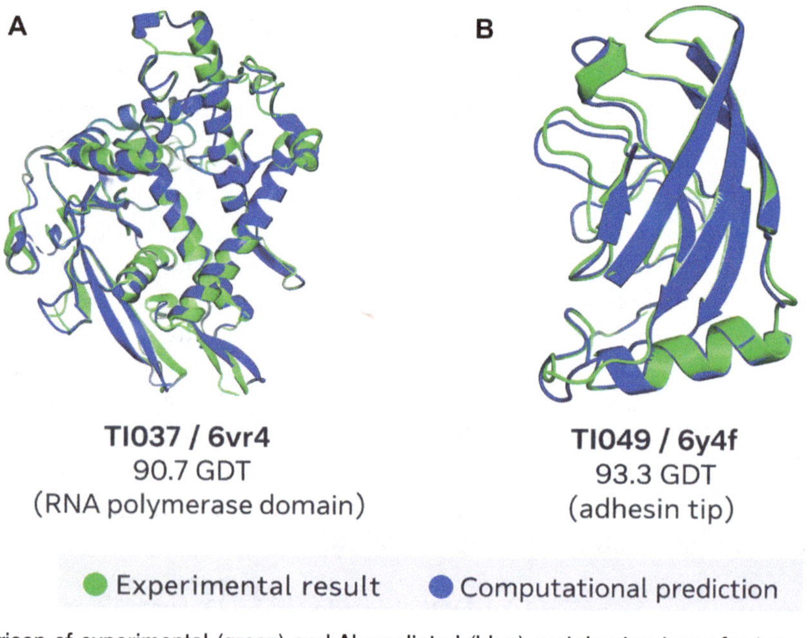

Fig. 2 Comparison of experimental (green) and AI-predicted (blue) protein structures for two representative CASP14 targets. (**A**) Target T1037 (GDT 90.7). (**B**) Target T1049 (GDT 93.3). The close agreement between experimental and predicted structures illustrates near-experimental prediction accuracy.

4 Expanding Beyond Proteins: Co-Folding Models

Released in 2024, AlphaFold 3 (AF3) represented a fundamental redesign. Unlike AF2, which was limited to predicting single protein chains, AF3 was built to model complete biomolecular assemblies and their interactions (see, e.g., Fig. 3).

Instead of representing proteins as sequences of amino acids, AF3 predicts the 3D coordinates of individual atoms directly. This enables the system to model proteins, RNA, DNA, small molecules, metals, and modified residues using a single, unified framework. In practice, AF3 delivered roughly 50% higher accuracy than traditional methods on the PoseBusters benchmark for drug-like interactions and became the first AI system to outperform physics-based tools (i.e., molecular docking) in biomolecular structure prediction. Unlike AF2, which was freely available for commercial use, AF3 is released only for noncommercial research under restricted terms.

Fortunately, other research teams have stepped in to democratize access by releasing co-folding models with comparable accuracy and commercially permissive licenses. **Boltz-1** (2024) introduced a

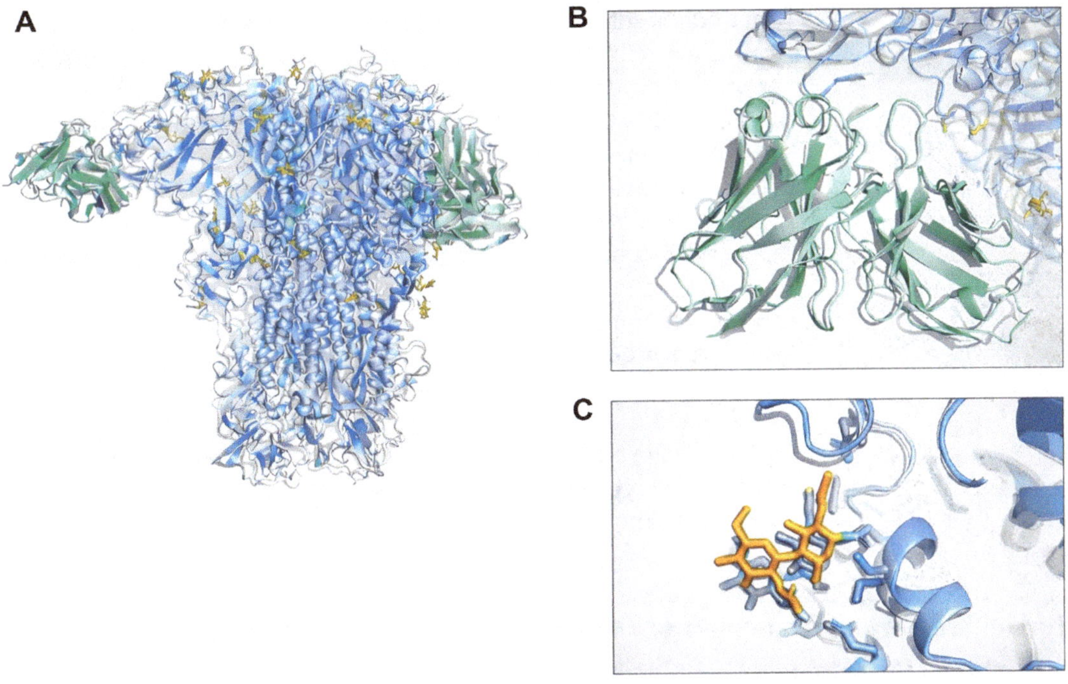

Fig. 3 AlphaFold 3 prediction of the Coronavirus OC43 spike protein complex (PDB: 7PNM). (**A**) Overall architecture of the spike protein (blue) in complex with antibodies (green) and glycans (yellow). (**B**) Close-up view of an antibody-binding region, showing accurate prediction of the protein–protein interface. (**C**) Detailed view of a glycan-binding site, illustrating precise placement of small-molecule components within the predicted structure

unified framework for predicting structures of proteins, nucleic acids, and small molecules within a single model [18]. Boltz-2 [19] extended this by incorporating binding-affinity prediction, allowing the system to estimate not only how molecules bind structurally but also their binding strength quantitatively of binding strength. More recently, **BoltzGen** [20] targets protein design by generating novel sequences and structures optimized for binding a specified target.

OpenFold3, developed by the OpenFold AI Research Consortium with industry partners, provides an open-source alternative aiming to match AF3's performance while retaining full commercial access (OpenFold Consortium (2025). Link).

5 Why These Methods Matter: Some Real-World Applications

Since AF2's release, structural coverage of the human proteome has expanded dramatically, bringing previously "undruggable" proteins into the spotlight. This created new opportunities to pursue orphan disease targets and challenging classes once considered beyond reach.

The impact of AlphaFold on drug discovery has been remarkable. For instance, researchers used AF2 to predict structures of CDK20 (an understudied cancer target) to guide the design of new inhibitors for hepatocellular carcinoma. Within just 30 days, after synthesizing only seven molecules, they obtained a compound with a binding constant K_d value of $9.2 \pm 0.5\,\mu M$; a second optimization iteration produced a nanomolar inhibitor (K_d of 566.7 ± 256.2 nM and IC_{50} of 33.4 ± 22.6 nM) [21].

Another group applied AlphaFold models of the trace amine-associated receptor 1 (TAAR1) to virtually screen 16 million compounds, achieving a striking 60% hit rate [22]. The resulting molecules included potent agonists (12–0.03 μM) that showed antipsychotic-like activity in animal studies, doubling the success rate achieved with traditional homology models. Similarly, an AlphaFold-based model of histone deacetylase 11 (HDAC11), a protein without an experimentally solved structure, enabled virtual screening that uncovered selective inhibitors with micromolar potency ($IC_{50} \approx 3.5\,\mu M$) and clear subtype selectivity [23].

6 Limitations and Challenges of Folding and Co-Folding Models

Despite remarkable progress in protein structure prediction using deep-learning-based folding/co-folding models, there are important limitations that must be acknowledged:

1. Models excel at recognizing structural patterns from training data. Performance correlates strongly with training similarity:

excellent for proteins with known homologs, often problematic for genuinely novel folds (Fig. 4) [24].

2. Models often predict **one dominant functional state** of a target protein, typically the most represented in the training data. A systematic study on 92-fold-switching proteins found only modest success in recovering both conformations: best-case ~35% (32/92) for proteins likely represented in training data, while baseline AF2 variants achieved ~12%–13%, and AF3 ~8% (7/92); for seven fold-switching proteins outside the likely training set, only 1/7 was recovered [25].
3. Membrane proteins remain challenging for current folding models: their structures depend on interactions with lipids or cofactors that are not encoded in their primary sequence [26]. For the substantial fraction of human proteins that are intrinsically disordered, models assign low confidence scores because they represent conformational ensembles rather than single conformations: thus, predictions in these regions are unreliable for functional interpretation [26].
4. While more recent co-folding models can predict 3D coordinates of ligands, a very recent 2025 Nature Communications study revealed that co-folding models (like AF3 and Boltz) fail

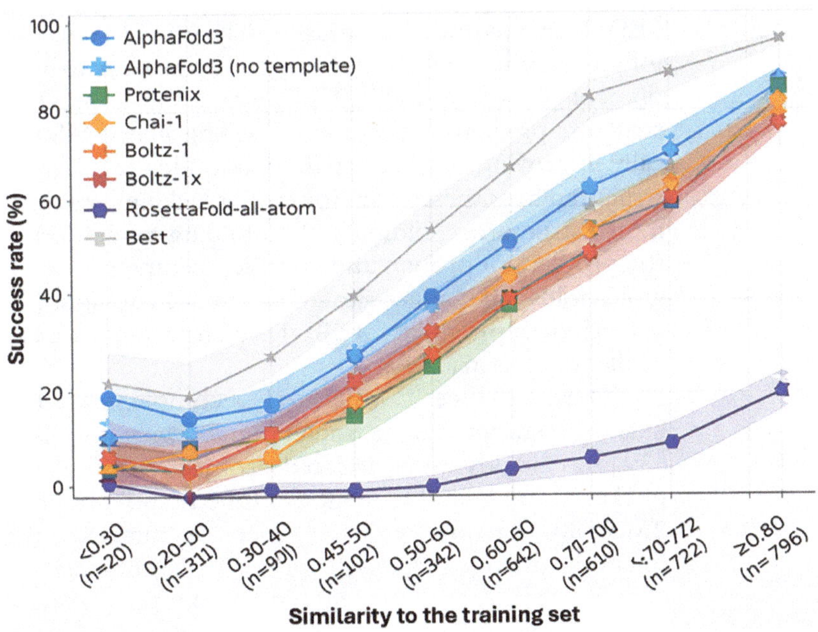

Fig. 4 Performance of protein–ligand co-folding methods as a function of similarity to the training set. Success rate (percentage of complexes with ligand RMSD < 2Å and LDDT-PLI > 0.8) is shown across bins of increasing similarity to training data for AlphaFold 3, Boltz-1, Chai-1, Protenix, RoseTTAFold-all-atom, and related variants. Performance improves markedly with higher training-set similarity, highlighting limited generalisation to low-similarity targets. (Adapted from [24])

to account for changes in binding-site physics (i.e., modifying binding site residues from charged to uncharged and vice versa, flooding the binding site with bulky hydrophobic residues, etc.) [27]. Despite these alterations, ligands were still placed with high confidence in energetically impossible poses. Even when the local chemistry was altered to physically hinder binding, the models reproduced memorized conformations from training data. By contrast, physics-based simulations correctly captured loss of binding. This demonstrates that while deep-learning models can generalize structural patterns, they still struggle to recognize when a predicted ligand pose is physically or chemically impossible.

5. Co-folding methods struggle with allosteric binding sites, consistently placing ligands in active site pockets. In a benchmark of 16 protein targets with known orthosteric and allosteric structures, methods like Boltz-1 successfully predicted most orthosteric poses but failed on allosteric sites (RMSD >10 Å in most cases) [28]. This bias toward orthosteric sites limits their utility for validating allosteric modulators.

7 Integrating AI-Based Protein Models with FMO for Structure-Based Drug Design

SBDD remains one of the most widely used and effective approaches for progressing compounds from the design stage to preclinical development (see Chap. 2) [1, 2]. Recent advances in AI-based protein structure prediction have further strengthened SBDD by dramatically improving access to structural models, enabling high-quality structural hypotheses even for targets lacking experimentally determined structures. Folding and co-folding models primarily address the problem of protein and complex geometry, reliably capturing global folds and binding-site architecture. In doing so, these methods extend the applicability of SBDD to targets that are experimentally challenging or inaccessible.

These advances can be further complemented by AI-driven ligand-design approaches. Integration of generative AI within SBDD [29] has already demonstrated promising results, particularly in pocket-aware ligand generation. One notable example is Pocket-2Mol [30], a geometry-driven generative framework that constructs ligand structures directly from the three-dimensional shape and chemical features of a protein binding pocket.

QM approaches, particularly the Fragment Molecular Orbital (FMO) method [31–33], further extend the power of SBDD by enabling accurate, physics-based analysis of protein–ligand and protein–protein interactions. In integrated workflows, AI-derived protein structures provide structurally plausible starting points, while

FMO calculations interrogate complexes at the electronic-structure level. Within the Sygnature integrated FMO platform (FMO-SP) (see Chap. 4), FMO yields quantitative interaction energies (kcal/mol), detailed residue-level interaction profiles, and decomposition into physically meaningful energy terms. These outputs can be exploited both for rational ligand optimization and for energetic validation or refinement of predicted binding modes. Full methodological and practical details of FMO-SP are described in Chaps. 4 and 6.

Crucially, FMO and its implementation within FMO-SP address key limitations of both co-folding models and ligand-focused AI by acting as a physics-based filter. While AI methods may generate geometrically plausible protein–ligand poses or chemically attractive ligand candidates, FMO calculations respond sensitively to changes in charge state, hydrogen bonding, metal coordination, and hydrophobic packing. This sensitivity enables discrimination between memorized or artefactual solutions and energetically viable binding modes, a capability that is particularly valuable when working with predicted structures in the absence of experimental validation.

From a practical SBDD perspective, combining AI-based structure prediction, ligand-design AI, and FMO-SP enables a robust hybrid workflow. Folding models provide protein structures, generative or docking-based approaches propose ligand hypotheses, and FMO-SP analysis quantifies interactions, identifies energetic hot spots, and guides ligand optimization. Rather than replacing experimental validation, this integrated strategy focuses experimental efforts on the most promising and physically sound design hypotheses.

More broadly, such AI–QM hybrid strategies represent a powerful paradigm for modern drug discovery (Fig. 5). They preserve the scalability and creative breadth of AI while anchoring decision-making in quantum-mechanical accuracy. By clearly separating the roles of structure generation, ligand generation, and energetic validation, and connecting them through physics-based methods such as FMO and its industrial implementation within FMO-SP, these workflows offer a principled and practical route toward next-generation SBDD.

8 Conclusions

SBDD remains the dominant computational paradigm for small-molecule discovery and continues to deliver the highest success rates in progressing compounds beyond the design stage [11, 13, 14]. Within this context, the distinction between protein folding and co-folding methods is critical for understanding how recent advances in AI-based structure prediction can be most effectively applied in practical drug discovery workflows. Deep-learning-based folding models, exemplified by AlphaFold, have clearly demonstrated their value in SBDD by providing reliable protein structures that can be readily integrated into established virtual screening and docking pipelines.

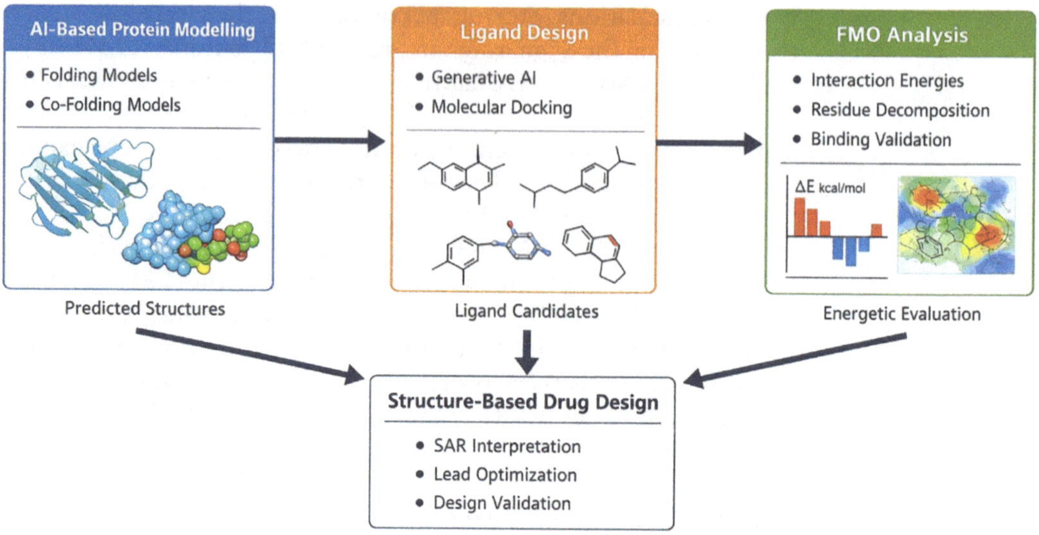

Fig. 5 Conceptual workflow integrating AI-based protein structure prediction, ligand-design approaches, and FMO analysis in structure-based drug design. AI-derived protein or protein–ligand geometries are combined with ligand-design methods to generate candidate molecules, while FMO provides quantitative interaction energies and energetic validation, enabling physically informed SAR interpretation and prospective ligand optimization

Most successful applications to date follow a common and effective strategy in which AI-based folding models are used to generate accurate apo protein structures that are subsequently combined with physics-based methods to model ligand binding. This separation of tasks leverages the respective strengths of each approach, with deep-learning models excelling at structure prediction and physics-based methods remaining competitive and well validated for modeling protein–ligand interactions. As a result, folding models have become a practical and impactful component of modern SBDD workflows.

Co-folding methods seek to address both structure and binding simultaneously by directly predicting protein–ligand complexes. While this represents a logical and ambitious extension of structure prediction, current implementations have not yet achieved the level of robustness observed for folding models alone. As discussed above, predicted binding poses frequently diverge from experimentally determined structures, and claims of reliable affinity prediction require further validation through systematic and prospective studies. At present, the most defensible application of co-folding approaches lies in hypothesis generation, where they can propose plausible binding modes or interaction patterns that are subsequently evaluated using physics-based methods or validated experimentally through established techniques such as site-directed mutagenesis or photoaffinity labeling.

The rapid progress in protein structure prediction over the past decade, from homology modeling to highly accurate folding and, more recently, to co-folding, has been substantial. Realizing the full potential of co-folding models will require the same rigorous validation and iterative refinement that established folding models as reliable tools for SBDD. This includes systematic benchmarking, careful assessment of failure modes, and close collaboration between method developers, computational chemists, and experimental scientists. While the conceptual foundations are promising, meaningful impact in drug discovery will depend on translating methodological advances into reproducible and quantitatively reliable performance.

AI-based protein structure prediction [29] has now become an integral component of SBDD, enabling the routine use of predicted structures to support compound design when combined with established physics-based and experimental approaches. Co-folding methods represent an important emerging direction but should currently be viewed as complementary tools for hypothesis generation rather than replacements for conventional virtual screening or experimental validation. Continued methodological development and validation are likely to further expand their role in structure-based drug design in the near future.

Recent work has also demonstrated that carefully filtered protein–ligand co-folding predictions can effectively substitute for experimentally determined structures when training machine-learning-based scoring functions [34]. Under stringent quality-control criteria, co-folded structures can achieve scoring and ranking performance comparable to experimental complexes. Importantly, indiscriminate expansion of training datasets with low-quality synthetic examples provides little benefit, underscoring the necessity of robust filtering strategies. Practical heuristics, including prioritization of single-chain complexes, confidence thresholds, and controlled train–test similarity, enable scalable and reliable data augmentation. Collectively, these findings establish co-folding as a viable approach for expanding training datasets, particularly for targets with limited structural coverage, and highlight its growing relevance for structure-based machine learning in drug discovery.

In this context, integration of AI-based protein structure prediction with quantum-mechanical approaches such as the FMO method provides a robust and practical path forward for SBDD. AI folding and co-folding models address the long-standing challenge of structural availability, while FMO delivers the energetic resolution and chemical interpretability required for informed medicinal chemistry decision-making. Together, these approaches form a complementary framework in which predicted structures are transformed into quantitatively interpretable models capable of supporting SAR analysis, binding-mode validation, and prospective ligand design.

In this broader context, integration of AI-based protein structure prediction with QM approaches provides a robust and practical path forward for SBDD. AI folding and co-folding models address the long-standing challenge of structural availability, while QM methods supply the electronic resolution and chemical rigor required to refine predicted structures and interpret molecular interactions [31]. Importantly, AI-based models and QM approaches address fundamentally different sources of uncertainty in structural modeling and are therefore highly complementary. AI methods substantially reduce conformational uncertainty at the level of global protein folds and binding-site architecture, providing physically plausible input geometries for downstream calculations. However, many chemically critical features remain poorly defined or inaccessible to current machine-learning models [31, 32]. QM and QM/MM methods [33, 34] enable accurate determination of protonation and tautomeric states in binding sites, explicit treatment of polarization and charge transfer between ligands and their protein environment, optimization of metal coordination geometries, and validation of hydrogen-bond networks. QM approaches further allow treatment of covalent mechanisms [35], transition states, nonclassical interactions, such as σ-hole-driven halogen bonding [36], and reaction pathways relevant to enzyme catalysis and covalent inhibitor design [32–37]. By refining these local electronic and chemical features, QM-based approaches transform AI-derived structures into chemically consistent models [33, 34] that can be reintegrated into SBDD workflows, improving the reliability of docking, molecular dynamics, free-energy calculations, and AI-driven scoring or generative models. In this way, QM methods beyond FMO serve as a critical refinement layer that complements AI structure prediction and enables more predictive, physics-grounded drug design.

References

1. Schneuing A, Harris C, Du Y et al. (2024) Structure-based drug design with equivariant diffusion models. Nat Comput Sci 4:899–909
2. Amzel LM (1998) Structure-based drug design. Curr Opin Biotechnol 9:366–369
3. Ouma RBO, Ngari SM, Kibet JK (2024) A review of the current trends in computational approaches in drug design and metabolism. Discov Public Health 21:108
4. Lin X, Li X, Lin X (2020) A review on applications of computational methods in drug screening and design. Molecules 25:1375
5. Neelam AKN (2025) Advancing drug discovery: the role of computer-aided design and development in modern pharmaceuticals. Discov Pharm Sci 1:8
6. Sadybekov AV, Katritch V (2023) Computational approaches streamlining drug discovery. Nature 616:673–685
7. Mihai DP, Nitulescu GM (2025) Computer-aided drug design and drug discovery. Pharmaceuticals 18:436
8. Wu Z, Chen S, Wang Y et al. (2024) Current perspectives and trend of computer-aided drug design: a review and bibliometric analysis. Int J Surg 110:3848–3878
9. Medina-Franco JL (2021) Grand challenges of computer-aided drug design: the road ahead. Front Drug Discov 1:728551
10. Sabe VT, Ntombela T, Jhamba LA et al. (2021) Current trends in computer aided drug design and a highlight of drugs discovered via

computational techniques: A review. Eur J Med Chem 224:113705
11. Dokholyan NV (2020) Experimentally-driven protein structure modeling. J Proteomics 220:103777
12. Timofeev V, Samygina V (2023) Protein crystallography: achievements and challenges. Crystals 13:71
13. Batool M, Ahmad B, Choi S (2019) A structure-based drug discovery paradigm. Int J Mol Sci 20:2783
14. Lounnas V, Ritschel T, Kelder J et al. (2013) Current progress in structure-based rational drug design marks a new mindset in drug discovery. Comput Struct Biotechnol J 5:e201302011
15. Balasco N, Esposito L, Vitagliano L (2025) Structural biology in the AlphaFold era: how far is artificial intelligence from deciphering the protein folding code?. Biomolecules 15:674
16. Jumper J, Evans R, Pritzel A et al. (2021) Highly accurate protein structure prediction with AlphaFold. Nature 596:583–589
17. Abramson J, Adler J, Dunger J et al. (2024) Accurate structure prediction of biomolecular interactions with AlphaFold 3. Nature 630:493–500
18. Wohlwend J, Corso G, Passaro S et al (2024) Boltz-1 democratizing biomolecular interaction modeling. http://biorxiv.org/lookup/doi/10.1101/2024.11.19.624167
19. Passaro S, Corso G, Wohlwend J et al (2025) Boltz-2: towards accurate and efficient binding affinity prediction. http://biorxiv.org/lookup/doi/10.1101/2025.06.14.659707
20. Stark H, Faltings F, Choi M et al (2025) Toward Universal Binder Design. bioRxiv. https://doi.org/10.1101/2025.11.20.689494
21. Ren Feng, Zhang Yu, Wang Y, Li X, Liu J, Chen Z, Liu H (2023) Structure-based discovery of cyclin-dependent kinase 20 inhibitors enabled by AlphaFold-predicted structures. J Med Chem 66:10945–10959
22. Díaz-Holguín A, Saarinen M, Vo DD et al. (2024) AlphaFold accelerated discovery of psychotropic agonists targeting the trace amine–associated receptor 1. Sci Adv 10:eadn1524
23. Baselious F, Hassan GS, El-Kashak WA, Abdelrahman MH, El-Shemy HA, Keri RS et al (2024) Structure-based discovery of selective histone deacetylase 11 inhibitors enabled by AlphaFold-predicted models. J Med Chem 67:7421–7436
24. Škrinjar P, Eberhardt J, Tauriello G et al (2025) Have protein-ligand cofolding methods moved beyond memorisation? http://biorxiv.org/lookup/doi/10.1101/2025.02.03.636309
25. Lee M, Schafer JW, Prabakaran J et al. (2025) Large-scale predictions of alternative protein conformations by AlphaFold2-based sequence association. Nat Commun 16: 5622
26. Fedorov DG, Kitaura K (2004) The importance of three-body terms in the fragment molecular orbital method. J Chem Phys 120:6832–6840
27. Fedorov DG, Kitaura K (2016) Subsystem analysis for the fragment molecular orbital method and its application to protein–ligand binding in solution. J Phys Chem A 120:2218–2231
28. Fedorov DG, Nagata T, Kitaura K (2012) Exploring chemistry with the fragment molecular orbital method. Phys Chem Chem Phys 14:7562
29. Hsu W-T, Grevtsev S, Herz AM et al. (2025) Can AI-predicted complexes teach machine learning to compute drug binding affinity?. J Chem Inf Model 65:13051–13056
30. Peng X, Luo S, Guan J, et al (2022) Pocket2Mol: efficient molecular sampling based on 3D protein pockets https://arxiv.org/abs/2205.07249
31. Yan Z, Wei D, Li X et al. (2024) Accelerating reliable multiscale quantum refinement of protein–drug systems enabled by machine learning. Nat Commun 15: 4181
32. Niazi SK, Mariam Z, Paracha RZ (2024) Limitations of protein structure prediction algorithms in therapeutic protein development. BioMedInformatics 4:98–112
33. Senn HM, Thiel W (2009) QM/MM methods for biomolecular systems. Angew Chem Int Ed 48:1198–1229
34. Van Der Kamp MW, Mulholland AJ (2013) Combined quantum mechanics/molecular mechanics (QM/MM) methods in computational enzymology. Biochemistry 52:2708–2728
35. Lodola A, Callegari D, Scalvini L et al. (2020) Design and SAR analysis of covalent inhibitors driven by hybrid QM/MM simulations. In: Heifetz A (Ed.)Eds. Quantum mechanics in drug discovery. Springer US, New York, NY, 307–337
36. Politzer P, Murray JS, Clark T (2013) Halogen bonding and other σ-hole interactions: a perspective. Phys Chem Chem Phys 15:11178
37. Warshel A, Sharma PK, Kato M et al. (2006) Electrostatic basis for enzyme catalysis. Chem Rev 106:3210–3235

Index

A

ABINIT-MP 2, 27, 57–58, 120, 136, 301–305, 307, 326
ADMET prediction 26
Affinity prediction 228–229, 338, 342
AI–QM integration 32–33
Allosteric modulation 340
AlphaFold 29, 32–33, 155, 334, 336, 338, 341
Amber10:EHT 55, 135, 305
Amide···π stacking 210–212
Anion–π 203–207
Antibody–drug conjugates (ADCs) 52, 64, 74–75, 92, 94–96
Artificial intelligence (AI) 1, 25–35, 46, 75–76, 224, 337–338, 340–344
Atropisomerism 225
Automated workflows 327
Automation 51, 75, 234–235, 238, 260, 280, 324, 326, 328–329

B

Binding affinity 40, 65, 84, 139, 179, 197, 201, 203, 208, 228–229, 243, 249, 255, 261
Binding energetics 56, 99, 273, 289, 335
Bioactive conformation 28, 84, 98, 109, 188, 203, 224, 244
Biologics 25, 92–95, 110–111
BRD4–MZ1 .... 106–108, 110

C

Cation–π interactions 144, 203–206, 272
Chalcogen bonding 200
Charge transfer 2–4, 9, 28, 50, 53–54, 96–97, 111, 120, 261, 272, 307, 344
Chemical factor 54, 58, 62, 83, 100, 105
Chlorine interactions 197
C–N fragmentation scheme 51, 55
Computer-assisted synthesis planning (CASP) 312, 316–317, 321
Conformational analysis 28, 224
Conformational sampling 33, 244, 260
Covalent inhibitors 28, 247, 344
CRBN 86, 101–103, 110
Cryo-electron microscopy (cryo-EM) 39, 42–43, 67–68, 70, 133, 142, 242, 248–249, 255, 333
CSV matrix 58

D

Danuglipron 70–71, 74, 101
Density functional theory (DFT) 3–4, 28, 33, 52, 192–193, 206, 208, 212–213, 224–225, 228–230, 233–234, 242, 246, 252, 255, 261, 265–267, 284, 287–289, 312–315, 319, 321
DFTB-PA 52, 55–57
DNA/RNA modulators 92, 94, 104
Docking 29, 32–33, 35, 41, 55, 75, 95–96, 111, 133, 197, 212–213, 241, 244, 246–247, 252, 254–255, 259–262, 264, 266–267, 280, 297, 326–327, 329, 337, 341, 344
Docking poses 28, 41, 52, 196, 247, 326
Dynamic simulations 64, 74

E

Electronic structure 228–229, 302
Electrostatic complementarity (ESC) 51–52, 63, 67–70, 74, 96
Electrostatic complementarity score (ESCS) 63
Energy decomposition analysis 15, 265
Enzyme inhibition 94, 246

F

Facio 2, 8–12, 15, 21
Fluorine interactions 195, 197
FMO database (FMODB) 39, 324, 328–329
FMO-PPI 83–84
FMO2 .... 2–3, 5–6, 12–14, 20–21, 136, 142, 146, 159, 302
FMO3 .... 2–3, 5–7, 143–144, 159, 302
Food and Drug Administration (FDA) 91–92, 94–96, 205
Fragmentation 1–2, 9, 12, 51–52, 57, 63, 73, 158, 225, 228, 238, 303, 324
Fragment molecular orbital (FMO) 1–2, 8–13, 15, 21–22, 26–29, 32–34, 39–44, 46, 50–59, 70, 82, 84–88, 96–97, 99, 107, 110, 118, 120,

Alexander Heifetz (ed.), *Quantum Mechanics Applications in Drug Design*, Methods in Molecular Biology, vol. 3061, https://doi.org/10.1007/978-1-0716-5420-0,

124–128, 134, 136–139, 141–145, 148–149, 158–159, 165, 169–170, 174, 179–180, 228, 233–235, 237–238, 272–274, 277–280, 301–304, 306–307, 323–329, 335, 340–344
Fragment molecular orbital-based molecular dynamics (FMO-MD)..301–308
Fragment molecular orbital drug design consortium (FMODD)......................52, 323–324, 326–329
Fragment molecular orbital sygnature platform (FMO-SP) 26, 50–59, 62–76, 96–106, 108–111, 341
Free energy..............28, 51, 56, 81–82, 155, 169, 171, 187–188, 196, 228–229, 242–243, 260, 262, 264–267, 297, 299, 314

G

GAMESS................2, 8–9, 12, 27, 53–54, 57, 97, 169, 233–238, 301, 304
GFN2-xTB....................................244–245, 247, 251
GLP-1R/Semaglutide......... 70–71, 73–74, 95, 99–101, 110
G-protein coupled receptors (GPCRs) 52, 70, 73, 84, 99, 110, 242, 252, 271–277, 279–280, 327

H

Halogen bonding ... 191–193, 195, 201–202, 261–262, 265, 267, 344
High-performance computing (HPC).........34–35, 149, 236, 238, 280, 324, 326
Hotspot residues 81, 86
Human ether-à-go-go–related gene (hERG) 52, 63, 66–70, 74, 93, 248–250
Hydrogen bonding....................42, 44, 104–105, 195, 212, 227, 265, 341
Hydrophilic 51, 58, 154
Hydrophobic................. 26, 50–51, 53–54, 59, 62, 65, 67–68, 71, 73–74, 83, 86, 97–100, 102–103, 105, 109–110, 138–139, 154, 188, 190, 249, 251–252, 272–273, 276–277, 279, 324, 340–341

I

Interaction energy... 4, 6, 16–20, 32, 40, 43, 56, 58–59, 73, 75, 97–98, 120–121, 125, 127–128, 139, 148–149, 158–160, 166, 188–189, 197, 200–201, 203, 238, 242–243, 265, 276–277, 341–342
Interaction networks......... 64, 104, 139–140, 143–144, 163–164, 179, 275–279
Inter-fragment interaction energies (IFIE) 39, 57, 134, 136–139, 141–143, 145–146, 148–149, 306–307
Ion channels..................................242, 248, 254–255

K

KNIME hub..30, 32, 59

L

Lead optimisation (LO)...........................26, 28, 30, 53
Ligand binding.............. 21, 40, 49–50, 56, 58–59, 64, 85–86, 104, 229, 241, 243, 248, 253, 255, 272–273, 275–276, 279, 334, 342
Lipophilicity 190, 243, 249
LogP 69, 243–245, 249

M

Machine learning (ML)............ 1, 29–33, 46, 155, 225, 227–228, 266–267, 280, 299, 329, 334–335, 343
Matched molecular pairs 196, 203, 244, 248, 251, 255
Medicinal chemistry........26, 28, 30, 35, 50–51, 59, 69, 73–75, 188, 192–193, 200, 205, 211, 272, 280, 324, 327, 334, 343
Metabolic GPCRs.................................70, 74, 99, 272
Molecular dynamics (MD) 7–8, 13, 20–22, 33, 40–44, 46, 52, 55, 62, 64–66, 73–75, 87, 96, 111, 118, 120–121, 124–125, 128–129, 133–136, 141, 145–150, 179, 208, 229, 234, 244, 264–267, 280, 283, 293–298, 302–305, 307, 327, 329, 344
Molecular glue degraders (MGDs)................64, 74–75, 94–96, 101–103
Molecular glues 52, 82, 85–86, 92, 94, 96–98, 101, 103, 110
Molecular mechanics (MM).......... 1, 27–28, 33, 49, 52, 55, 63, 73, 75, 82, 95, 225, 227, 229, 264–267, 272–273, 278, 283–285, 287–288, 290–292, 295–296, 298, 303–305, 307, 323, 325
Molecular operating environment (MOE).....28, 50, 55, 59, 63, 86, 134–135, 145–146, 305
Møller–Plesset perturbation theory (MP2).............. 3–4, 8, 13, 16–17, 20–21, 52, 56–58, 120, 136, 138, 142, 146, 193, 200, 208, 211–212, 233–234, 302, 308
Mutual polarization 18, 56

N

Non-covalent interactions...............................156, 227
Nonequilibrium simulation (NES) 62, 64–66, 73–75
Nontraditional stacking...210
Nucleic acids.................... 44, 46, 64, 94–95, 104–105, 110–111, 116–119, 124, 127–129, 153, 206, 284–285, 338

O

OpenFMO 2, 58
Orthogonal multipolar interactions 195, 197
Ozempic® 70, 99

P

P38α MAP kinase 62, 64
PAICS 2, 58
Pair interaction energy (PIE) 2–3, 7, 26–27, 33, 50–51, 54, 58–59, 62, 65, 68, 72, 83–86, 96, 98, 100–101, 103, 105, 107–108, 158–159, 165, 177, 272–274
Pair interaction energy decomposition analysis (PIEDA) 3, 9, 11–17, 20–21, 27, 41, 50, 53–54, 56, 58, 96–97, 120, 129, 144, 272, 274–275
Peptide-drug conjugates (PDCs) 94
Peptides 12, 51, 70–71, 73, 84, 93–95, 99–100, 110, 153, 189, 197, 210, 212, 235
PF-06882961 .... 70, 101
Pimozide 67–68
Poisson–Boltzmann equation 63
Polarisation..... 28, 51, 53–54, 56, 63, 96–97, 159, 165, 272, 326
Polarized C–H bonds 189–190
Protein/ligand deformation 56
Protein–ligand interactions........27, 56, 62, 87, 97, 194, 198, 201, 340, 342
Protein–protein interactions (PPIs) ... 39, 46, 52–53, 64, 74–75, 81–82, 84–88, 93–99, 101, 106–108, 110–111, 133–134, 136–138, 141, 143, 145, 149–150, 154–155, 163–164, 166–170, 172, 176, 179–180, 195, 312, 323, 340
Protein structure prediction 29, 334–335, 338, 340, 342–344
Proteolysis-targeting chimeras (PROTACs)........52, 64, 74, 76, 82, 85–87, 92, 94–96, 101, 106–107, 109–110
Putative directional interactions:Bürgi–Dunitz and tetrel 198
PyMol 2, 50, 59, 165, 170, 179, 244
Python......... 11, 55, 59, 135, 146, 174–176, 236, 284, 298

Q

Quantum chemistry 35, 230, 233, 239
Quantum computing (QC) 22, 34, 224, 229–230
Quantum mechanics (QM) 1, 3, 8, 20, 22, 25–29, 32–35, 50, 53, 56, 73, 75, 82, 86–87, 95, 97, 100, 104, 109–110, 189, 192–193, 197, 212, 223–225, 227–230, 237, 241–244, 246–247, 261–264, 266–267, 272–273, 278–280, 283–293, 296–299, 302, 311–312, 315–316, 319–321, 323–324, 326–327, 340–341, 344
Quantum mechanics/molecular mechanics (QM/MM) 28, 58, 224, 229, 246, 295–296, 344

R

Reactivity prediction 227
Residue interaction networks 2, 144, 156–160, 162–165, 167–168, 170, 173–174, 176
RNA/DNA modulators 92
RNA-SS0 complex 104–105
RNA therapeutics 104
Rybelsus® 99

S

Scoring functions 26, 32–33, 53, 57, 75, 133, 212, 224, 260–261, 264–265, 324, 326, 343
Semi-empirical methods 225, 266
Singular value decomposition (SVD) 134, 136, 138–139, 143–144, 150, 161, 163, 166, 169, 173, 176
Solvation 4, 18, 51, 54, 56, 74, 104, 136, 159, 165, 187, 229, 238, 242–243, 245, 260–262, 265, 267, 272, 315, 326
Structure–activity relationship (SAR) 26–27, 34, 50–51, 54, 56–59, 63, 66, 71, 73–74, 97, 110, 188–189, 197–198, 202, 212–213, 272, 326, 334, 342–343
Structure-based drug design (SBDD) 26–29, 32–35, 49, 53, 63, 75, 81, 85, 109, 111, 212, 272, 323–324, 326–327, 333–335, 340–344
Subsystem analysis (SA) 5, 11–14, 18–21, 51, 56, 58, 62–63, 97, 170
SVL 55, 99–100
Synthetic feasibility 27, 29, 311–312, 319
S–π interactions 201

T

Theoceptor 241–255
Torsion barriers 224, 230
Torsion profiles 224–226
Transition states (TS) 227, 312, 318–319, 344

U

Undruggable 93, 101, 106, 338

V

Virtual screening....... 26–27, 33, 35, 52, 56–57, 75–76, 84–85, 212, 243, 261–262, 338, 341, 343
Visualization 9, 15, 83, 104, 234, 237, 324

W

Wegovy® 70, 99
Workflow automation 75

FSC
www.fsc.org
MIX
Papier | Fördert gute Waldnutzung
FSC® C083411

Kolibri 360 GmbH
Ferdinand-Jühlke-Straße 7
99095 Erfurt, Deutschland
produktsicherheit@kolibri360.de